I0476967

Impacts of Climate Change on Biodiversity, Ecosystems, and Ecosystem Services

Technical Input to the 2013 National Climate Assessment

July 2012

Recommended Citation for overall document:
Michelle D. Staudinger, Nancy B. Grimm, Amanda Staudt, Shawn L. Carter, F. Stuart Chapin III, Peter Kareiva, Mary Ruckelshaus, Bruce A. Stein. 2012. *Impacts of Climate Change on Biodiversity, Ecosystems, and Ecosystem Services: Technical Input to the 2013 National Climate Assessment.* Cooperative Report to the 2013 National Climate Assessment. 296 p.

STEERING COMMITTEE

Shawn L. Carter	U.S. Geological Survey
F. Stuart Chapin III	University of Alaska, Fairbanks
Nancy B. Grimm	National Science Foundation; Arizona State University; National Climate Assessment
Peter Kareiva	The Nature Conservancy
Mary Ruckelshaus	NaturalCapital Project
Michelle D. Staudinger	USGS National Climate Change & Wildlife Science Center; University of Missouri
Amanda Staudt	National Wildlife Federation
Bruce A. Stein	National Wildlife Federation

ACKNOWLEDGMENTS

We would like to thank the U.S. Geological Survey (USGS) for providing financial resources to support the development of this technical input including personnel, travel and accommodations for workshop participants. We thank the Gordon and Betty Moore Foundation for use of their facilities and catering services during the Steering Committee meeting and plenary workshop. We thank the U.S. Global Change Research Program (USGCRP) for providing technical support during phone and web-based meetings, as well as for a shared online workspace. We are grateful to Kathy Jacobs (OSTP), Anne Waples (NOAA), Robert Bradburne (UK National Ecosystem Assessment), Robin O'Malley (USGS), Doug Beard (USGS), Walt Reid (Packard Foundation), Joel Smith (Stratus Consulting), Richard Moss (University of Maryland), Steve McCormick (Moore Foundation), and Guillermo Castilleja (Moore Foundation) for giving presentations during webinars and the plenary workshop. A special thanks to Joanna Whittier and the University of Missouri for managing workshop participant travel and expenses. We are also appreciative to Martin Saunders (Santa Clara University), Michael Bernstein (Arizona State University), Jennifer Riddell (EPA/AAAS), Elda Varela-Acevedo (Michigan State University), Laura Thompson (USGS), and Daniel Nover (EPA/AAAS) for their help as recorders at the plenary workshop.

This cooperative agency report underwent a formal USGS peer review process following the Fundamental Science Practices requirements and has Bureau approval. We thank Virginia Burkett (USGS) and Susan Aragon-Long (USGS) for administrative support and coordination of this review. We appreciate the constructive comments received from J.P. Schmidt (University of Georgia), Craig Allen (USGS; Nebraska Cooperative Fish and Wildlife Research Unit), and Jayne Belnap (USGS) on the entire technical input, as well as the comments received from several external reviewers on individual chapters during the preparation of this report.

AUTHOR TEAMS FOR THIS REPORT BY CHAPTER

Executive Summary
Lead Authors: Amanda Staudt, Michelle D. Staudinger, Mary Ruckelshaus, Peter Kareiva, Nancy B. Grimm, Shawn L. Carter, Bruce A. Stein, F. Stuart Chapin III

Chapter 1: Introduction
Lead Authors: Michelle D. Staudinger, Nancy B. Grimm, Amanda Staudt, Shawn L. Carter, Peter Karieva, Mary Ruckelshaus, F. Stuart Chapin III, Bruce A. Stein

Chapter 2: Impacts of Climate Change on Biodiversity
Convening Lead Authors: Michelle D. Staudinger, Shawn L. Carter
Lead Authors: Molly S. Cross, Natalie S. Dubois, J. Emmett Duffy, Carolyn Enquist, Roger Griffis, Jessica Hellmann, Josh Lawler, John O'Leary, Scott A. Morrison, Lesley Sneddon, Bruce A. Stein, Laura Thompson, Woody Turner
Contributing Author: Elda Varela-Acevedo, Walt Reid

Chapter 3. Impacts of Climate Change on Ecosystem Structure and Functioning
Convening Lead Authors: Nancy B. Grimm, F. Stuart Chapin III
Lead Authors: Britta Bierwagen, Patrick Gonzalez, Peter M. Groffman, Yiqi Luo, Forrest Melton, Knute Nadelhoffer, Amber Pairis, Peter Raymond, Josh Schimel, Craig E. Williamson
Contributing Author: Michael J. Bernstein

Chapter 4. Impacts of Climate Change on Ecosystem Services
Convening Lead Authors: Peter Kareiva, Mary Ruckelshaus
Lead Authors: Katie Arkema, Gary Geller, Evan Girvetz, Dave Goodrich, Erik Nelson, Virginia Matzek, Malin Pinsky, Walt Reid, Martin Saunders, Darius Semmens, Heather Tallis

Chapter 5: Impacts of Climate Change on Already Stressed Biodiversity, Ecosystems, and Ecosystem Services
Convening Lead Authors: Amanda Staudt, Allison K. Leidner
Lead Authors: Jennifer Howard, Kate A. Brauman, Jeff Dukes, Lara Hansen, Craig Paukert, John Sabo, Luis A. Solórzano,
Contributing Author: Kurt Johnson

Chapter 6: Adaptation to Impacts of Climate Change on Biodiversity, Ecosystems, and Ecosystem Services
Convening Lead Authors: Bruce A. Stein, Amanda Staudt
Lead Authors: Molly S. Cross, Natalie Dubois, Carolyn Enquist, Roger Griffis, Lara Hansen, Jessica Hellman, Josh Lawler, Eric Nelson, Amber Pairis
Contributing Authors: Doug Beard, Rosina Bierbaum, Evan Girvetz, Patrick Gonzalez, Susan Ruffo, Joel Smith

Chapter 7. Proposed Actions for the Sustained Assessment of Biodiversity, Ecosystems, and Ecosystem Services
Lead Authors: Shawn L. Carter, Amanda Staudt, Gary Geller, Woody Turner

TABLE OF CONTENTS

Executive Summary S-1

Chapter 1. Introduction 1-1

   1.1. Federal mandate for the National Climate Assessment 1-1
   1.2. Building on past assessments... 1-1
   1.3. Technical input process.. 1-2
   1.4. Report approach and organization 1-3
   1.5. Literature cited 1-4

Chapter 2. Impacts of Climate Change on Biodiversity 2-1

   2.1. Introduction.. 2-1
      2.1.1. What is biodiversity and why does it matter?... 2-1
      2.1.2. Threats to biodiversity 2-2
      2.1.3. Biodiversity at risk in the United States 2-7
      2.1.4. Objectives 2-10
   2.2. Observed impacts of recent climate change on biodiversity 2-11
      2.2.1. Genetic diversity, traits, and phenotypic plasticity.............................. 2-12
      2.2.2. Phenological shifts 2-13
      2.2.3. Shifts in species distributions 2-17
      2.2.4. Shifts in biotic interactions and species assemblages 2-18
   2.3. How will climate change affect biodiversity in the coming century? 2-20
      2.3.1. Introduction... 2-20
      2.3.2. Projected impacts on organisms, species and populations.................... 2-21
      2.3.3. Projected impacts on communities, ecosystems, and biomes................ 2-23
   2.4. Vulnerabilities and risks: Why impacts of climate change on biodiversity matter 2-25
      2.4.1. Vulnerability and risk to climate change 2-25
      2.4.2. What types of ecosystems and species are most vulnerable? 2-26
      2.4.3. Policy implications for vulnerable species 2-28
   2.5. What human response strategies could address the most harmful impacts of climate
      change on biodiversity and what barriers and opportunities exist to their
      implementation? 2-30
      2.5.1. Climate change adaptation for biodiversity conservation.................... 2-30
      2.5.2. Techniques and approaches for understanding the impacts of climate
      change and human actions on biodiversity 2-33
   2.6. Synthesis of impacts on biodiversity 2-37
   2.7. Critical gaps in knowledge, research, and data needs...................................... 2-39
   2.8. Literature cited 2-41

Chapter 3. Impacts of Climate Change on Ecosystem Structure and Functioning............. 3-1

   3.1. Introduction.. 3-1
      3.1.1. Ecosystem impacts in context 3-2
      3.1.2. A conceptual framework for ecosystem change 3-3

3.2. Historical changes, current status, and projected changes in ecosystem structure
and functioning 3-5
    3.2.1. Biome shifts 3-5
    3.2.2. Ecosystem state transitions 3-9
    3.2.3. Forest growth, mortality, pests, and fire 3-12
    3.2.4. Changes in winter have surprising impacts 3-15
    3.2.5. Intensification of the hydrologic cycle 3-17
    3.2.6. Ecosystem effects from physical changes in lakes and oceans..................... 3-19
    3.2.7. Feedbacks from ecosystem functioning to climate 3-23
    3.2.8. Resource management in the context of climate change 3-24
3.3. Synthesis: likelihood and consequence of the key impacts of climate change
on ecosystems 3-29
3.4. Societal response: Managing change—it can be done. Adaptation and mitigation
responses for ecosystems 3-35
3.5. Critical gaps in knowledge, research, and data needs.. 3-36
    3.5.1. Observational networks for documenting ecosystem change 3-37
3.6. Literature cited 3-39

Chapter 4. Impacts of Climate Change on Ecosystem Services **4-1**

4.1. Introduction: What are ecosystem services and why do they matter?........................... 4-2
4.2. What are observed impacts of recent climate change on ecosystem services
and their value? 4-3
    4.2.1. Marine fishery yields 4-3
    4.2.2. Nature-dependent tourism... 4-6
    4.2.3. Hazard reduction: Coastal protection services 4-7
    4.2.4. Fire regulation 4-8
    4.2.5. Carbon storage and sequestration 4-11
4.3. How will climate change affect ecosystem services and human well being
over the next 50 to 100 years? 4-13
    4.3.1. Marine fishery yields 4-13
    4.3.2. Nature-based recreation and tourism 4-13
    4.3.3. Hazard reduction by coastal habitats 4-15
    4.3.4. Water supply and water quality under future climate................................... 4-18
4.4. What response strategies could address the most harmful impacts of climate
change on ecosystem services? 4-20
4.5. Critical gaps in knowledge, research, and data needs for climate impacts
on ecosystem services.. 4-27
4.6. Literature cited 4-28

**Chapter 5. Impacts of Climate Change on Already Stressed Biodiversity, Ecosystems,
and Ecosystem Services** **5-1**

5.1. Introduction... 5-1
5.2. Conceptual framework of climate change interactions with other stressors................... 5-2
5.3. Interactions between stressors and climate change are already being observed 5-5
    5.3.1. Land use and land cover change 5-5

5.3.2. Extraction of natural resources 5-6

5.3.3. Biological disturbance 5-7

5.3.4. Pollution 5-8

5.4. Anticipated interactions of climate change with other stressors 5-12

5.4.1. Projections for climate change interactions with land use and land
cover change 5-13

5.4.2. Projections for climate change interactions with water extraction 5-14

5.4.3. Projections for climate change interactions with biological disturbances 5-15

5.4.4. Projections for climate change interactions with pollutants 5-15

5.5. Case study: water use in California's Central Valley 5-16

5.6. Integrating climate adaptation and conservation strategies for other stressors 5-18

5.6.1. Habitat connectivity imperative for addressing habitat loss 5-20

5.6.2. Managing harvest must contend with new situations 5-20

5.6.3. Best practices for managing invasive species, pest, and disease outbreaks
need to be reconsidered 5-21

5.6.4. Pollution regulations can be undermined by climate change effects 5-21

5.7. Concluding thoughts 5-24

5.8. Literature cited 5-25

**Chapter 6. Adaptation to Impacts of Climate Change on Biodiversity, Ecosystems, and
Ecosystem Services 6-1**

6.1. Responding to climate change 6-1

6.1.1. What is adaptation? 6-2

6.1.2. Managing change 6-4

6.1.3. Targets of adaptation: From species to services 6-6

6.2. Dramatically increased interest in adaptation 6-8

6.2.1. Trends in adaptation attention 6-8

6.2.2. Adaptation at the Federal level 6-9

6.2.3. Adaptation at the State level 6-11

6.2.4. Adaptation by tribal governments 6-12

6.2.5. Challenges 6-12

6.3. The need to reconsider conservation goals 6-13

6.3.1. Forward-looking goals: A key to successful adaptation 6-13

6.3.2. Reconsidering goals 6-14

6.3.3. Challenges 6-15

6.4. Convergence on adaptation principles and strategies 6-16

6.4.1. Overarching principles for adaptation 6-16

6.4.2. Key adaptation strategies 6-19

6.4.3. Advances in adaptation planning 6-22

6.4.4. Challenges 6-25

6.5. Mainstreaming adaptation 6-27

6.6. Illustrative case study: Climate adaptation in the San Francisco Bay 6-28

6.7. Into the future: Adaptation and its limits 6-29

6.8. Literature cited 6-31

Chapter 7. Proposed Actions for the Sustained Assessment of Biodiversity, Ecosystems, and Ecosystem Services **7-1**

7.1. Proposed actions for the sustained assessment 7-1

7.2. Literature cited 7-5

Appendix A. List of Contributors to the National Climate Assessment Technical Input on Biodiversity, Ecosystems, and Ecosystem Services **A-1**

Appendix B. Plenary Workshop Agenda **A-3**

FIGURES

Figure 2.1. Map of observed and projected biological responses to climate change across the
United States 2-3

Figure 2.2. Distribution of climate related threats evaluated by the IUCN on 4,161 terrestrial and
aquatic species globally 2-7

Figure 2.3. Terrestrial and aquatic species at risk by State in the United States 2-8

Figure 2.4. Aquatic and terrestrial biodiversity hotspots across the United States..................... 2-9

Figure 2.5. Changes in plant phenology across an elevation gradient near Tucson, Arizona.
Ninety-three species (26 percent) showed change in flowering range with elevation with
warmer summers. (A) 12 species exhibited flowering range shift upslope. (B) 34 species
exhibited flowering range expansion upslope. (C) 23 species exhibited flowering range
contraction upslope 2-16

Figure 2.6. Physiological and life history traits of species and populations that influence
vulnerability or resilience in response to climate-related disturbance.......................... 2-26

Figure 2.7. A process for using the results of a vulnerability assessment to identify potential
adaptation strategies is illustrated for a limited set of factors affecting the Atlantic salt
marsh snake (*Nerodia clarkii taeniata*) throughout its range in Florida. 2-28

Figure 3.1. Distribution of the world's biomes with respect to mean annual temperature and total
annual precipitation 3-2

Figure 3.2. Biome shifts. (A) Observed linear temperature trend 1901-2002 (^{o}C century^{-1}) and
field sites of detected shifts. (B) Potential vegetation under observed 1961–90 climate.
(C) Potential vegetation under projected 2071–2100 climate where any of nine GCM–
emissions scenario combinations project change. (D) Vulnerability of ecosystems to
biome shifts based on historical climate and projected vegetation................................. 3-8

Figure 3.3. Hypothesized changes in soil moisture in xeric, mesic, and hydric systems with less
frequent but larger precipitation events 3-11

Figure 3.4. An example of long-term trends in DOC concentration for a lake in northern
Pennsylvania. As DOC concentration increases, lake transparency decreases, with
consequences for primary productivity, the base of lake food webs 3-22

Figure 3.5. Relative vulnerability quadrants resulting from degree of exposure and
sensitivity 3-27

Figure 3.6. Relative vulnerability of catchments to low flow events and warming
temperatures.. 3-27

Figure 3.7. Locations in the United States of major historical changes at the ecosystem level
attributed to climate change, including bark beetle infestations, biomes shifts, increased
forest growth, forest mortality, stream intermittency, increased streamflow and
accelerated nutrient flushing, thermal stratification, wildfire 3-30

Figure 3.8. Fraction of net anthropogenic nitrogen inputs (NANI) exported in riverine nitrogen flux as a function of (A) discharge, (B) precipitation and (C) temperature for watersheds greater than 250 km^2. The strongest relationship is with discharge in individual watersheds 3-33

Figure 3.9. Ecological production functions linking nitrogen and ecosystem services 3-33

Figure 4.1. Winners and losers as a result of lobster range shifts 4-5

Figure 4.2. Vulnerability of Gulf coastal counties based on physical and social indicators and their integration into place vulnerability 4-17

Figure 4.3. Location of the 184 of 213 Alaska Native villages already affected by flooding and erosion 4-18

Figure 4.4. The number of U.S. counties with water sustainability risk by 2050 with and without climate change 4-19

Figure 4.5. A map of soil classes 4-22

Figure 4.6. Estimated average corn yield from 2000-2008 by soil class 4-22

Figure 5.1. Conceptual diagram of climate change interaction with a single other environmental stress. 5-3

Figure 5.2. Conceptual diagram illustrating the multiple interacting environmental stresses that can affect natural systems, including climate change 5-4

Figure 5.3. Conceptual diagram of the impacts of climate change interaction with multiple environmental stressors on salmon and their aquatic habitat in California. 5-17

Figure 5.4. An environmental stressor, coupled with climate change, may have a larger consequence for the condition of biodiversity than an individual stressor alone. Adaptation actions can be implemented to ameliorate the impact of these combined stressors 5-19

Figure 6.1. Scope and scale of adaptation efforts 6-5

Figure 6.2. Generalized framework for climate change adaptation planning and implementation 6-23

Figure 6.3. Framework for assessing the relative intensity of conservation interventions based on the vulnerability of a species or ecosystem 6-26

Figure 7.1. An operational process for conducting regular assessments of biodiversity and ecosystem services 7-2

TABLES

Table 1.1. Metrics used to assess and communicate confidence levels and uncertainties in key findings 1-3

Table 2.1. Overview of physical changes associated with climate change and examples of the potential ecological consequences associated with these changes 2-11

Table 2.2. Examples of observed phenological change across geographical regions of the United States 2-14

Table 2.3. Examples of facilities and networks that organize and archive observations of biodiversity on national and global scales for public use 2-36

Table 4.1. Current status, and projected future impacts of climate on ecosystem services 4-42

Table 4.2. Factors affecting adaptation responses to climate change impacts.......................... 4-58

Table 4.3. Predicted annual corn yield from 2000 to 2008... 4-21

Table 4.4. Acres available for cropping on the best soils as of 2001 4-22

Table 4.5. Contemporaneous yield impact of marginal soil improvement................................. 4-23

Table 4.6. Change in average annual corn GDD and growing season precipitation between the periods of 1950–1958 and 2000–2008.. 4-23

Table 4.7. Predicted average corn yield in the period 2050–2058 assuming that average annual GDD and growing season precipitation increase 10 percent between the periods of 2000-2008 and 2050–2058 across the entire study area 4-24

Table 4.8. Predicted average corn yield in the period 2050 – 2058 assuming that average annual GDD and growing season precipitation increase 10 percent between the periods of 2000-2008 and 2050–2058 across the entire study area but technological improvements in corn farming occur at half the rate that they did in the past... 4-24

Table 4.9. Predicted average corn yield in the period 2050–2058 assuming that average annual GDD and growing season precipitation increase 20 percent between the periods of 2000-2008 and 2050–2058 across the entire study area but technological improvements in corn farming occur at half the rate that they did in the past... 4-25

Table 4.10. Potential improvements by improving marginal corn soils 4-25

Table 5.1. Interacting stressors and non-speculative examples of their effects on biodiversity, ecosystems and ecosystem services when combined with climate change 5-9

Table 5.2. Example strategies for conserving and managing natural resources and ways that the strategies have been modified to integrate climate change adaptation 5-22

Table 6.1. Change continuum and strategic responses 6-6

BOXES

Box E.1. Case study of the 2011 Las Conchas, New Mexico wildfire..S-4

Box 2.1. Examples of observed and projected biological response to climate change across the
    United States 2-3

Box 2.2. Case study: Recent and projected changes in plant communities in the Sky Islands
    region of the Southwest 2-16

Box 2.3. Impacts of climate change on marine ecosystems 2-21

Box 2.4. Integrating vulnerability assessments into adaptation planning: Updating the Florida
    State Wildlife Action Plan 2-27

Box 2.5. Recent advances in the genetics and evolution of climate responses 2-34

Box 3.1. A resilience-based framework for considering climate change impacts on ecosystem
    structure and functioning 3-4

Box 3.2. Climate change and ecosystem disruption: Whitebark pine and mountain
    pine beetles...3-13

Box 3.3. Ominous signals from ocean ecosystems...3-21

Box 3.4. An example of vulnerability assessment approaches applied to northeastern stream
    and river ecosystems 3-26

Box 3.5. Nitrogen regulation for rivers and the coastal zone 3-32

Box 4.1. Climate impacts on New England groundfish fisheries 4-4

Box 4.2. Climate impacting fire risk, water supply, recreation, and flood risk in western U.S.
    forests...4-9

Box 4.3. Climate impacts on coastal hazards in the Gulf of Mexico.......................................4-16

Box 4.4. Adapting to climate change by maximizing a supporting service: Soil quality..........4-22

Box 5.1. National fish, wildlife and plant climate adaptation strategy draft: Goal 7 5-18

Executive Summary

Lead Authors: Amanda Staudt, Michelle D. Staudinger, Mary Ruckelshaus, Peter Kareiva, Nancy B. Grimm, Shawn L. Carter, Bruce A. Stein, F. Stuart Chapin III

Ecosystems, and the biodiversity and services they support, are intrinsically dependent on climate. During the twentieth century, climate change has had documented impacts on ecological systems, and impacts are expected to increase as climate change continues and perhaps even accelerates. This technical input to the National Climate Assessment synthesizes our scientific understanding of the way climate change is affecting biodiversity, ecosystems, ecosystem services, and what strategies might be employed to decrease current and future risks.

Building on past assessments of how climate change and other stressors are affecting ecosystems in the United States and around the world, we approach the subject from several different perspectives. First, we review the observed and projected impacts on biodiversity, with a focus on genes, species, and assemblages of species. Next, we examine how climate change is affecting ecosystem structural elements—such as biomass, architecture, and heterogeneity—and functions—specifically, as related to the fluxes of energy and matter. People experience climate change impacts on biodiversity and ecosystems as changes in ecosystem services; people depend on ecosystems for resources that are harvested, their role in regulating the movement of materials and disturbances, and their recreational, cultural, and aesthetic value. Thus, we review newly emerging research to determine how human activities and a changing climate are likely to alter the delivery of these ecosystem services.

This technical input also examines two cross-cutting topics. First, we recognize that climate change is happening against the backdrop of a wide range of other environmental and anthropogenic stressors, many of which have caused dramatic ecosystem degradation already. This broader range of stressors interacts with climate change, and complicates our abilities to predict and manage the impacts on biodiversity, ecosystems, and the services they support. The second cross-cutting topic is the rapidly advancing field of climate adaptation, where there has been significant progress in developing the conceptual framework, planning approaches, and strategies for safeguarding biodiversity and other ecological resources. At the same time, ecosystem-based adaptation is becoming more prominent as a way to utilize ecosystem services to help human systems adapt to climate change.

In this summary, we present key findings of the technical input, focusing on themes that can be found throughout the report. Thus, this summary takes a more integrated look at the question of how climate change is affecting our ecological resources, the implications for humans, and possible response strategies. This integrated approach better reflects the impacts of climate in the real world, where changes in ecosystem structure or function will alter the viability of different species and the efficacy of ecosystem services. Likewise, adaptation to climate change will simultaneously address a range of conservation goals. Case studies are used to illustrate this complete picture throughout the report; a snapshot of one case study, *2011 Las Conchas, New Mexico Fire*, is included in this summary.

KEY FINDINGS

Biodiversity and ecosystems are already more stressed than at any comparable period of human history. Climate change almost always exacerbates the problems caused by other environmental stressors including: land use change and the consequent habitat fragmentation and degradation; extraction of timber, fish, water, and other resources; biological disturbance such as the introduction of non-native invasive species, disease, and pests; and chemical, heavy metal, and nutrient pollution. As a corollary, one mechanism for reducing the negative impacts of climate change is a reduction in other stressors.

Climate change is causing many species to shift their geographical ranges, distributions, and phenologies at faster rates than previously thought. Changes in terrestrial plant and animal species ranges are shifting the location and extent of biomes, and altering ecosystem structure and functioning. These rates vary considerably among species. Terrestrial species are moving up in elevation at rates 2 to 3 times greater than initial estimates. Despite faster rates of warming in terrestrial systems compared to ocean environments, the velocity of range shifts for marine taxa exceeds those reported for terrestrial species. Species and populations that are unable to shift their geographic distributions or have narrow environmental tolerances are at an increased risk of extinction.

There is increasing evidence of population declines and localized extinctions that can be directly attributed to climate change. Ecological specialists and species that live at high altitudes and latitudes are particularly vulnerable to climate change. Overall, the impacts of climate change are projected to result in a net loss of global biodiversity and major shifts in the provision of ecosystem services. For example, the range and abundance of economically important marine fish are already changing due to climate change and are projected to continue changing such that some local fisheries are very likely to cease to be viable, whereas others may become more valuable if the fishing community can adapt.

Range shifts will result in new community assemblages, new associations among species, and promote interactions among species that have not existed in the past. Changes in the spatial distribution and seasonal timing of flora and fauna within marine, aquatic, and terrestrial environments can result in trophic mismatches and asynchronies. Novel species assemblages can also substantially alter ecosystem structure and function and the distribution of ecosystem services.

Changes in precipitation regimes and extreme events can cause ecosystem transitions, increase transport of nutrients and pollutants to downstream ecosystems, and overwhelm the ability of natural systems to mitigate harm to people from these events. Changes in extreme events affect systems differentially, because different thresholds are crossed. For example, more intense storms and increased drought coupled with warming can shift grasslands into shrublands, or facilitate domination by other grass types (for example, mixed grass to C-4 tallgrass). More heavy rainfall also increases movement of nutrients and pollutants to downstream ecosystems, restructuring processes, biota, and habitats. As a consequence, regulation of drinking water quality is very likely to be strained as high rainfall and river discharge lead to higher levels of nitrogen in rivers and greater risk of waterborne disease outbreaks.

Changes in winter have big and surprising effects on ecosystems and their services. Changes in soil freezing, snow cover, and air temperature have affected carbon sequestration, decomposition, and carbon export, which influence agricultural and forest production. Seasonally snow-covered regions are especially susceptible to climate change as small changes in temperature or precipitation may result in large changes in ecosystem structure and function. Longer growing seasons and warmer winters are enhancing pest outbreaks, leading to tree mortality and more intense and extensive fires. For winter sports and recreation, future economic losses are projected to be high because of decreased or unreliable snowfall.

The ecosystem services provided by coastal habitats are especially vulnerable to sea-level rise and more severe storms. The Atlantic and Gulf of Mexico coasts are most vulnerable to the loss of coastal protection services provided by wetlands and coral reefs. Along the Pacific coast long-term erosion of dunes due to increasing wave heights is projected to be an increasing problem for coastal communities. Beach recreation is also projected to suffer due to coastal erosion. Other forms of recreation are very likely to improve due to better weather, and the net effect is likely a redistribution of the industry and its economic impact, with visitors and tourism dollars shifting away from some communities in favor of others.

Climate adaptation has experienced a dramatic increase in attention since the last National Climate Assessment and become a major emphasis in biodiversity conservation and natural resource policy and management. Federal and State agencies are planning for and integrating climate change research into resource management and actions to address impacts of climate change based on historical impacts, future vulnerabilities, and observations on the ground. Land managers have realized that static protected areas will not be sufficient to conserve biodiversity in a changing climate, requiring an emphasis on landscape-scale conservation, connectivity among protected habitats, and sustaining ecological functioning of working lands and waters. Agile and adaptive management approaches are increasingly under development, including monitoring, experimentation, and a capacity to evaluate and modify management actions. Risk-based framing and stakeholder-driven scenario planning will be essential in enhancing our ability to respond to the impacts of climate change.

Climate change responses employed by other sectors (for example, energy, agriculture, transportation) are creating new ecosystem stresses, but also can incorporate ecosystem-based approaches to improve their efficacy. Ecosystem-based adaptation has emerged as a framework for understanding the role of ecosystem services in moderating climate impacts on people, although this concept is currently being used more on an international scale than within the United States.

Ecological monitoring efforts need to be improved and better coordinated among Federal and State agencies to ensure that the impacts of climate change are adequately observed as well as to support ecological research, management, assessment, and policy. As species and ecosystem boundaries shift to keep pace with climate change, improved and better-integrated research, monitoring, and assessment efforts will be needed at national and global scales. Existing monitoring networks in the United States are not well suited for detecting and attributing the impacts of climate change to the wide range of affected species at the appropriate spatio-temporal scales.

Box E.1. Case Study of the 2011 Las Conchas, New Mexico Fire

In the midst of severe drought in the summer of 2011, Arizona and New Mexico suffered the largest recorded wildfires in their history, affecting more than 694,000 acres. For example, the Las Conchas fire in northern New Mexico burned 63 residences, 1100 archeological sites, more than 60 percent of Bandelier National Monument (BNM), and more than 80 percent of the forested lands of the Santa Clara Native American Pueblo. Some rare threatened and endangered species were devastated by the fire. For instance, the major canyon systems of Bandelier National Monument experienced extensive, to near complete mortality, of all tree and shrub cover, which represents a total loss of nesting and roosting habitat for Mexican Spotted Owls (*Strix occidentalis lucida*; NPS 2011). The Jemez salamander (*Plethodon neomexicanus*) is another endangered species whose population was put in further danger by this fire.

Following the fire, heavy rainstorms led to major flooding and erosion, including at least ten debris flows originating from the north slopes of a single canyon in Bandelier National Monument. Popular recreation areas in the Monument were evacuated for four weeks and the flash floods damaged the newly renovated multi-million dollar U.S. Park Service Visitor Center. Sediment and ash eroded by the floods were washed downstream into the Rio Grande, which supplies 50 percent of drinking water for Albuquerque, the largest city in New Mexico. Water withdrawals by the city from the Rio Grande were stopped entirely for a week and reduced for several months, due to the increased cost of treatment.

These fires provide an example of how forest ecosystems, biodiversity, and ecosystem services are affected by the impacts of climate change, other environmental stresses, and past management practices. Warmer temperatures, reduced snowpack, and earlier onset of springtime are leading to increases in wildfire in the western United States (Westerling and others, 2006); while extreme droughts are becoming more frequent (Williams and others, 2011). In addition, climate change is affecting naturally occurring bark beetles: warmer winter conditions allow these pests to breed more frequently and successfully (Jönsson and others, 2009; Schoennagel and others, 2011). The dead trees left behind by bark beetles make crown fires more likely (Hoffman and others, 2010; Schoennagel and others, 2011). Forest management practices also have made the forests more vulnerable to catastrophic fires. In New Mexico, even-aged, second-growth forests were hit hardest because they are much denser than naturally occurring forest and consequently consume more water from the soil and increase the availability of dry above-ground fuel.

Looking to the future, the National Research Council (2011) projects that for every 1°C warming across the West there will be a 2- to 6-fold increase in area burned by wildfire. Potential impacts include: reduced provisioning of timber, large-scale terrestrial-atmospheric carbon fluxes, increased water scarcity, loss of homes, and increases in homeowner's insurance prices. Some of the adaptation solutions being considered include forest restoration activities such as non-commercial, mechanical thinning of small-diameter trees; controlled burns to reintroduce the low-severity ground fires that historically maintained forest health; and comprehensive ecological monitoring to determine effects of these treatments on forest and stream habitats, plants, animals, and soils.

LITERATURE CITED

Hoffman C, Parsons R, Morgan P, Mell R. 2010. Numerical simulation of crown fire hazard following bark beetle-caused mortality in lodgepole pine forests. *In* Wade DD, Robinson ML (Eds), Proceedings of 3rd Fire Behavior and Fuels Conference; 25-29 October 2010; Spokane, WA. Birmingham, AL: International Association of Wildland Fire. 1 p.

Jönsson AM, Appelberg G, Harding S, and Bärring L. 2009. Spatio-temporal impact of climate change on the activity and voltinism of the spruce bark beetle, *Ips typographus*. *Global Change Biology* **15**: 486–499.

NPS (National Park Service). 2011. Las Conchas Post-Fire Response Plan. Available at: http://www.nps.gov/fire/

NRC (National Research Council). 2011. Climate stabilization targets: Emissions, concentrations, and impacts over decades to millennia. The National Academies Press, Washington, DC: 286 p.

Schoennagel T, Veblen TT, Negron JF, and Smith JM. 2012. Effects of mountain pine beetle on fuels and expected fire behavior in lodgepole pine forests, Colorado, USA. *PLoS ONE* **7**(1): e30002.

Westerling AL, Hidalgo HG, Cayan DR, and Swetnam TW. 2006. Warming and earlier spring increase western U.S. forest wildfire activity. *Science* **313**(5789): 940-943.

Williams P, Meko D, Woodhouse C, Allen CD, Swetnam T, Macalady A, Griffin D, Rauscher S, Jiang X, Cook E, Grissino-Mayer H, McDowell N, and Cai M. 2011. 1,100 years of past, present, and future forest response to drought in the North American Southwest. American Geophysical Union Fall Meeting, December 5-9, 2011 San Francisco, CA. Poster presentation.

Impacts of Climate Change on Biodiversity, Ecosystems, and Ecosystem Services |
Technical Input to the 2013 National Climate Assessment`

Chapter 1
Introduction

Chapter 1. Introduction

Lead Authors: Michelle D. Staudinger, Nancy B. Grimm, Amanda Staudt, Shawn L. Carter, Peter Kareiva, Mary Ruckelshaus, F. Stuart Chapin III, Bruce A. Stein

This report assesses how climate change has affected biodiversity, ecosystems, and ecosystem services, the projected impacts during the coming century, and potential response options. It has been produced as part of the technical input process for the 2013 National Climate Assessment (NCA), with primary support from the U.S. Geological Survey. Drawing upon an extensive review of the available literature, the report focuses on advances in our understanding since about 2008.

The primary intended audience for this report is the NCA Development and Advisory committee (NCADAC) and the lead authors of the 2013 NCA report. In addition, we hope that this technical input report will be a useful resource for the community of scholars, resource managers, and decision makers who are concerned with safeguarding our nation's natural assets.

As documented in this report, climate change is already markedly altering biodiversity and ecosystems in the United States. These impacts are expected to increase as climate change continues during the coming decades. As natural resource managers grapple with the challenges posed by climate change, systematic efforts to assess our knowledge, such as this technical input and the overall NCA process, are essential for informing their decision making.

1.1. FEDERAL MANDATE FOR THE NATIONAL CLIMATE ASSESSMENT

The Global Change Research Act (GCRA) of 1990, Section 106, requires that an assessment be conducted not less frequently than every four years, which: 1) integrates, evaluates, and interprets findings of the United States Global Change Research Program (USGCRP) and discusses uncertainties; 2) analyzes the effects of global change on various sectors; and 3) analyzes current trends in global change and projects change for the future 25–100 years (GCRA 1990). The Act specifically calls for an analysis of global change impacts on the natural environment and biological diversity, among several other sectors. For the 2013 report, the NCADAC has recommended that the assessment of the natural environment and biodiversity also include consideration of ecosystem services.

Several sectors and cross-cutting themes impinge on the ecosystems and biodiversity sector, including water, forestry, agriculture, land use and cover change, urban infrastructure and vulnerability, coastal systems, and interactions of biogeochemical cycles and climate change. Many of these topics are being addressed by separate technical input teams.

1.2. BUILDING ON PAST ASSESSMENTS

This technical input builds on several previous assessment efforts. These include assessments that have focused on the United States—in particular the previous National Climate Assessments (NAST, 2001; USGCRP, 2009), the *State of the Nation's Ecosystems* reports (for example, Heinz Center, 2002), and the *Report to the President on Sustaining Environmental Capital: Protecting Society and the Economy* (PCAST, 2011)—as well as international assessment efforts, such as the *Millennium Ecosystem Assessment* (MA, 2005), the

Intergovernmental Panel on Climate Change reports (for example, IPCC, 2007), and the ongoing *Intergovernmental Platform on Biodiversity and Ecosystem Services* (IPBES) (Larigauderie and Mooney, 2010). Some of these past assessment efforts focused on ecosystems and biodiversity, and considered climate change as one of many factors affecting ecosystem health (for example, MA, 2005; PCAST, 2011). Others were focused on climate change, with the impacts on ecosystems or biodiversity being one of many different sectors addressed (for example, NAST, 2001; USGCRP, 2009; IPCC, 2007).

1.3. TECHNICAL INPUT PROCESS

In August 2011, a steering committee comprised of Federal agency, academic, and non-governmental organization participants was assembled to direct the development of this technical input.[1] The steering committee developed an outline for the report and a strategy for soliciting input from the broader expert community. Approximately 60 scholars were invited to contribute to writing the report, participate in a series of conference calls and webinars, and attend a 3-day workshop. The steering committee held one in-person meeting, on December 12-14, 2011, to begin synthesizing the materials gathered to date, discuss the draft findings and potential cross-cutting topics, and conduct additional planning.

The contributors to this report (full list and affiliations in Appendix A) represented diverse expertise and perspectives, and included scholars engaged in relevant research, and expert stakeholders involved in developing response strategies to address the impacts of climate change on ecosystems. Specifically, the workshop included 25 participants from academic institutions, 20 from Federal agencies (USGS, NOAA, EPA, NSF, NASA, USDA, NPS, USFS, and FWS), 13 from non-governmental organizations, and 2 from State fish and wildlife agencies. These invited participants were encouraged to reach out to other colleagues to fill any gaps in expertise. Additional contributors are recognized in the author lists and acknowledgments for each chapter.

Four working groups were established in November 2011 to lead the authorship of the chapters on biodiversity, ecosystems, ecosystem services, and other environmental stressors. These working groups held multiple conference calls and developed preliminary drafts in advance of the January workshop. They started this process by identifying the most important contributions to the literature since the last National Climate Assessment, which was published in 2009. Then, they developed draft key findings and chapter outlines to specifically highlight these recent advances. A fifth working group was created at the workshop to develop the adaptation chapter, which similarly focuses on advances in scholarship and practice during the last few years.

The plenary workshop was held on January 17-19, 2012 at the Gordon and Betty Moore Foundation in Palo Alto, California. The objectives of the workshop were to 1) continue developing content for individual chapters, 2) identify ways to integrate across the chapters, 3) strengthen and expand key findings, 4) determine the level of confidence in the key findings of the report, and 5) discuss how best to sustain the assessment process towards the 2017 NCA. Several presentations were made during plenary sessions, primarily to help guide participants to develop findings that would be useful for the continuing NCA process, inform policy and resource management decisions, and build on best practices from past assessments (See Appendix B for the workshop agenda). During break-out sessions, workgroups used guidance

---

[1] Steering committee members are listed as the authors of this chapter.

Impacts of Climate Change on Biodiversity, Ecosystems, and Ecosystem Services |
Technical Input to the 2013 National Climate Assessment`
Chapter 1
Introduction

provided by the NCADAC on how to characterize and communicate certainty in key findings; levels of confidence used by author teams were based on the quality of evidence, and the level of agreement among experts with relevant knowledge and experience (Moss and Yohe, 2011). Confidence ratings and probabilistic estimates of uncertainty used to craft the key findings of the overall report and within individual chapters are presented in **Table 1.1**. Furthermore, key uncertainties as well as critical gaps in research, knowledge, and data identified by the authors of this report are summarized at the end of each chapter.

Table 1.1. *Metrics used to assess and communicate confidence levels and uncertainties in key findings (adapted from Moss and Yohe, 2011).*

Confidence level	Factors used to evaluate confidence ratings
High	Strong evidence (established theory, multiple sources, consistent results, well documented and accepted methods); high consensus
Moderate	Moderate evidence (several sources, some consistency, methods vary and/or documentation limited); medium consensus
Fair	Fair evidence (a few sources, limited consistency, models incomplete, methods emerging); competing schools of thought
Low	Inconclusive evidence (limited sources, extrapolations, inconsistent findings, poor documentation and/or methods not tested); disagreement or lack of opinions among experts
Subjective likelihood level	**Corresponding range of probability events**
Very likely	Greater than 9 in 10
Likely	Greater than 2 in 3
As likely as not	Approximately 1 in 2
Unlikely	Less than 1 in 3
Very unlikely	Less than 1 in 10

1.4. REPORT APPROACH AND ORGANIZATION

This report examines climate change effects on ecological resources from three perspectives: (1) the effects on biodiversity with a focus on genes, species, and assemblages of species (*Chapter 2*); (2) the effects on ecosystem structure and functioning (*Chapter 3*); and (3) the effects on ecosystem services (*Chapter 4*). It is important to note that there are not distinct boundaries between biodiversity, ecosystem structure and functioning, and ecosystem services. Climate change impacts on ecosystem structure and functioning are experienced by people as changes in ecosystem services. Changes to biodiversity in the form of species loss or homogenization can be seen by the everyday observer and often have impacts on ecosystem services.

This technical input also examines two cross-cutting topics. First, we recognize that climate change is happening against the backdrop of a wide range of other environmental stressors, many of which have caused dramatic ecosystem degradation already. *Chapter 5* examines the key environmental stressors that interact with climate change and that complicate our abilities to predict and respond to climate change. The second cross-cutting topic is the rapidly advancing field of climate adaptation. *Chapter 6* reviews the significant progress in developing the conceptual framework, planning approaches, and strategies for safeguarding biodiversity and other ecological resources as the climate changes. It also addresses how ecosystem-based adaptation is becoming more prominent as a way to utilize ecosystem services to help human systems adapt to climate change. We conclude with a short discussion of ways to expand and sustain assessment activities addressing biodiversity, ecosystems, and ecosystem services in *Chapter 7*.

1.5. LITERATURE CITED

GCRA (Global Change Research Act). 1990. Public Law 101-606(11/16/90) 104 Stat. 3096-3104. Available at: http://www.gcrio.org/gcact1990.html

Heinz Center (The H. John Heinz III Center for Science, Economics and the Environment). 2002. The State of the Nation's Ecosystems: Measuring the Lands, Waters, and Living Resources of the United States: Cambridge University Press.

IPCC (Intergovernmental Panel on Climate Change). 2007. Climate Change 2007: The Physical Science Basis. Contribution of Working Group I to the Fourth Assessment Report of the Intergovernmental Panel on Climate Change, Cambridge, UK.

Larigauderie A, and Mooney HA. 2010. The Intergovernmental science-policy platform on biodiversity and ecosystem services: moving a step closer to an IPCC-like mechanism for biodiversity. *Current Opinion in Environmental Sustainability* **2**: 9-14.

MA (Millennium Ecosystem Assessment). 2005. Ecosystems and Human Well-being: Biodiversity Synthesis. World Resources Institute, Washington, D.C.

Moss RH, and Yohe G. 2011. Assessing and Communicating Confidence Levels and Uncertainties in the Main Conclusions of the NCA 2013 Report: Guidance for Authors and Contributors. National Climate Assessment Development and Advisory Committee (NCADAC). Available at: http://www.globalchange.gov/images/NCA/Draft-Uncertainty-Guidance_2011-11-9.pdf

NAST (National Assessment Synthesis Team). 2001. Climate Change Impacts on the United States: The Potential Consequences of Climate Variability and Change, Report for the U.S. Global Change Research Program. Cambridge University Press, Cambridge, UK.

PCAST (President's Council of Advisors on Science and Technology). 2011. Sustaining environmental capital: protecting society and the economy. 145 p. Available at: http://www.whitehouse.gov/administration/eop/ostp/pcast/docsreports

USGCRP (U. S. Global Change Research Program). 2009. Global Climate Change Impacts in the United States. Cambridge University Press, Cambridge, UK.

Chapter 2. Impacts of Climate Change on Biodiversity

Convening Lead Authors: Michelle D. Staudinger, and Shawn L. Carter
Lead Authors: Molly S. Cross, Natalie S. Dubois, J. Emmett Duffy, Carolyn Enquist, Roger Griffis, Jessica Hellmann, Josh Lawler, John O'Leary, Scott A. Morrison, Lesley Sneddon, Bruce A. Stein, Laura Thompson, Woody Turner
Contributing Authors: Elda Varela-Acevedo, Walt Reid

Key Findings

- Climate change is causing many species to shift their geographical ranges, distributions, and phenologies at faster rates than were previously thought; however, these rates are not uniform across species.

- Increasing evidence suggests that range shifts and novel climates are very likely to result in new community assemblages, new associations among species, and promote interactions that have not existed in the past.

- Differences in how organisms respond to climate change determine which species or populations will benefit (winners), and which will decline and possibly go extinct (losers) in response to climate change.

- The potential for biodiversity to respond to climate change over short (for example, plasticity) and long (for example, evolutionary) time scales is enhanced by increased genetic diversity; however, the rate of climate change may outpace species' capacity to adjust to environmental change.

- Identifying highly vulnerable species and understanding why they are vulnerable are critical to developing climate change adaptation strategies and reducing biodiversity loss in the coming decades.

- As species shift in space and time in response to climate change, effective management and conservation decisions require consideration of uncertain future projections as well as historic conditions.

- Broader and more coordinated monitoring efforts across Federal and State agencies are necessary to support biodiversity research, management, assessment, and policy.

2.1. INTRODUCTION
2.1.1. What is biodiversity and why does it matter?

More than two decades ago, E.O. Wilson (1988) warned that global *biodiversity*, defined as the variation of all life on earth and the ecological complexes in which they occur (Leadley and others, 2010), faced an unprecedented threat from habitat loss and other anthropogenic stressors. Thus humankind was in a race to describe, classify, and preserve global biodiversity before it was lost forever through extinction. Since then, advances in scientific research and technology have greatly improved our knowledge of the vast array of animals, plants, fungi, invertebrates, and microorganisms that comprise the earth's ecosystems; yet threats to biodiversity resulting from a suite of human activities including habitat loss and degradation, introduction of non-native species, overexploitation, pollution, and disease have only accelerated since Wilson's call to arms (Williams, 1989; Flather and others, 1997, Wilcove and others, 1998; Purvis and others, 2000; Butchart and others, 2010; Leadley and others, 2010).

Biodiversity is fundamental to ecosystem structure and function, and underpins the broad spectrum of goods and services that humans derive from natural systems (Chapin and others, 1997; Walther and others, 2002; MA, 2005, Naeem, 2009; Mace and others, 2012). Declines or loss of any aspect of biodiversity can have direct or indirect impacts on ecosystem function, persistence, and services (Hooper and others, 2005). Keystone or foundation species play a central role in ecosystems, either through trophic processes (for example, as dominant primary producers; major predators or prey), by providing structure (for example, habitat forming), or as ecological engineers (for example, by moderating the availability of resources to other species). Many such species also provide beneficial services to humans in the form of food (for example, fisheries), storm and flood protection (for example, mangroves), and/or maintenance of water quality (MA, 2005; Leadley and others, 2010). However, in many cases there is limited understanding of the functional or interactive role a species or group plays in a system, which in turn limits our ability to predict how the system will respond to changes in climate and other anthropogenic stressors, and ultimately affect the societal benefits they support.

2.1.2. Threats to biodiversity

Climate change is having widespread impacts across multiple scales of biodiversity including genes, species, communities, and ecosystems (Parmesan, 2006; Bellard and others, 2012). Biological responses to climate change vary widely among species and populations; some responses are positive, leading to increased growth rates or range expansions, while others are negative, resulting in localized or widespread declines (Miller-Rushing and others, 2010; Montoya and Raffaelli, 2010; Dawson and others, 2011; Geyer and others, 2011). Many species have already shifted their geographical ranges, generally poleward, towards higher elevations, or to deeper depths in marine environments (Nye and others, 2010; Burrows and others, 2011; Chen and others, 2011; Doney and others, 2012). Species have altered the temporal patterns of seasonal migrations and other life cycle events (phenology), showed changes in population demographics (Doak and Morris, 2010; Miller-Rushing and others, 2010; Dawson and others, 2011), or in some cases are adapting in place to the new environmental conditions (Bellard and others, 2012). These shifts will likely bring about new assemblages of species (Williams and Jackson, 2007), cause novel interspecific interactions (Suttle and others, 2007), and in worst case scenarios result in some extinctions (Butchart and others, 2010; Barnosky and others, 2011) (**Box 2.1**).

Although there are currently few direct examples of climate-induced extinctions (Monzón and others (2011) lists 19 known species extinctions due to climate change), current trends in the velocity and magnitude of climate change will likely exceed many species' abilities to adjust to new environmental conditions thus leading to increased extinction rates (Loarie and others, 2009; Bellard and others, 2012). Estimations of extinction rates are uncertain and expert opinion differs as to what the magnitude of loss will be (He and Hubbell, 2011). Predictions are complicated in part due to the great deal of uncertainty regarding the number of species that exist on earth (May, 2011; Mora and others, 2011). Approximately 1.4 million species have been catalogued to date, and the most recent estimations for the total number of species on earth (described and unknown) is about 8.7 million species (Mora and others, 2011). All estimates have some degree of uncertainty, yet the common conclusion among studies is that the majority of life on earth has yet to be catalogued (Erwin, 1982, 1991; Bouchet, 2006; May, 1988, 2011).

Impacts of Climate Change on Biodiversity, Ecosystems, and Ecosystem Services |
Technical Input to the 2013 National Climate Assessment

Chapter 2
Biodiversity

**Box 2.1. Examples of Observed and Projected Biological Responses to
Climate Change across the United States**

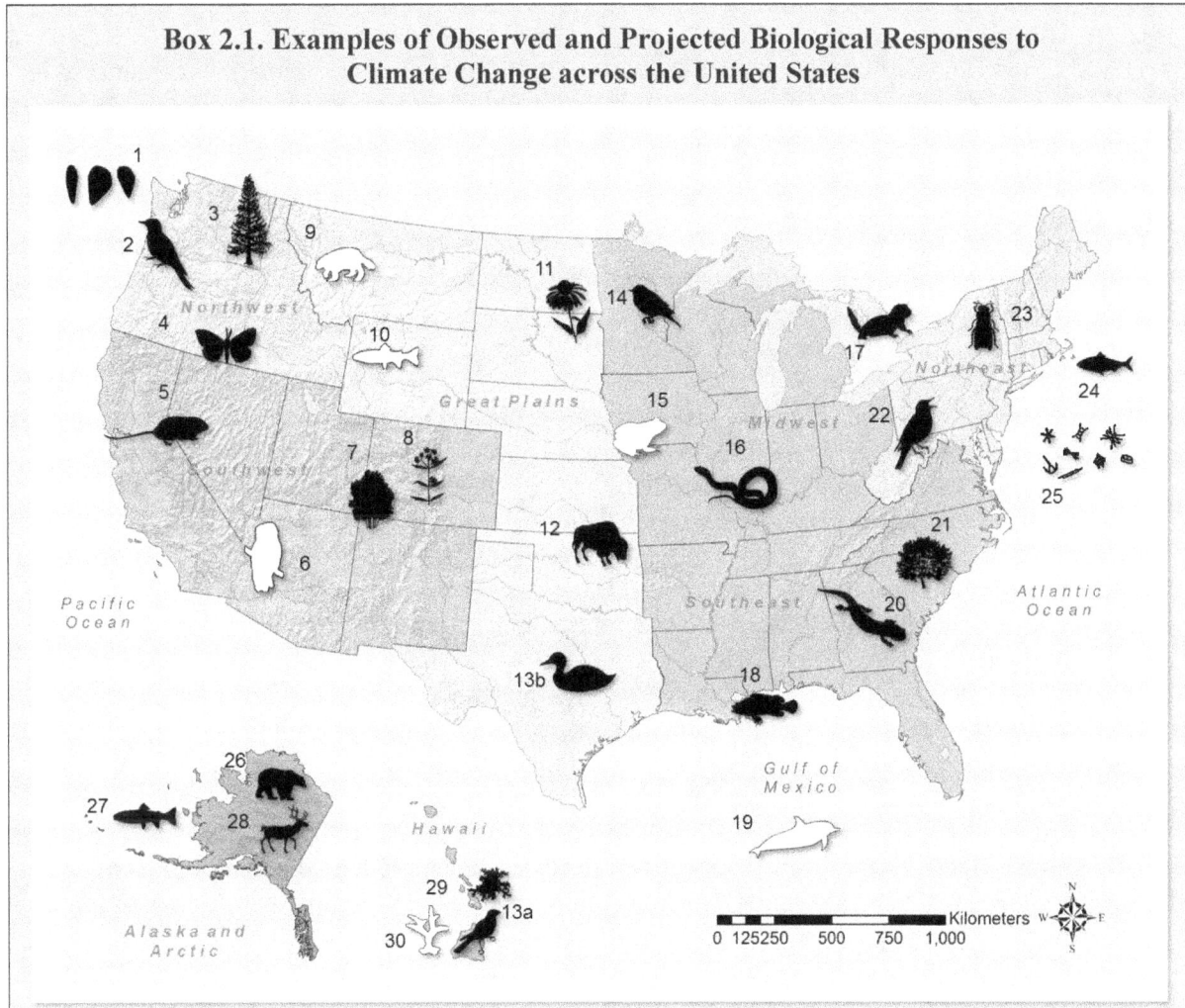

Figure 2.1. *Map of observed and projected biological responses to climate change across the
United States. Case studies listed below correspond to observed (black icons on map) and
projected (white icons on map; italicized statements (below)) responses.*

1. Time series data and experiments demonstrated that mussel and barnacle beds have declined
   or disappeared along northwest coast rocky shores (the Strait of Juan de Fuca) due to
   warming temperatures; these hotter, drier conditions have compressed the habitable
   intertidal space and decreased or eliminated refuge from predators (Harley, 2011).
2. In a 12 year study of Northern Flickers (*Colaptes auratus*) in the Pacific Northwest, birds
   arrived at breeding sites earlier when temperatures along their migration routes were
   warmer, and temperatures at breeding sites correlated significantly with initiation of egg
   laying (Wiebe and Gerstmar, 2010).
3. Climate-induced changes in pests and pathogens have been blamed for recent mortality events
   in conifer forests across western North America including the Pacific Northwest, California,
   and the Rocky Mountains (van Mantgem and others, 2009).

Box 2.1, continued.

4. Larval growth and survival of an oak specialist butterfly (*Erynnis propertius*) was examined in Oregon and California to determine if the insect could change tree hosts (to different *Quercus* sp.) after moving to a new area under climate change; findings suggest local adaptation of butterflies to specific oak species often precluded their populations from colonizing new areas under climate change (Pelini and others, 2010).

5. Of 28 small mammal species monitored during the past century, 50 percent showed substantial upward changes in elevation limits (~500 meters); formerly low-elevation species expanded their ranges, while high-elevation species contracted theirs. These shifts resulted in major changes in community composition at mid- and high-elevations in Yosemite National Park (Moritz and others, 2008).

6. *Population dynamics and extinction risk were projected for 3 populations of northern spotted owls (Strix occidentalis) in the southwestern United States. Owl populations in Arizona and New Mexico are projected to decline during the next century and are at high risk for extinction due to future climatic changes, while the southern California population is projected to be insensitive to future climatic changes* (Peery and others, 2012).

7. Quaking aspen-dominated systems (*Populus tremuloides*) are experiencing declines in the western United States after stress due to climate-induced drought conditions during the last decade (Anderegg and others, 2012).

8. Warmer and drier conditions during the early growing season in high elevation habitats in Colorado are disrupting flowering phenology across meadow habitats, and resulting in mid-season declines in flower resources that may affect pollinator populations, particularly with limited foraging ranges (Aldridge and others, 2011). Earlier spring growth in high altitude perennial plant species have made flower buds more susceptible to late season frost events that hinder flowering or result in increased mortality (Inouye, 2008). In addition, meadow plants in the west Elk Mountains responded more readily to springtime warming than some insects thus increasing the potential for decoupling important plant-pollinator relationships (Forrest and Thomson, 2011).

9. *Population fragmentation of wolverines (Gulo gulo luscus) in the northern Cascades and Rocky Mountains is expected to increase as spring snow cover retreats (33 percent to 63 percent) over the coming century* (McKelvey and others, 2011).

10. *Cutthroat trout (Oncorhynchus clarkii) populations are forecasted to decline by up to 58 percent due to increasing temperatures, seasonal shifts in precipitation, and biotic interactions with non-native species in the western United States (Columbia, and Colorado River Basins; Yellowstone and Lahontan lineages* (Wenger and others, 2011).

11. Comparisons of historical (1910-1961) and recent (2007-2010) first flowering dates (FFD) in 178 plant species from North Dakota revealed significant shifts have occurred in over 40 percent of all species examined; the greatest changes were observed during the two warmest years of the study (Dunnell and Travers, 2011).

12. Variation in the timing and magnitude of precipitation was found to impact weight gain of bison (*Bison bison*) in the Konza Prairie in Kansas and the Tallgrass Prairie Preserve in Oklahoma. Late-summer precipitation was related to increased weight gain, whereas midsummer precipitation was related to decreased weight gain due to reduced nutritional quality of grazing biomass (Craine and others, 2009).

Box 2.1, continued.

13. Increased environmental variation has been shown to influence mate selection and increase the probability of infidelity in birds that are normally socially monogamous to increase the gene exchange and the likelihood of offspring survival. Examples include infidelity in palilas (*Loxioides bailleui*) in Hawaii, and avian divorce in black-bellied whistling ducks (*Dendrocygna autumnalis*) in Texas (Botero and Rubenstein, 2012).

14. Of 44 species of migratory birds monitored in Minnesota over a 40 year period, 36 percent showed significantly earlier arrival dates, particularly in short-distance migrants, due to increasing winter temperatures (Swanson and Palmer, 2009).

15. *The northern leopard frog (Rana pipiens) is projected to experience poleward and elevational range shifts in response to climatic changes in the latter quarter of the century* (Lawler and others, 2010).

16. Seasonal activity patterns were similar among black ratsnake (*Elaphe obsoleta*) populations found at different latitudes in Canada, Illinois, and Texas. Although climate conditions differed among sites, snake activity was not found to vary as a function of temperature; findings suggest that snake populations, particularly in the northern part of their range, should be able to adjust and possibly benefit from warming temperatures if there are no negative impacts on their habitat and prey (Sperry and others, 2010).

17. Climate-induced hybridization was detected between southern and northern flying squirrels (*Glaucomys volans* and *Glaucomys sabrinus*, respectively) in the Great Lakes region of Ontario Canada, and Pennsylvania as a result of increased sympatry after a series of warm winters (Garroway and others, 2010).

18. Increased temperatures are believed to have caused northward shifts of warm-water fishes and additions of tropical and subtropical fishes (for example, butterfly (*Chaetodon* sp.) and surgeon (*Acanthurus* sp.) fishes) to temperate seagrass meadow assemblages in the northern Gulf of Mexico (Fodrie and others, 2010); similar shifts and invasions have been documented in Long Island Sound and Narragansett Bay in the Northeast Atlantic (Wood and others, 2009).

19. *Global marine mammal diversity is projected to decline by as many as 11 species by mid-century, particularly in coastal habitats, due to climatic change* (Kaschner and others, 2011).

20. Warmer night-time temperatures and cumulative seasonal rainfalls were correlated with changes in the arrival times of amphibians to wetland breeding sites in South Carolina over a 30 year time period (1978-2008). Autumn-breeding amphibians arrived later (for example, up to 76 days in the dwarf salamander (*Eurycea quadridigitata*)), while winter-breeding species arriving earlier. Overall rates of change ranged from 5.9 – 37.2 days/decade, and are representative of some of the fastest rates of phenological change observed to date (Todd and others, 2011).

21. Seedling survival for nearly 20 species of trees decreased during years of lower rainfall in the Southern Appalachians and the Piedmont areas (Ibáñez and others, 2008).

22. Widespread declines in body size of resident and migrant birds at a bird-banding station in western Pennsylvania were documented over a 40-year period; body sizes of breeding adults were negatively correlated with mean regional temperatures from the preceding year (Van Buskirk and others, 2010).

Box 2.1, continued.

23. Over the last 130 years (1880-2010), native bees have advanced their spring arrival in the northeastern United States by an average of 10 days, primarily due to increased warming. Plants have also showed a trend of earlier blooming thus helping preserve the synchrony in timing between plants and pollinators (Bartomeus and others, 2011).

24. In the Northwest Atlantic, 24 out of 36 commercially exploited fish stocks showed significant range (latitudinal and depth) shifts between 1968–2007 in response to increased sea surface and bottom temperatures (Nye and others, 2009).

25. Increases in maximum and decreases in the annual variability of sea surface temperatures in the North Atlantic Ocean have promoted growth of small phytoplankton and led to a reorganization in the species composition of primary (phytoplankton) and secondary (zooplankton) producers (Beaugrand and others, 2010)

26. Changes in female polar bear (*Ursus maritimus*) reproductive success (decreased litter mass, and numbers of yearlings) along the north Alaska coast have been linked to changes in body size and/or body condition following years with lower availability of optimal sea ice habitat (Rode and others, 2010).

27. Water temperature data and observations of migration behaviors over a 34 year time period showed that adult pink salmon (*Oncorhynchus gorbuscha*) migrated earlier into Alaskan creeks, and fry advanced the timing of migration out to sea. Shifts in migration timing may increase the potential for a mismatch in optimal environmental conditions for early life stages, and continued warming trends will likely increase pre-spawning and egg mortality rates (Taylor, 2008).

28. Warmer springs in Alaska have caused earlier onset of plant emergence, and decreased spatial variation in growth and availability of forage to breeding caribou (*Rangifer tarandus*). This trophic and spatial asynchrony ultimately reduced calving success in caribou populations (Post and others, 2008).

29. Hawaiian mountain vegetation including subalpine and alpine shrubland, montane cloud forest, and epiphytic communities (lichens and bryophytes), were found to vary in their sensitivity to changes in moisture availability; consequently, climate change will likely influence vegetation patterns (distribution, assemblage, and turnover) along east-west and elevation gradients in this region (Crausbay and Hotchkiss, 2010).

30. *A 0.5-1.0 meter sea-level rise in Hawaiian waters is projected to increase wave heights, the duration of turbidity, and the amount of re-suspended sediment in the water; consequently, this will change the amount of light available for photosynthesis and create potentially stressful conditions for coral reef communities* (Storlazzi and others, 2011).

The International Union for Conservation of Nature (IUCN) Red List of Threatened Species provides one of the most comprehensive evaluations of the conservation status of global populations of plants and animals. Of the 4,161 terrestrial and aquatic species currently recognized by the IUCN as being threatened by climate change, 33 percent are at risk from habitat shifts and alteration due to climate change, 29 percent due to temperature extremes, and 28 percent due to drought (**Figure 2.2**). Evaluating species at risk due to climate change is a relatively new effort by the IUCN, and as is true for other assessments, it is difficult to disentangle the impacts of climate change from other anthropogenic stressors for a range of species. Consequently, evaluations may provide insight into the relative distribution of climate threats to global biodiversity, but may not fully capture the cumulative impacts and synergistic interactions with other anthropogenic stressors (for example, habitat loss) (IUCN, 2010).

In many cases, these other stressors (See *Chapter 5*: *Multiple stressors*) currently act as the primary drivers of biodiversity loss (Jetz and others, 2007), and systems that are already stressed from human activities are likely more sensitive to the impacts of climate change. In addition, biodiversity loss is expected to accelerate as climate change interacts with and exacerbates the impacts of other anthropogenic stressors (Brook and others, 2008; Walther and others, 2009; Hoffmann and Sgró, 2011; Bellard and others, 2012). Consequently, conservation and management efforts should take a multidimensional approach that simultaneously addresses climate change and other anthropogenic impacts on biodiversity now and over the long-term (Game and others, 2010; Glick and others, 2011).

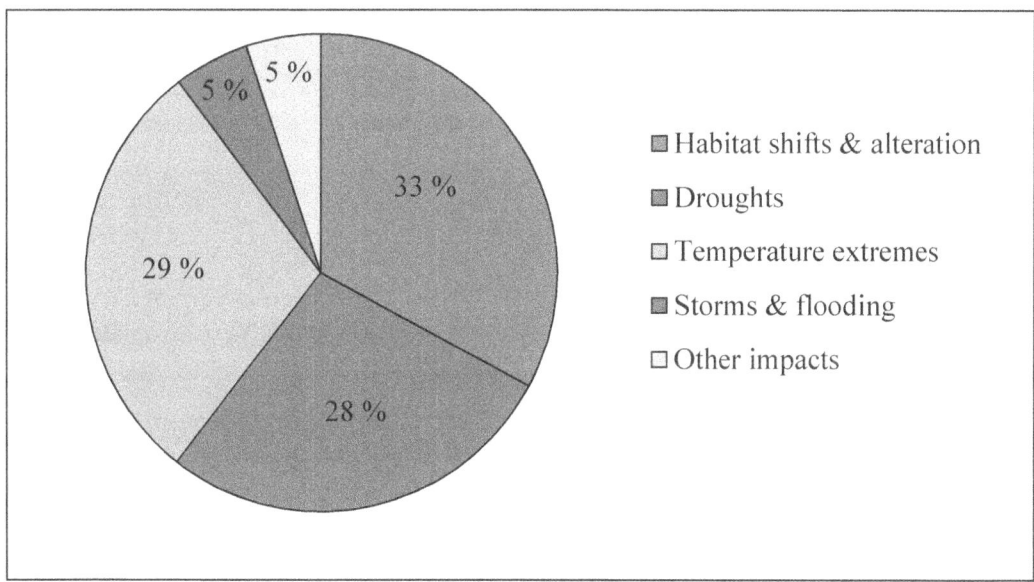

Figure 2.2. *Distribution of climate related threats evaluated by the IUCN on 4,161 terrestrial and aquatic species globally. Note there may be overlaps in species evaluated among categories (IUCN 2010).*

2.1.3. Biodiversity at risk in the United States

Knowledge of different taxonomic groups and their conservation status varies widely based on body size, distribution, perceived economic importance, and the environments in which they live (Fautin and others, 2010). Vertebrate animals and vascular plants are some of the best described taxonomic groups, while invertebrates, microorganisms, and fungi are some of the least described biota in the United States and globally. Recent advances in molecular techniques are improving our ability to detect and describe organisms, and some of the greatest advances in the past decade have been in marine environments (Heidelberg and others, 2010).

NatureServe and its network of State agency-based natural heritage programs have assessed the conservation status of nearly 25,000 of the approximately 205,000 species of plants, vertebrates and invertebrates (not including microorganisms) in the United States (**Figure 2.3**).

Of these, 20 percent are ranked in categories of high extinction risk (ranks of presumed extinct [GX], possibly extinct [GH], critically imperiled [G1], and imperiled [G2]), including 6 percent of butterflies, 7 percent of mammals, and up to 61 percent of freshwater snails (Wilcove and Master, 2005). In addition, the U.S. Fish and Wildlife Service has formally listed 1,387 species of plants and animals as endangered or threatened according to the criteria in the Endangered Species Act (USFWS, 2012). A consistent finding is that most aquatic animals, particularly amphibians, and freshwater mollusks, are more at-risk than terrestrial groups (Stein and others, 2000; Wilcove and Master, 2005; Heinz Center, 2008; NatureServe, 2011). Extinction risks across marine biota are more uncertain than on land, and evaluations of marine animals have often been skewed towards large vertebrates, leaving the status and protection of many marine invertebrates deficient in comparison (McClenachan and others, 2012). Marine taxonomic groups known to be most at risk include reef-building corals, sea turtles, marine mammals, sharks, and apex predatory finfishes such as tunas, and billfishes (Roberts and Hawkins, 1999; Wallace and others, 2011, Collette and others, 2011).

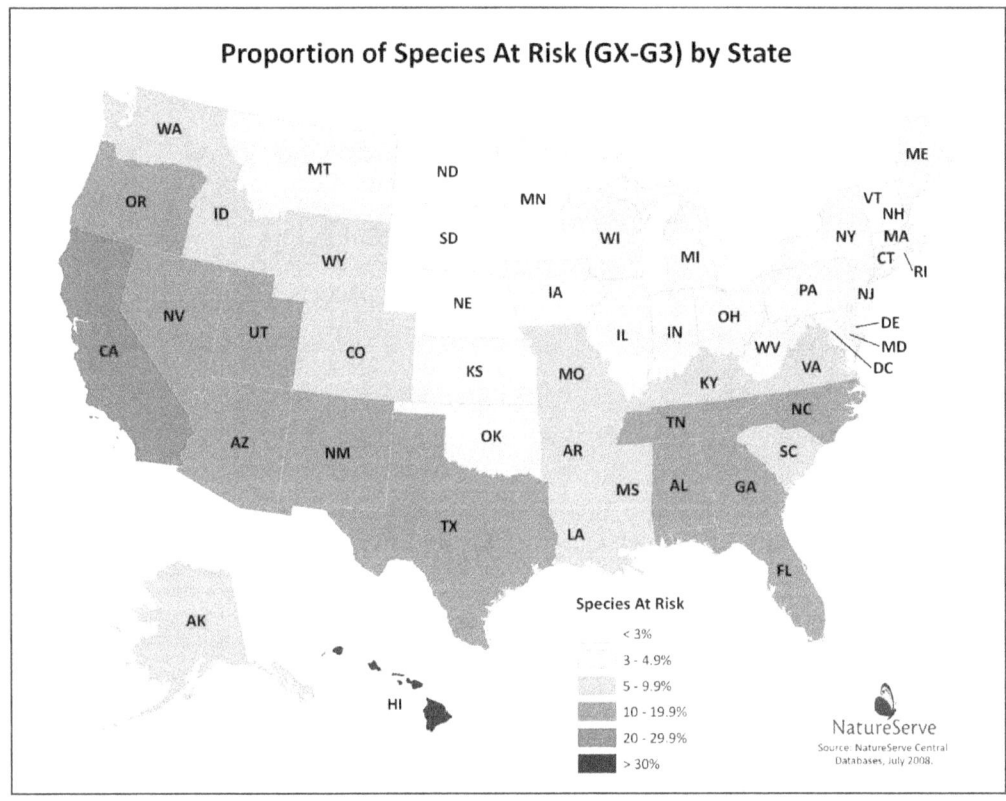

Figure 2.3. *Terrestrial and aquatic species at risk by State in the United States. NatureServe ranks species conservation status as presumed extinct [GX], possibly extinct [GH], critically imperiled [G1], and imperiled [G2], Vulnerable [G3], Apparently Secure [G4], Secure [G5], Unrankable [GU], and Not Yet Ranked [GNR]. Figure was prepared by NatureServe, and values reflect over three decades of data collection and assessment up to 2008.*

Patterns of biodiversity are not uniform across the United States nor globally (**Figure 2.4**). Areas of exceptionally high levels of endemic species, known as "biodiversity hotspots" are contained within a relatively small proportion of the earth's surface (Myers, 1988, 1991, 2000; Tittensor and others, 2010). Some of the best known examples include tropical rain forests, coral reefs, and the Hawaiian Islands (Myers and others, 2000). Protected areas have been the principal approach to providing refuge to biodiversity from previous anthropogenic threats (for example, land use change, exploitation), and have achieved some success. However, fixed boundary preserves will be ineffective against climate change as species move into unprotected areas to follow favorable climate conditions (Monzón and others, 2011). Species that have restricted ranges are particularly vulnerable to climate change since the environmental conditions and vegetation types that have supported these unique complexes may not be suitable or exist in the coming decades (Malcolm and others 2006). Future conservation planning will need to evaluate not only where species currently reside in the landscape, but also where they are most likely to move and persist. Conservation in a changing environment will require creative and flexible real-time planning that maintains environmental heterogeneity, and increases protected area coverage and connectivity as species responses are realized (Monzón and others, 2011).

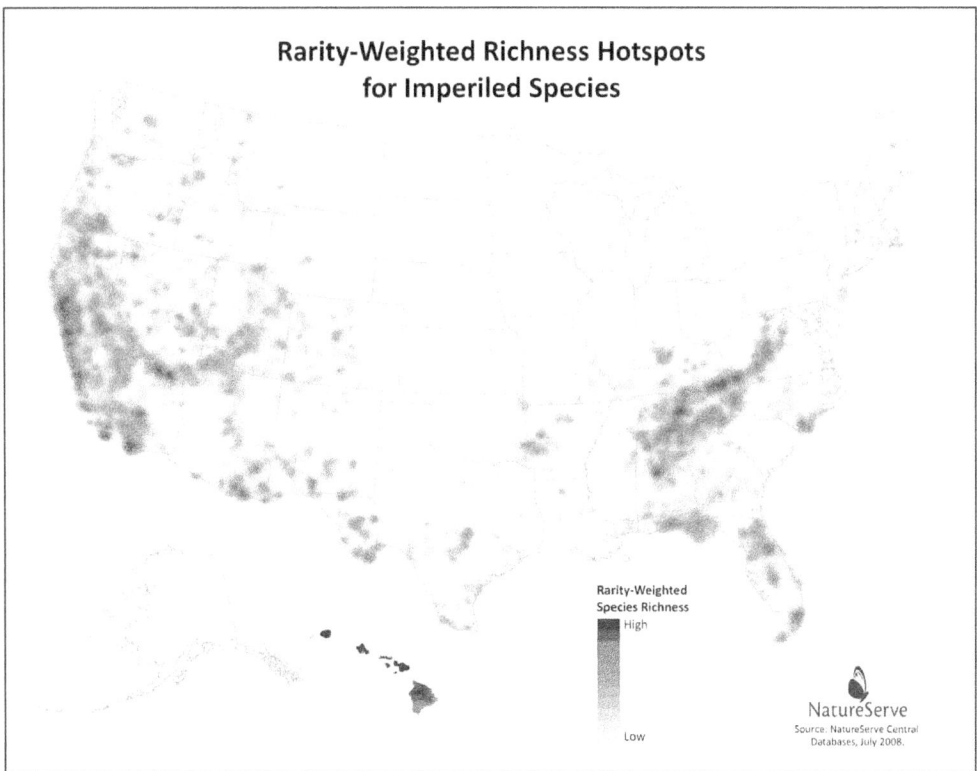

Figure 2.4. Aquatic and terrestrial (marine species not shown) biodiversity hotspots across the United States. Figure was originally published by NatureServe on LandScope America (www.landscope.org), and reflects data collected from 1970 - 2008. Analysis excludes extirpated or historic populations.

2.1.4. Objectives

In this chapter we provide an assessment and synthesis of recent advances in our knowledge of the major biological responses to key climate variables across multiple levels of biodiversity. Because many of the studies we reviewed were conducted on individual species or species groups, our framework and conclusions are largely focused on this component of biodiversity; nonetheless, our findings are broadly applicable across other elements of biodiversity (for example, genetic and ecosystem diversity). Examples are presented from a range of systems and taxonomic groups representative of biodiversity within the United States and globally. In addition, case studies of specific biota or habitats are presented to highlight the range of known possible outcomes including species that may benefit (that is, winners) and those that may decline (i.e. losers) in response to climate change. The primary goal of this report is to provide guidance to natural resource managers and decision makers on critical issues related to the impacts of climate change on all components of biodiversity; to support this mission we have organized our findings using the following four policy relevant questions:

1. Can we currently observe the impact of recent climate change on biodiversity?
2. How will climate change affect biodiversity in the coming century?
3. What impacts of climate change on biodiversity are likely to pose particularly high costs or risks?
4. What human response strategies could address the most harmful impacts of climate change on biodiversity, and what barriers and opportunities exist to their implementation?

The impacts of climate change on ecosystems and processes (*Chapter 3*), ecosystem services (*Chapter 4*), as well as evaluations of climate interactions with other anthropogenic stressors (*Chapter 5*), and how humans are responding to climate change through adaptation (*Chapter 6*) are addressed in forthcoming chapters of this technical input.

2.2. OBSERVED IMPACTS OF RECENT CLIMATE CHANGE ON BIODIVERSITY

There is unequivocal evidence that climate change is occurring and having impacts on biodiversity (IPCC, 2007). The combined impacts of climate change and other anthropogenic stressors are believed to be responsible for extinction rates that are 2-5 orders of magnitude above historical rates, and are leading to the sixth mass extinction of global biodiversity (Pimm and others, 1995; Dirzo and Raven, 2003; Pimm, 2008; Butchart and others, 2010; Pereira and others, 2010; Barnosky and others, 2011). What makes the current rates of decline different from historical events is the prominent role human activities (for example, exploitation, habitat degradation) are playing relative to natural agents of change (Vitousek and others 1997; Sala and others, 2000; deBuys, 2008; Barnosky and others, 2011). Attribution of climate change specifically to human activities (for example, greenhouse gas emissions) has been advocated in other assessments, and is important to mitigating anthropogenic impacts in the future (Rozenzweig and others, 2008). Regardless of the source of climate change, biological systems respond to fluctuations in local climate conditions, and do not partition their responses to anthropogenic and natural climate variation (Parmesan and others, 2011).

Climate change is altering the abiotic conditions that influence biological systems and processes (**Table 2.1**); biological responses to climate change depend on a number of factors, including the rate, magnitude, and character of the change, ecological sensitivity, and adaptive capacity to environmental change. The combination of these factors is affecting all levels of biodiversity, such that the distribution, organization, and interactions among biota are shifting over spatial and temporal scales (Walther, 2010).

Table 2.1. Overview of physical changes associated with climate change and examples of the potential ecological consequences associated with these changes.

Observed or projected physical change	Examples of potential impacts on biodiversity
Increased ambient temperature	Species and population range shifts and/or changes in phenology leading to alteration or loss of biotic interactions
Changes in annual and seasonal precipitation	Changes in community composition
Increased frequency of extreme events	Mortality resulting from flooding after storms or drought events; damage or mortality resulting from deep freezes or heat waves
Changes to hydrologic regimes	Reduced stream flow affecting population persistence and community composition
Changes to fire regimes	Changes in community composition
Ocean acidification	Change in water chemistry affecting calcification rates of marine organisms
Sea level rise	Habitat loss and fragmentation affecting population persistence
Increases in ocean stratification	Reduced productivity of pelagic ecosystems
Changes in coastal upwelling	Changes in productivity of coastal ecosystems and fisheries

2.2.1. Genetic diversity, traits, and phenotypic plasticity

Genetic variation is fundamental to biodiversity and helps populations respond to changes in environmental conditions across short and long-term time scales. As the rate and magnitude of projected climate change continue to increase and exert selective pressure on populations, natural selection will favor genes that increase species survival in new environments, and may lead to the decline of genes that were dominant under previous conditions (Hoffmann and Sgró, 2011). Evolutionary responses to climate change are less likely when genetic diversity is absent or beneficial alleles occur at a low frequency within a population (Lynch and Lande, 1993; Bürger and Lynch, 1995; Hoffmann and Sgró, 2011). In those cases, evolution will depend on new genes arising from mutation or gene shuffling. Under strong selection pressure (for example, rapid climate change) such populations risk going extinct before beneficial genes have a chance to increase population fitness (Hellmann and Pineda-Krch, 2007). However, there is increasing evidence that rates of evolution can be rapid when genetic variation for differing environmental tolerances already exists in populations. For example, the annual plant, turnip mustard (*Brassica rapa*), was found to evolve in just a few generations to climate induced drought by shifting the timing of first flowering by up to 8 days earlier in the year compared to previous generations (Franks and others, 2007; Franks and Weis, 2008).

Our understanding of the capacity of species and populations to adapt to climate change remains one of the least understood aspects of climate science despite its importance in predicting persistence under future environmental conditions (Donelson and others, 2011). We are limited in part because observed evolutionary changes in response to modern climate change are relatively rare, even though analogs have been shown repeatedly under laboratory conditions (for example, Sørensen and others, 2009). It is difficult to determine if observed phenotypic changes are induced, plastic responses, or changes in the gene pool (Hoffman and Sgró, 2011). In addition, it is often unknown which genes (and combinations of genes) are responsible for allowing populations to adapt to the new conditions imposed by climate change. It is therefore of critical importance to maintain genetic diversity to prevent unintended losses of traits that can enhance population survival, and increase the potential for adaptive evolution (Lynch and Lande, 1993; Bürger and Lynch 1995; Hoffmann and Sgró, 2011; Schwartz and others, 2012).

Observed changes in phenotype in response to environmental change may be due to evolutionary adaptation (that is, genetic change across generations as discussed above), phenotypic plasticity (that is, direct influences of environment on phenotype within a generation) or a combination of both (Price and others, 2003). Although potentially more rapid, plasticity does not ensure that a population can persist under rapid rates of climate change. In some cases, the rate and magnitude of climate change will exceed the limits of phenotypic plasticity, at which point adaptive evolution or gene flow from other populations provide the only mechanisms for population persistence (Reed and others, 2011). Physiological responses are mediated by mechanisms involving biochemical adaptations, changes in membrane properties, and molecular adaptations such as shifts in gene expression in response to environmental fluctuations (Pörtner and Farrell, 2008; Fuller and others, 2010; Hofmann and Todgham, 2010). However, physiological responses are limited in some populations, leading to sub-lethal impacts on fitness by increasing energetic demands on individuals (for example, Gibbs, 2011; Kelly and others, 2012; Williams and others, 2012). Ectotherms in particular, are often sensitive to changes in temperature, which effects metabolic rates, and has been shown to influence life-history traits and foraging behavior (for example, Le Lann and others, 2011). Heat shock protein synthesis is protective against temperature stress (Feder and Hofmann, 1999), but varies widely among organisms from different thermal environments. For example, species that live in either very stable (for example, tropical) or highly variable thermal environments (for example, intertidal zones) are often more sensitive to climate change than those that live in more moderately variable thermal environments (Tomanek, 2010).

A growing number of studies have associated observed phenotypic changes in morphology, behavior, and other life history traits with changing environmental conditions. For example, increasing evidence supports a general trend towards smaller body sizes with increases in temperature (Daufresne and others, 2009; Van Burskirk and others, 2010; Gardner and others, 2011; Sheridan and Bickford, 2011), though notable exceptions exist (for example, Yom-Tov and Yom-Tov, 2005, 2008). Changes in morphology and physiology can influence a number of life history characteristics such as age at first reproduction, number and size of offspring, and reproductive lifespan. We are just beginning to understand the complex ways in which these underlying mechanisms are affecting individual fitness and population dynamics in response to climate change; there are numerous examples from a range of taxa demonstrating that biological responses to climate change vary widely with positive, negative or uncertain effects (Steigenga and Fischer, 2007; Massot and others, 2008; Forster and others, 2011; Le Lann and others, 2011; Olsson and others, 2011).

Shifts in precipitation and temperature can also affect food availability and indirectly lead to changes in body size. In Arctic environments, female polar bear (*Ursus maritimus*) reproductive success (decreased litter mass, and numbers of yearlings) has been linked to changes in body size and/or body condition following years with lower availability of optimal sea ice habitat (Rode and others, 2010). Alternatively, examination of museum specimens has shown that several small mammals including ground squirrels (*Callospermophilus lateralis, Urocitellus beldingi*) (Eastmann and others, 2012), masked shrews (*Sorex cinereus*) (Yom-Tov and Yom-Tov, 2005), and American martens (*Martes americana*) (Yom-Tov and others 2008) have increased in size over the past century possibly in response to milder winters, which could have improved food availability and lowered metabolic demands.

Demographic responses (for example, survival, population growth rates or recruitment) are often the result of local conditions operating directly or indirectly on behavior. As with other organismal-level responses to climate change, it is how these responses affect individual fitness and population dynamics (for example, age structure, sex ratio, and abundance) under climate change that will ultimately determine species viability (Tuomainen and Candolin, 2011). In a recent study conducted by Botero and Rubenstein (2012), increased environmental variation was shown to influence mate selection and increase the probability of infidelity in birds that are normally socially monogamous such as owls, warblers, gulls, and cranes. This mate-swapping behavior is believed to function to increase genetic diversity among offspring and increase the likelihood of offspring survival under variable environmental conditions.

2.2.2. Phenological shifts

Changes in phenology, or the seasonal timing of life events, have been observed in response to variations in temperature, precipitation, and photoperiod in terrestrial and aquatic habitats, as well as temperature-driven patterns in ocean currents in marine environments (Parmesan and Yohe, 2003; Root and others, 2003; Parmesan, 2007). Phenological events include changes in leaf out, flowering and blooming in plants, and shifts in the timing of spawning and migrations in animals (Walther and others, 2002). A prominent example revealed through long-term observations of lilac flowering indicates that the onset of spring has advanced one day earlier per decade across the Northern Hemisphere in response to increased winter and spring temperatures (Schwartz and others, 2006) and by 1.5 days per decade earlier in the western United States (Ault and others, 2011). Among animals, changes in the timing of springtime bird migrations are among the most well recognized biological responses to warming, and have been documented in the western (MacMynowski and others, 2007), mid-western (MacMynowski and Root, 2007) and eastern (Miller-Rushing and others, 2008; VanBuskirk and others, 2009) United States. Other taxonomic groups have demonstrated a range of phenological shifts that are linked to recent changes in regional climates (**Table 2.2**).

Table 2.2. Examples of observed phenological change across geographical regions of the United States. Compiled by S. Leicht-Young and C. Enquist, written communication, 2012.

U.S. Region	Taxonomic Group	Observed Changes in Phenology	Reference
Alaska & The Arctic	Fish	In a 34-year study of an Alaskan creek, fry of pink salmon (*Oncorhynchus gorbuscha*) migrated increasingly earlier over time.	Taylor 2008
Great Plains	Plants	In a study of first flowering dates (FFD) for 178 species of plants from 1910-1961 and 2007-2010 in North Dakota, over 40 percent of plants showed a change in FFD when compared to flowering data for 2007-2010. Most species showed a difference in FFD during the two warmer years (2007 and 2010) of this study.	Dunnell and Travers 2011
Midwest & Great Lakes	Birds	Of 44 species of birds monitored in Minnesota over a 40 year period, 36 percent showed significantly earlier arrival dates. Increasing winter temperatures correlated with the earlier arrival of birds, particularly for short-distance migrants.	Swanson and Palmer 2009
Northeast	Insects	Native bees have appeared in spring an average of 10 days earlier over the last 130 years (1880-2010). The majority of this advance occurred in the last 40 years in parallel to increasing warming trends. Plants with bee visitation also showed a trend of earlier blooming, helping preserve synchrony in timing between the observed plant species and their insect pollinators.	Bartomeus and others 2011
Northwest	Birds	In a 12 year study of Northern Flickers (*Colaptes auratus*), the birds arrived at breeding sites earlier when temperatures along their migration routes were warmer. Temperatures at the breeding site correlated significantly with initiation of egg laying.	Wiebe and Gerstmar 2010

Impacts of Climate Change on Biodiversity, Ecosystems, and Ecosystem Services |
Technical Input to the 2013 National Climate Assessment

Chapter 2
Biodiversity

Southeast	Reptiles	Two species of autumn-breeding amphibians arrived later, and two winter-breeding species arrived earlier to their breeding area in a 30 year study (1978-2008). Arrival times were up to 76 days later for the autumn-breeding dwarf salamander (*Eurycea quadridigitata*). Overall, rates of change ranged from 5.9 – 37.2 days/decade, and are representative of some of the fastest rates of phenological change observed to date. Increasing overnight temperatures during the breeding season and amount of cumulative rainfall were correlated with these changes.	Todd and others 2011
Southwest	Mammals	Yellow-bellied marmots (*Marmota flaviventris*) observed over a 33 year period (1976-2008) in Colorado emerged earlier from hibernation, and gave birth earlier in the season; this gave marmots more time to grow before the end of the season. Consequently, marmots tended to have larger body sizes at the beginning of hibernation, which ultimately led to lower mortality rates and higher population sizes.	Ozgul and others 2010

In the more arid regions of the southwestern United States, the timing of precipitation has a greater influence on plant phenology than temperature, particularly in the form of available soil moisture (Crimmins and others, 2011a; **Box 2.2**). At a field site in the Sonoran Desert, mean annual precipitation decreased while mean annual temperatures increased during the winter growing season (September –May) over a 25-year period (Kimball and others, 2010). A study of desert annuals at this site found that the timing of germination occurred under colder conditions (-0.4°C decrease/year) due to delays in the occurrence of winter rains, which now peak in December rather than October. This shift in the timing of rainfall led to an increase in abundance of cold-adapted plant species as they were able to successfully germinate in cooler conditions with improved soil moisture conditions (Kimball and others, 2010).

Recent studies suggest that phenotypic plasticity increases the probability of population persistence compared to those of species with relatively fixed phenotypes. For example, in Concord, Massachusetts, an evaluation of 150 years of floral data showed that species that were able to track short-term seasonal temperature variation were more likely to persist, while species whose flowering time did not track seasonal temperature declined (Willis and others, 2008). In addition, Willis and others (2008) provided some of the first evidence that there may be a phylogenetically selective pattern in climate change induced extinction risk among species with shared traits. Consequently, climate induced changes in phenology have the potential to alter the abundance of entire clades, and effect patterns in community composition.

**Box 2.2. Case Study: Recent and Projected changes in Plant Communities
in the Sky Islands region of the Southwest**

In this case study of the Sky Islands Region, a series of studies demonstrate how plant communities have already been observed to respond to changing climatic conditions, and provide some projections of potential future impacts on this system.

In the southwestern U.S., it is expected that the climate will become warmer and drier (USGCRP, 2009). Researchers examined a 20-year data set in southeastern Arizona across a 1200 m elevation gradient to determine whether local plant communities have changed over time (Crimmins *and others* 2009). Out of 363 plant species, 93 (25.6 percent) showed a significant upward shift in flowering range. Furthermore, there was an expansion in flowering range of some species in higher elevations, a pattern consistent with expectations under increased summer warming. In a related study, Crimmins and others (2010) found that only 10 percent of the total species examined exhibited a trend toward earlier spring blooming. The drivers of bloom time were diverse, with a general trend of plants at lower elevations showing a delay of spring flowering when insufficient chilling or moisture occurred the previous autumn, and plants at higher elevations blooming earlier with warmer spring temperatures. With future warmer and drier conditions, plants at lower elevations are predicted to experience delayed flowering if the timing or amount of rain is altered, or if it is too warm for plants to experience sufficient chilling, whereas plants at higher elevations are predicted to advance blooming with increased temperatures (Crimmins and others, 2010). In contrast, additional research found that onset of flowering in summer is strongly linked to the amount and timing of July 'monsoon' rains across elevations and plant life forms (Crimmins and others, 2011a). As a result, with projected future drying in the Southwest, species across elevations are predicted to flower later in summer due to decreased soil moisture conditions resulting from increased summer temperatures.

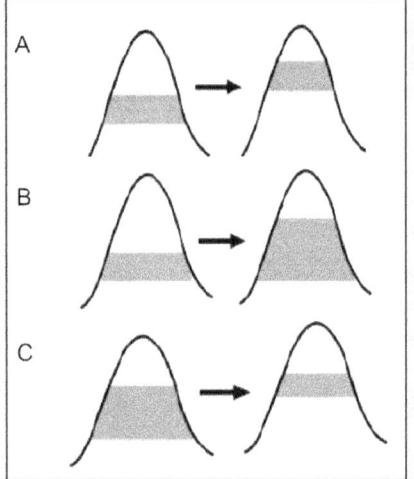

Figure 2.5. Changes in plant phenology across an elevation gradient near Tucson, Arizona. Ninety-three species (26 percent) showed change in flowering range with elevation with warmer summers. (A) 12 species exhibited flowering range shift upslope. (B) 34 species exhibited flowering range expansion upslope. (C) 23 species exhibited flowering range contraction upslope. Adapted with permission from Crimmins and others, 2009.

2.2.3. Shifts in species distributions

Many species are shifting their geographic ranges in response to rapid changes in temperature and precipitation regimes. Often populations track temperature gradients by moving poleward, up in elevation, or to increased depths in the oceans. For example, an analysis of four decades of Christmas bird counts revealed significant northward movement among 58 percent (177 of 305) of species tracked in the survey; on average species moved 35 miles northward, and more than 60 species moved in excess of 100 miles north (National Audubon Society, 2009). In addition, 14 out of 28 species of small mammals evaluated in the Sierra Nevada region showed substantial upward changes in elevation limits (about 500 meters up in elevation) during the past century; formerly low-elevation species were found to expand their ranges while high-elevation species contracted theirs (Moritz and others, 2008). However, not all species movements have been poleward or up in elevation. For example, a study in California showed that numerous vascular plants have exhibited a significant downward shift in altitude since the 1930s due to regional changes in climatic water balance rather than temperature (Crimmins and others, 2011b).

Recent analyses suggest that range shifts examined in 23 taxonomic groups were approximately 2-3 times greater than reported previously (Chen and others, 2011). In terrestrial and aquatic environments, plants and animals moved to higher elevations at a median rate of 0.011 kilometers per decade, and to higher latitudes at a median rate of 16.9 kilometers per decade. Despite faster rates of warming on land (0.24°C per decade) compared to the ocean (0.07°C per decade), geographic shifts of isotherms and the onset of spring have occurred 2.08 days per decade earlier in ocean environments of the Northern Hemisphere compared to on land (Burrows and others, 2011). These changes in climate threaten biodiversity if organisms are unable to track their optimal thermal conditions through range shifts and/or phenology.

Establishment of new populations at the leading edge of a range expansion may increase the potential for genetic bottlenecks, potentially reducing the ability of the population to adapt to future environmental change (Nei and others, 1975). Similarly, range contractions may also decrease genetic diversity across a species' range as populations along the declining edge go extinct (for example, Rubidge and others, 2012). To help populations become established and be successful in new regions, natural resource managers will need to consider how best to preserve high diversity among individuals in newly established or founding populations, maintain population viability in historical regions where range contractions have occurred, and how to sustain gene flow with historic populations (Marsico and others, 2009; Lawton and others, 2011).

Interspecific hybridization may be another mechanism that allows species to persist in marginal habitats as environmental conditions change, and may become more prevalent as new combinations of species occur over spatial and temporal scales (Edwards and others, 2011; Vonlanthen and others, 2012). A large proportion of biodiversity is of recent evolutionary origin; these relatively young species occur as a result of divergent adaptation to differences in environmental conditions. However, as landscapes become less heterogeneous due to the effects of climate change and other anthropogenic activities (for example, land use and land cover change), the incidence of hybridization among closely related species has the potential to increase (Seehausen and others, 2008). Some recent examples of climate-induced hybridization have been detected between southern and northern flying squirrels (*Glaucomys volans* and *Glaucomys sabrinus*, respectively) in Ontario, Canada (Garroway and others, 2010), and the spread of a hybrid zone between two salamander species (*Plethodon teyahalee* and *P. shermani*) in the Nantahala Mountains of North Carolina (Walls, 2009). It has also been suggested that brown bears (*Ursus arctos*) and polar bears have hybridized multiple times throughout the last 100,000 years, and the dynamics of their dispersal events have been mostly climate-driven (Edwards and others, 2011). Although this mechanism has the potential to help species persist as environmental conditions change, an increase in interspecific hybridization could reduce species richness and diversity, and has important consequences for ecosystem function (Seehausen and others, 2008).

Overall, range shifts may raise the probability of persistence of species and populations; however the ability to disperse or migrate to new areas does not guarantee survival as there are additional factors such as species interactions and land use change that may influence populations (Hoffmann and Sgró, 2011). Species-specific differences in physiological, behavioral, and morphological plasticity may allow individuals and populations to respond *in situ* and delay or eliminate the need for range shifts (Doak and Morris, 2010); however, in many cases, these responses may be difficult to predict.

2.2.4. Shifts in biotic interactions and species assemblages

Climate change is having both direct and indirect effects on the way species interact over spatial and temporal scales, with sometimes profound impacts on ecosystem structure and function (Walther, 2010; Singer and Parmesan, 2010; Yang and Rudolf, 2010). Higher temperatures can affect food-web interactions by increasing vital rates such as growth and consumption. For example, warming-mediated increases in consumption and interaction strengths of marine consumers, both herbivores and carnivores, have been documented in a range of intertidal, benthic, and pelagic habitats (Sanford, 1999; Philippart and others, 2003; O'Connor and others, 2009). Observations and experiments have provided evidence that, on average, consumer pressure tends to be stronger at low latitudes (Bertness, 1981; Bolser and Hay, 1996; Pennings and Silliman, 2005). Thus, one likely outcome of poleward range expansions of lower-latitude species might be a strengthening of predatory (top-down) control. A dramatic example involves the recent establishment of a lithodid ("king") crab (*Neolithodes yaldwyni*) population on the Antarctic shelf as waters there have warmed; these generalist predators are reducing diversity and abundance of the formerly luxuriant Antarctic benthic invertebrate community (Smith and others, 2012). Studies have also shown that in some systems such as aquatic habitats and terrestrial grasslands, higher trophic levels are often less able to vary their responses to environmental variability, leading to increases in population dynamics and extinction risk (Petchey and others, 1999; Voigt and others, 2003). For example, when the match between food availability and demand decreased over time for several insectivorous birds and raptors, consumer level responses to warming were weaker than those of their food sources (Both and others, 2009).

Since physiological responses to climate change are highly species-specific, changes in the timing of phenological events will often differ among interacting species as well as across trophic levels, and lead to trophic mismatches (Brander, 2010; Yang and Rudolf, 2010). The complex networks of interactions among species makes it difficult to predict climate-mediated changes in communities, and sometimes the responses are counterintuitive. Nonetheless, many researchers agree that an increased frequency of ecological mismatches across trophic levels will increase the potential for population declines, local extirpations, and novel species interactions (Visser and Both, 2005; Parmesan, 2007; Miller-Rushing and others, 2010). Although there is

Impacts of Climate Change on Biodiversity, Ecosystems, and Ecosystem Services |
Technical Input to the 2013 National Climate Assessment

Chapter 2
Biodiversity

currently a paucity of studies documenting observed negative effects of trophic mismatch, there are noteworthy exceptions. In Alaska, warmer springs have caused earlier onset of plant emergence, and decreased spatial variation in growth and availability of forage to breeding caribou (*Rangifer tarandus*). This trophic and spatial asynchrony ultimately reduced calving success in caribou (Post and others, 2008). In another example from Lake Washington, algal blooms have advanced by as much as 27 days in synchrony with warming air and water temperatures; however, the algae predator, *Daphnia* spp., has not tracked this change in food source and thus has experienced decreases in population size (Winder and Schindler, 2004). In addition, changes in temperature and precipitation during the early growing season in high elevation habitats are disrupting flowering phenology across meadow habitats, and resulting in a mid-season decline in flowering resources that may act to decouple important plant-pollinator relationships (Aldridge and others, 2011; Forrest and Thomson, 2011).

Changes in environmental conditions have led to shifts in species dominance, and community composition in a range of ecosystem types (Collie and others, 2008; Moritz and others, 2008; Beaugrand and others, 2010; Dijkstra and others, 2011). In some cases, these shifts are leading to associations and assemblages among organisms that have only occurred in rare instances or have not occurred in the past. For example, in marine habitats such as Long Island Sound, Narragansett Bay, and the Gulf of Mexico there has been an increase in tropical and subtropical species in historically temperate habitats (Wood and others, 2009; Fodrie and others, 2010). Juvenile and larval life stages of warm-water marine fishes such as butterfly (*Chaetodon* sp.) and surgeon (*Acanthurus* sp.) fishes were previously absent or considered seasonal visitors in northern habitats. The presence of new species could lead to local increases in species diversity, but also have the potential to impact food-web dynamics, and productivity thus effecting valuable ecosystem services.

Some of the most dramatic examples of changing species assemblages involve climate-mediated disease outbreaks. Warming ocean waters have aggravated the spread and prevalence of diseases in marine organisms (Harvell and others, 2002), including microbial disease outbreaks in reef-building corals (Bruno and others, 2007), and pathogens of the eastern oyster (Ford and Smolowitz, 2007), and contributed to widespread loss of these important habitat-forming species. Across western North America, climate-induced changes in pests and pathogens have also been blamed for recent mortality events in conifer forests (van Mantgem and others, 2009) and quaking aspen-dominated systems (*Populus tremuloides*) after being stressed by drought (Anderegg and others, 2012).

In general, biotic interactions are complex, and there is much uncertainty in the greater ecological consequences that climate-mediated changes in abundance and distribution will have at the ecosystem-level (for example, Harley, 2011). Although it is inevitable that there will be surprising outcomes brought about by changes in interspecific interactions, baseline data on existing species relationships will alleviate this to some degree by allowing up us to recognize and track novel interactions as they develop, and when necessary take action to minimize loss.

2.3. HOW WILL CLIMATE CHANGE AFFECT BIODIVERSITY IN THE COMING CENTURY?

2.3.1. Introduction

Climate change impacts on biodiversity are projected to increase in magnitude and pervasiveness as CO_2 levels and temperatures continue to rise, and extreme events (for example, heat and storms) increase in frequency and intensity (IPCC, 2007). A range of methods and approaches are being employed to predict the impacts of climate change on biodiversity including historical trends and relationships, experiments, and model projections. As these methods have become more sophisticated, so has our understanding of projected impacts, particularly on aquatic and marine systems, which have been less well studied compared to terrestrial systems.

Although there are many ways to categorize the modeling approaches used to assess potential impacts of climate change on biodiversity, most can be described as either empirical (correlative) or process-based (mechanistic). Projections of species distributions often come from empirical models that relate observed occurrences to current or historical climate conditions, and predict future distributions using projected changes in the geographic distributions of abiotic variables (for example, temperature). These models, which are often referred to as climate envelope, niche, or species distribution models, have the advantage of being relatively easy to apply to large numbers of species, and have been used to project potential shifts in areas of climatic suitability for plants and animals at varying scales across the United States (Matthews and others, 2004; Iverson and others, 2008; Stralberg and others, 2009; Wegner and others, 2011). Similar empirical models have been used to project changes in biomes and vegetation types (Rehfeldt and others, 2012). Despite their flexibility, empirical models generally do not directly model biotic interactions (for example, competition), account for evolution, or address dispersal; these limitations can lead to an overestimation of the ability of species to track climatic changes (Schloss and others, 2012), and an underestimation of extinction rates (Maclean and Wilson, 2011; Urban and others, 2012).

In contrast, process-based models are designed to specifically account for a number of the mechanisms that determine species distributions or vegetation patterns, and can simulate physiological responses, population processes, dispersal, ecosystem functions, and plant growth. Some examples include spatially explicit, individual-based population models, dynamic global vegetation models (DGVM), and forest gap models (for example, Battin and others, 2007; Carroll, 2007; Buckley and others, 2010; Rogers and others, 2011).

There have been several recent efforts to integrate individual empirical and process-based models to ensembles of such models to improve projections of the impacts of climate change on various components of biodiversity (Araujo and New, 2007). These include efforts to 1) account for dispersal or movement in empirical models of changes in species or population distributions (Iverson and others, 2004; Early and Sax, 2011), 2) combine metapopulation models with projected shifts in climatic suitability (Keith and others, 2008), 3) integrate multiple mechanisms into projected changes in the distribution of marine species (Cheung and others, 2009), and 4) combine niche models with physiological mechanistic models (Kearney and Porter, 2009). These new and increasingly sophisticated methods of linking niche models, trophic models, dynamic vegetation models, and global climate models with socioeconomic scenarios are increasing our abilities to predict and evaluate future impacts of climate change on biodiversity (Pereira and others, 2010).

2.3.2. Projected impacts on organisms, species and populations

Forecasting biological responses to the impacts of climate change requires identifying the climate drivers (for example, temperature, precipitation, ocean current patterns) responsible for instigating change, and the temporal and spatial scales at which they are influential (Pau and others, 2011). In addition, biological responses to abiotic drivers are idiosyncratic and are based on morphological, physiological, and behavioral traits of individual organisms (Bellard and others, 2012; **Box 2.3**). Consequently, research and monitoring of how all components of biodiversity have tracked changes in environmental conditions in the past informs our understanding of future changes.

Box 2.3. Impacts of Climate Change on Marine Ecosystems

Marine biodiversity face two primary threats due to climate change: increasing temperature, and acidity due to the absorption of CO_2 by the oceans (Doney and others, 2010). As the physical and chemical conditions in global oceans continue to change, marine biodiversity is predicted to respond idiosyncratically; some species and populations will respond positively and adjust to these changes, while others will respond negatively and decline. The examples below highlight just a few of the direct and indirect effects of climate change on marine biodiversity and the potential consequences for population dynamics, survival, and species interactions.

Since the mid-1950s, ocean temperatures have risen by an average of 0.2° Celsius in the upper 700 meters of water (Bindoff and others, 2007; Levitus and others, 2009). Warming temperatures have already caused numerous shifts in the distribution, abundance, and phenology of marine organisms (Nye and others, 2010; Chen and others 2011; Smith and others 2011), and are expected to increasingly affect population connectivity, community structure, and spatial patterns of biodiversity in the future. In particular, temperature has been shown to have a strong and negative effect on the duration of the larval period across a diverse range of species, body sizes, and habitats (O'Connor and others, 2007). This consistent finding is important because for many marine species, the planktonic larval phase is the only life stage in which individuals disperse away from the parental population, and planktonic larval duration is correlated with dispersal distance. This suggests that larvae of marine animals may disperse over systematically shorter distances in a warming climate, potentially reducing the connectivity among populations, and increasing the vulnerability of isolated populations to extinction. These results have important implications for how we conserve and manage marine populations, particularly though the use of Marine Protected Areas. Shifts in life history due to reduced larval duration will also likely alter population growth rates and interactions among species with community-level consequences that are difficult to predict.

Acidification results from the absorption of CO_2 by the oceans and chemical reactions that lead to decreases in ocean pH. Sea-surface pH has declined by an estimated 0.1 units over the last 150 years causing a 26 percent increase in acidity, most of which has occurred in the past several decades. Projections suggest further declines of 0.2-0.3 pH units during the coming century (Feely and others, 2009). By lowering carbonate ion levels and increasing carbonate

---

Box 2.3, continued.

solubility, ocean acidification is thought to increase the energetic cost of calcification (Fabry and others, 2008). Polar regions may be especially sensitive to ocean acidification because of an expected transition to undersaturated conditions of aragonite in surface waters within the next several decades (Steinacher and others, 2009).

This hypothesis is supported by observations of reduced calcification rates by a variety of marine invertebrates under elevated CO_2 conditions (Kroeker and others, 2009; Hofmann and others, 2010). Acidification is thus likely to have major negative impacts on biogenic habitats, including 1) coral reefs and oyster beds, 2) food webs, notably high-latitude pelagic systems where pteropod mollusks are important prey for fishes, and 3) planetary geochemical cycles due to changes in $CaCO_3$ flux by pelagic coccolithophorid algae.

Elevated CO_2 concentrations have also been shown to have chemosensory, auditory, and neurological effects that impair behavioral activities in non-calcifying animals (Domenici and others, 2012) including predatory and antipredator behaviors of marine fishes (Ferrari and others, 2010, 2012; Nowicki and others, 2012). For example, juvenile anemone fish, *Amphiprion melanopus,* exhibit decreased food consumption and foraging abilities when exposed to higher water temperatures and CO_2 levels projected in the coming century (Nowicki and others, 2012). Experimental exposure to elevated CO_2 representative of levels projected by 2100 also caused juvenile damsel fish (*Pomacentrus* sp.) to be more susceptible to predation and to a different set of predators than when they were exposed to present day CO_2 levels. In addition, larger individuals tended to be less impacted by CO_2 induced changes in predation, thus younger fishes may suffer greater climate-impacts compared to larger life stages and species (Ferrari and others, 2011).

Recent estimates of the velocity of climate change, described by Loarie and others (2009) as the speed and residence time of temperature change over space and time, are faster than was previously thought; this has raised concern as to whether species migration rates will be fast enough to track future environmental conditions (Pearson, 2006). Projected shifts in species distributions are generally poleward and upward in elevation, although forecasts can be more complex and sometimes counterintuitive (Lawler and others, 2010). Forecasts of terrestrial biomes estimate that the velocity of global temperature shifts will be lowest in topographically complex montane systems (0.08 km per year) whereas systems with much less topography such as flooded grasslands and deserts, will show higher rates of change (1.26 km per year) (Loarie and others, 2009).

In a recent study by Cheung and others (2009), marine species were projected to shift their distributions at rates of 45-59 km/decade. These shifts represent data on over 1,000 species of fishes and invertebrates, and far exceed those observed in terrestrial species over the past several decades (Parmesan, 2007; Burrows and others, 2011). The consequences of such dramatic shifts could lead to numerous local extinctions particularly in sub-polar regions, the tropics, and semi-enclosed seas, and a reshuffling of global marine floras and faunas in the majority (greater than 60 percent) of the world's oceans (Cheung and others, 2009). In situations where species cannot adapt or move quickly enough, populations face extinction.

To date, only a few species of amphibians, birds, fishes, and gastropods are known to have gone extinct due to the impacts of climate change (Monzón and others, 2011); however, widespread losses of global biodiversity are projected by numerous studies using a range of modeling approaches and climate scenarios. Thomas and others (2004) projected extinction rates for a variety of birds, mammals, frogs, reptiles, invertebrates, and plants in areas around the world using a climate envelope modeling approach. Extinction rates ranged from 11-34 percent for a 0.8°C increase in temperature, and from 33-58 percent for greater than 2°C change in temperature depending on assumptions about species dispersal abilities (Thomas and others, 2004). Using generalized linear models, Sekercioglu and others (2008) projected that extinction rates for the majority (87 percent) of terrestrial bird populations in the Western Hemisphere would range from 1.3 percent for a 1.1°C to 30 percent for a 6.4°C increase in temperature. Sinervo and others (2010) projected a 20 percent loss of global lizard species by 2080 by employing an empirically-validated physiological-based modeling approach. By applying IUCN Red List Categories and Criteria, Carpenter and others (2008) determined that 30 percent of reef-building corals (845 species) are at risk for extinction in the coming decades. In addition, Malcolm and others (2006) projected extinction rates among endemic species in global biodiversity hotspots would range from less than 1 to 43 percent using a combined approach of global vegetation models (GVMs) and general circulation models (GCMs).

Projections of climate-driven extinction rates vary widely among species largely due to the limitations and assumptions associated with different modeling approaches (Botkin and others, 2007; He and Hubbell, 2011; Bellard and others, 2012). Nonetheless, conclusions among projected studies agree that as the impacts of climate change continue to grow, we risk widespread declines and loss of global biodiversity in the coming century (Harte and Kitzes, 2011).

2.3.3. Projected impacts on communities, ecosystems, and biomes

It is expected that tropical and subtropical ecosystems will be some of the first to experience extreme temperatures, and up to 86 percent of global terrestrial and 83 percent of freshwater ecoregions will experience extreme (greater than 2 standard deviations (SDs) of 1961-1990) temperatures by 2070 (Beaumont and others, 2011). These and other projections suggest that the majority of the world's biodiversity will experience potentially stressful environmental conditions in the coming century. In addition, the disappearance of existing and creation of novel climate conditions will likely alter the magnitude and direction of existing interspecific relationships, reshuffle community compositions, bring about new combinations of species, and change the flow of materials and energy through food-webs (Walther and others, 2002; Williams and others, 2007; Hobbs and others, 2009; Harley, 2011; Wenger and others, 2011; Urban and others, 2012).

Several studies have projected climate-mediated turnover (that is, the replacement of one species by another) in species composition for particular areas and taxa resulting from the combination of local species losses and invasions from other regions. Based on projections from empirical bioclimatic models, terrestrial vertebrates are estimated to turnover at rates of 25-38 percent across the Western Hemisphere by the end of the century, with some areas in the United States experiencing a greater than 90 percent change in species composition (Lawler and others, 2009). Marine animal species are projected to experience mean invasion intensities (the number of invasions relative to current species richness) that would reach a global average of 55 percent by 2050, with invasions concentrated at high latitudes (Cheung and others, 2009)). In addition, novel bird communities are projected across 70 percent of the State of California by 2070

Impacts of Climate Change on Biodiversity, Ecosystems, and Ecosystem Services |
Technical Input to the 2013 National Climate Assessment

Chapter 2
Biodiversity

(Stralberg and others, 2009). Changes in community composition and species turnover have important implications for how ecosystems are structured and function, particularly because extinctions and species invasions are often biased by species' functional traits such as life history and trophic level (Byrnes and others, 2007).

Biotic interactions have received relatively less attention than studies of responses to direct (abiotic) effects in part because indirect effects are harder to quantify and predict. Efforts to model climate change effects on species interactions are increasing and will be helpful to understand how the increasing presence (or absence) of different species or populations (for example, due to invasions) can affect the greater community. Such efforts will be useful for identifying species that may be better able to adapt and successfully compete for resources as environmental conditions change (for example, Wenger and others, 2011). The impacts of climate change on communities and ecosystems can be complicated and even reversed when species interactions are considered in models or experiments, leading to sometimes counterintuitive outcomes. For example, a five-year experimental manipulation of seasonality and intensity of rainfall in a California grassland community showed that altering the availability of water strongly affected individual species, but the direction of effects were reversed over the long term as feedbacks and species interactions overrode the direct physiological responses of individual species (Suttle and others, 2007).

Climate-driven changes in the functioning of ecosystems are very likely to result in altered vegetation communities, shifts in major biome boundaries, and changes in habitat for animal species (IPCC, 2007; Alo and Wang, 2008; Bergengren and others, 2011; Gonzalez and others, 2010; Sitch and others, 2008). Overall, model projections generally agree on extensive poleward shifts of vegetation, although spatial distributions vary slightly from model to model due to differences in the model parameters, emissions scenarios, and vegetation models used in each analysis. Future climate warming is likely to exacerbate trophic mismatch in a wide range of ecosystems, disrupt food-web interactions, and ultimately impact ecosystem functions and services that are mediated by these functions (for example, fishery production, biological control of pests, and pollination). A further discussion of impacts at the ecosystem and biomes levels can be found in *Chapter 3: Ecosystems and Processes,* and impacts on ecosystem services will be addressed in *Chapter 4: Ecosystem Services.*

2.4. VULNERABILITIES AND RISKS: WHY IMPACTS OF CLIMATE CHANGE ON BIODIVERSITY MATTER

2.4.1. Vulnerability and risk to climate change

Climate change is already affecting biodiversity in myriad ways and impacts are anticipated to increase in the century ahead; consequently, evaluation of the risks and vulnerabilities of biodiversity is needed to inform decision-making and where and how to most effectively allocate scarce resources.

The vulnerability of biodiversity to climate change is dependent on the character, magnitude, and rate of changes experienced by a species or system (exposure), the degree to which they are, or are likely to be, affected by or responsive to those changes (sensitivity), and the ability to accommodate or cope with impacts with minimal disruption (adaptive capacity) (IPCC, 2007; Williams and others, 2008; Glick and others, 2011). Each of these factors is difficult to measure due to uncertainties in climate change projections in the coming decades, and gaps in our knowledge of biological and ecological responses to these changes (Glick and others, 2011). In addition, biodiversity is already impacted by a range of anthropogenic stressors including land use change, exploitation, pollution, non-native invasive species, and disease. In many cases, these other stressors have been, are currently (Flather and others, 1997, Wilcove and others, 1998; Jetz and others, 2007, Master and others, 2009), or are expected to be the primary drivers of biodiversity loss (Clavero, 2011). Overall, it is anticipated that the impacts of climate change will become increasingly pervasive and influential in the coming decades, and interact synergistically with existing stressors to affect biodiversity's vulnerability (Brook and others, 2008; Barnosky and others, 2011; Mantyka-Pringle and others, 2011). For example, new bioclimatic conditions and altered community compositions may enable invasions by non-native species, thus further stressing biological systems (Walther and others, 2009). Although the net effect on biodiversity globally is expected to be markedly negative (Bellard and others, 2012), an increasing number of studies shows that a range of species and populations may experience local benefits and thrive under the changing climate conditions (Schmidt and others 2009; Hare and others, 2010; Schmidt and others, 2011).

It is also important to note that our understanding of species' vulnerability to climate change is often based on basic biological research and theory that is not necessarily climate-based in application, or may have been derived from a relatively small set of studies that have specifically tested species' responses to climate change. In some cases, the observed biological responses differ from what was expected, thus revealing limitations in our ability to predict vulnerability to climate change. This was demonstrated by Angert and others (2011) who showed that species' traits were not always good predictors of geographic range changes and suggested that we need a better understanding of the process of range shifts to improve model forecasts and vulnerability assessments. Nonetheless, it is widely believed that the sensitivity of biodiversity to climate change is largely determined by intrinsic biological attributes (**Figure 2.6**), and the degree to which their viability has already been compromised by other anthropogenic factors (for example, habitat degradation) (Glick and others, 2011; Williams and others, 2008).

Historically, conservation status ranks of species have been among the key criteria for setting priorities for conservation action (Master and others, 2009). However, these ranks are not necessarily indicators of the vulnerability of species to climate change. It may not be prudent to

Impacts of Climate Change on Biodiversity, Ecosystems, and Ecosystem Services | Technical Input to the 2013 National Climate Assessment

Chapter 2
Biodiversity

assume, for example, that common species will remain common in light of climate change. Conservation practitioners therefore require additional tools to determine how vulnerable species are to climate change regardless of whether they are rare or common. Vulnerability assessments provide such a tool and are increasingly being used to inform climate change adaptation strategies (for example, U.S. EPA, 2009; Young and others, 2009; Bagne and others, 2011; Glick and others, 2011; Rowland and others, 2011) (**Box 2.4**). Vulnerability assessments require knowledge of the observed and projected biological responses to climate change, and are often based on species-specific factors such as dispersal ability, physical habitat specificity, and genetic factors (Glick and others, 2011). However, vulnerability assessments have the ability to group species based on similar drivers of vulnerability, allowing for greater efficiency in planning.

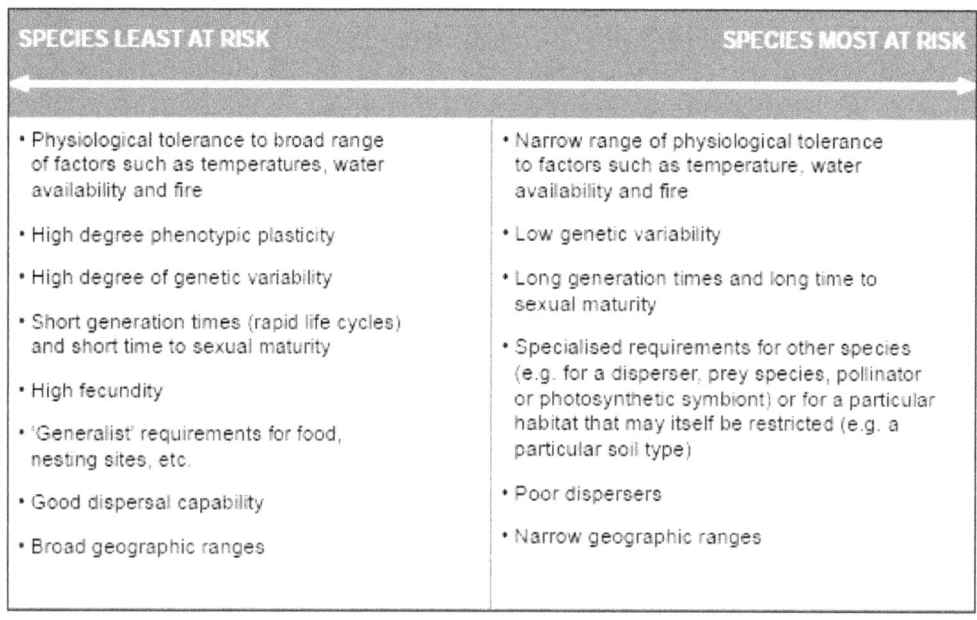

Figure 2.6. *Physiological and life history traits of species and populations that influence vulnerability or resilience in response to climate-related disturbance. Figure was reproduced with permission from the Department of Climate Change and Energy Efficiency.*

2.4.2. What types of ecosystems and species are most vulnerable?

There is increasing evidence that species will vary in their vulnerability to climate change due to biological and ecological factors such as physiology (Feder, 2010; Dawson and others, 2011), genetic diversity (Lawton and others, 2011), resource specialization (Munday, 2009; Yang and others, 2011), competitive abilities (Montoya and Raffaelli, 2011), interspecific (Wegner and others, 2011) and intraspecific interactions (Best and others, 2007). Species and populations that live at high latitudes, at or near their thermal limits (for example, in deserts, coral reefs and other tropical environments), and early-season species are likely to be most vulnerable to climate-mediated changes in phenology that disrupt trophic links (Deutsch and others, 2008; Moore and Jarvis, 2008; Eakin and others, 2010; Pau and others, 2011).

Ecological specialists may also be particularly vulnerable and react strongly to changing resources due to their relatively narrow dietary, thermal, and habitat niche breadths (Altermatt, 2010; Clavel and others, 2011; Lawton and others, 2011; Montoya and Raffaelli, 2011) and association with relatively stable environments (for example, tropical coral reefs) (Walther and others, 2002). To date, habitat degradation and introductions of non-native species by humans have been the primary drivers of specialist declines (Durak, 2010; Devictor and others, 2008), but projected increases in climatic variability (IPCC, 2007) could exacerbate the loss of native specialists if they are not able to adjust (Clavel and others, 2011; Montoya and Raffaelli, 2011). However not all specialists are affected in the same way (Buisson and Grenouillet, 2009) and some may benefit due to climate change (Altermatt, 2010). For example, "hidden" specialists may emerge when environmental conditions become optimal for their survival as was recently discovered in *Montastraea* corals, which make up approximately 25 percent of all corals globally (van Woesik and others, 2010). Under extreme environmental (or biological) stress, corals expel their zooxanthellae and bleach. Past bleaching events, have been precipitated by increases in sea surface temperatures of only a few degrees Celsius, and resulted in mass mortalities of corals worldwide (Loya and others, 2001). *Montastraea* corals host multiple species of symbiotic zooxanthellae (*Symbiodinium*), some of which were revealed to be high temperature specialists, and after a triggering event, were able to reshuffle and recolonize their coral hosts thus increasing the coral's ability to adjust to changing conditions (van Woesik and others, 2010).

Although there is still much uncertainty in how specialists (and non-specialists) will respond to changes in environmental conditions, increased knowledge of traits and interspecific relationships improves our ability to recognize and predict why different species respond in varying ways to climate change (Diamond and others, 2011).

Box 2.4. Integrating Vulnerability Assessments into Adaptation Planning: Updating the Florida State Wildlife Action Plan

In 2005, the Florida Fish and Wildlife Conservation Commission (FWC) released Florida's Wildlife Legacy Initiative, the State's wildlife action plan, which identified conservation threats impacting species of high conservation need, their associated habitats, and actions proposed to mitigate those threats. Like many other States, Florida is actively expanding efforts to address new threats emerging due to climate change. In 2011, the agency worked with partners to explore two complementary approaches to vulnerability assessment as part of a process to identify potential adaptation strategies for several priority species (FWC, 2011). The first approach utilized an existing vulnerability assessment tool, the *NatureServe* Climate Change Vulnerability Index (CCVI) (Young and others, 2010) to identify factors contributing to vulnerability to climate change for a set of species occurring in Florida (Dubois and others, 2011). Twenty-three species and sub-species were included in this assessment, generating vulnerability scores ranging from "Not Vulnerable" to "Extremely Vulnerable (**Figure 2.7.**). Species associated with coastal habitats and that are susceptible to sea level rise and changes in hydrology, tended to rank among the most vulnerable. Species with more generalized habitat and/or ecological requirements were considered less vulnerable to climate change. The second approach was a spatially explicit vulnerability analysis which estimated potential future habitat under land use scenarios incorporating sea level rise, public policy options, and financial

Impacts of Climate Change on Biodiversity, Ecosystems, and Ecosystem Services |
Technical Input to the 2013 National Climate Assessment

Chapter 2
Biodiversity

Box 2.4, continued.

conditions and was applied to a subset of species assessed with the CCVI (Flaxman and Vargas-Moreno, 2011).

The results of these assessments were integrated into a process for identifying potential adaptation strategies. The CCVI provided a framework for identifying the factors contributing to vulnerability to climate change. These factors were translated into anticipated causal relationships between potential climate-related threats, and the resulting biophysical impacts on the biodiversity target (in this case species). Existing synergistic threats were then integrated into the threat assessment and actions were identified that could be implemented to improve the condition of the biodiversity target. Finally, the spatially-explicit vulnerability analysis provided the landscape context in which to relate these actions to the species' geography.

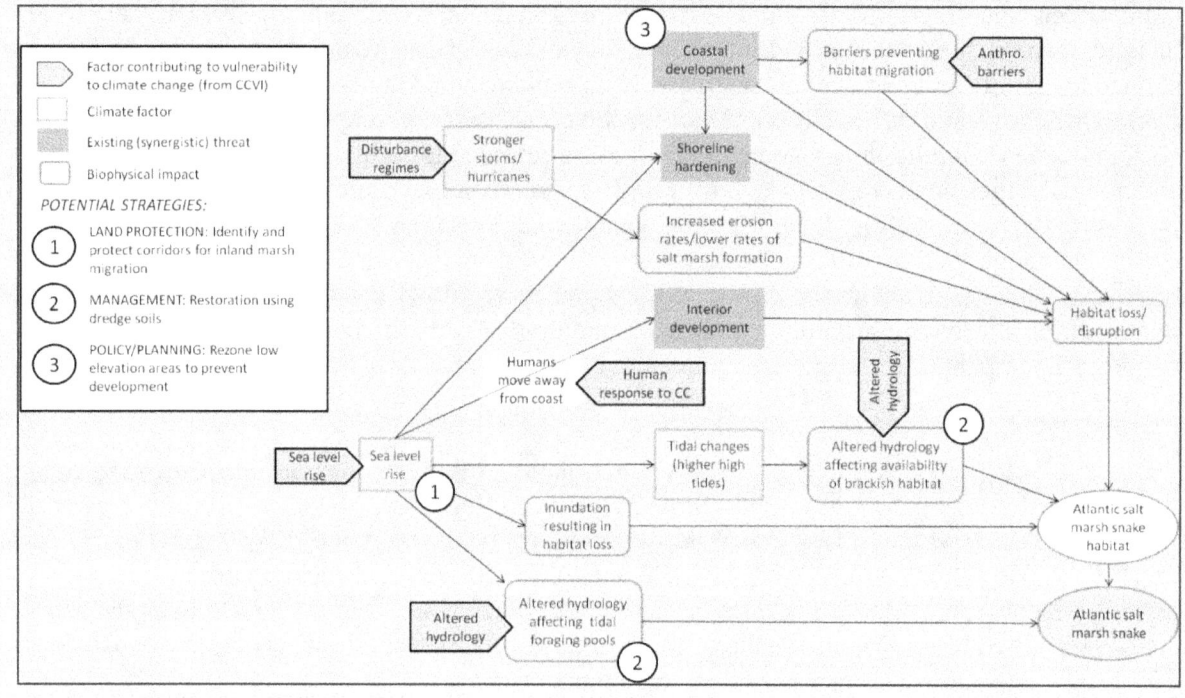

Figure 2.7. *A process for using the results of a vulnerability assessment to identify potential adaptation strategies is illustrated for a limited set of factors affecting the Atlantic salt marsh snake (Nerodia clarkii taeniata) throughout its range in Florida. Figure used with permission from Defenders of Wildlife.*

2.4.3. Policy implications for vulnerable species

The ultimate threat to biodiversity is total loss through global extinction, for which losses on local or regional scales can often provide early warning signs. For example, species or populations that are exploited (for example, overfished) to the point where they are no longer profitable are deemed commercially extinct (Sala and Knowlton, 2006; Ehrhardt and Deleveaux,

2007), and species or populations that become so scarce that they no longer perform their ecological function are said to be ecologically (or functionally) extinct. Determining when a species has become ecologically extinct is often more difficult in comparison to when a species is commercially extinct, but often commercially extinct species are also ecologically extinct, and species that are listed as critically imperiled (for example, by State, Federal or international bodies) are more likely ecologically extinct. Some species, particularly, rare and cryptic species may have surprising effects when they are eradicated from a system (Lyons and Schwartz, 2001; Lyons and others, 2005). Although many species have close relatives or functional equivalents that could fill their ecological roles with relatively little change to ecosystem dynamics, system resilience may still be compromised; the only way to know the full impact of species loss is through experimental removal from a system or extinction (Díaz and others, 2003).

Observed and projected ecological responses to climate change suggest that extinction risk will increase for many species (Maclean and Wilson, 2011). Indeed, a number of species have already been evaluated for possible listing under the U.S. Endangered Species Act (ESA) partially or wholly due to the impacts of climate change. Well known examples include, Arctic species such as the polar bear, and ringed seal (*Phoca hispida*), which are threatened by declines in critical sea ice habitats (USFWS, 2008; NOAA, 2010), and elkhorn (*Acropora palmata*) and staghorn (*A. cervicornis*) corals due to their high sensitivity to sea level rise and warming sea temperatures. Delisting species from ESA protection has been halted even when recovery targets were achieved (Goldstein, 2010; 9th Circuit Court Decision, 2011[2]). This was the case for the Yellowstone population of grizzly bears (*Ursus arctos horribilis*) when it was determined that the U.S. Fish and Wildlife Service (USFWS) did not adequately consider the potential impacts of global warming on the population and its primary prey, the whitebark pine (*Pinus albicaulis*). Yet the potential impacts of climate changes do not always warrant ESA listing (or prevent delisting) as was seen for the American pika (*Ochotona princeps*). Following a review, the USFWS declined to list either the full species or any of the five subspecies of American pika. It was determined there would be sufficient high elevation habitat to ensure pika persistence for the foreseeable future despite higher temperatures across much of its range due to climate change (USFWS, 2010); however it was noted that "there are no known existing regulatory mechanisms currently in place at the local, State, national, or international level that effectively address climate induced threats to pika habitat" (USFWS, 2010).

A species can be listed under the ESA if there is 1) present or threatened destruction, modification, or curtailment of its habitat or range, 2) over utilization for commercial, recreational, scientific, or educational purposes, 3) disease or predation, 4) an inadequacy of existing regulatory mechanisms, or 5) other natural or manmade factors affecting its continued existence (ESA, 1973). As climate change effects accumulate, more species will become "conservation reliant"; these are species for which it may not be possible to fully abate the threats to their persistence (Scott and others, 2010). Consequently, managers must have in place monitoring systems and a readiness to respond to population problems, perhaps in perpetuity (Scott and others, 2010).

Protected areas have historically provided refuge to biodiversity (for example, from anthropogenic stressors); however, with the disappearance of some habitats and the emergence of novel climates, these refugia may not contain the range of climatic conditions needed to

---

[2] *Greater Yellowstone Coalition, Inc. versus Servheen*, --- F.3d ----, 2011 WL 5840646, *7 (9th Cir. 2011).

support previously and newly vulnerable species in the future (Wiens and others, 2011). Schloss and others, (2012) projected that climate change impacts will decrease the amount of spatial refuge provided by national parks and protected areas in the Western Hemisphere by an average of 46 percent, and impact up to 85 percent of the mammals living there. In U.S. National Parks, projections of a doubling of atmospheric CO_2 are estimated to result in a loss of an average 8.3 percent of mammalian species diversity (individual parks range from 0-20 percent losses) (Burns and others, 2003). These types of changes are projected to result in a net emigration of mammals from 70 percent of the largest reserves and decrease biodiversity within reserves by an average of 6 species per area (Schloss and others, 2012). Natural resource managers who are tasked with managing a fixed place on the ground (for example, protected areas) or an individual species or population may need to look beyond the places where their management efforts are currently focused, to places that will become suitable in the future; these future climate refugia may often occur outside of the species' or population's current range (Ashcroft, 2010).

2.5. WHAT HUMAN RESPONSE STRATEGIES COULD ADDRESS THE MOST HARMFUL IMPACTS OF CLIMATE CHANGE ON BIODIVERSITY AND WHAT BARRIERS AND OPPORTUNITIES EXIST TO THEIR IMPLEMENTATION?

2.5.1. Climate change adaptation for biodiversity conservation

Developing successful human response strategies to climate change would include adopting approaches that benefit biodiversity and enhance the ability of its elements to adapt to change. Adaptation strategies meant to address the impacts of climate change to biota and ecosystems are becoming increasingly more common, though their effectiveness is often untested (Poiani and others, 2011). As humans respond to the impacts of climate change, a primary concern is to do no greater harm to already stressed natural systems that are also seeking to adapt to change (see *Chapter 5: Multiple stressors*). The potential for unintended negative impacts of human responses on biodiversity is large (Turner and others, 2010), and has strong implications for society because of the many ecosystem services that biodiversity supports including those that counter the deleterious effects of climate change.

Climate change adaptation seeks to reduce key vulnerabilities of natural and human systems against actual or expected climate change effects, and when possible, take advantage of beneficial opportunities (IPCC, 2007), after structured consideration of climate change impacts and associated uncertainties (Chester and others, In Press). Climate change adaptation strategies will often draw from past and existing conservation planning and tools; however, climate change adaptation may require re-prioritizing where, when and how we apply those tools, while also being on the lookout for when we need new tools (Chester and others, In Press).

It is very likely that current biodiversity conservation goals will need to be reconsidered, and framed to focus on an uncertain future, rather than past, climatic and ecological conditions (Pearson and Dawson, 2005; Simenstad and others, 2006; Hobbs and Cramer, 2008; Julius and West, 2008; Lawler, 2009; Game and others, 2010; Glick and others, 2011b, Cross and others, Accepted). Although embracing forward-looking goals may prove to be difficult in light of existing legislation, regulations, institutional cultures, and other barriers, historical reference points should be used with caution in the goal setting process (Millar and others, 2007). Ideally, climate-informed management goals and priorities would incorporate a needed flexibility to manage for change over the long-term (decades to centuries), yet also account for near-term conservation challenges and transition periods.

Although there is a growing consensus on several general adaptation principles (*Chapter 6: Adaptation*), capacity for taking action continues to lag (National Research Council 2010). While practical guidance for adaptation planning has been limited (Heller and Zavaleta, 2009), several emerging approaches to adaptation planning are being tested and refined in landscapes across the United States (for example, Peterson and others, 2011; Halofsky and others, 2011; Poiani and others, 2011; Weeks and others, 2011; Cross and others, Accepted). These efforts illustrate the value of constructive dialogue between scientists and managers, on local climate change projections and ecological responses, for producing science-based strategies for climate change adaptation (Halofsky and others, 2011; Cross and others, Accepted). Chapter 6 (Adaptation) provides an in-depth examination of climate change adaptation issues for biodiversity, ecosystems and ecosystem services; here we highlight and reiterate a few key points as they specifically relate to the conservation of biodiversity.

Adaptation for specific species vs. biodiversity in the aggregate

There are a number of adaptation strategies aimed at protecting specific species and populations including 1) habitat manipulations, 2) conserving populations with higher genetic diversity or more plastic behaviors or morphologies, 3) changing seed sources for re-planting to introduce species or ecotypes that are better suited for future climates, 4) assisted migration to help move species and populations from current locations to those areas expected to become more suitable in the future, and 5) ex-situ conservation such as seed banking and captive breeding. Some of these strategies rely on model projections of how climate may change within a place, or where climate conditions suitable for a specific species may be found in the future. Because of the associated uncertainties with modeling future climate changes and the response of specific species and biodiversity in general to those changes, some argue in favor of more coarse-filter approaches to conservation (Hunter and others, 1988; Anderson and Ferree, 2010; Beier and Brost, 2010; Groves and others, In Press). Proponents of a coarse-filter approach advocate for the inclusion of conservation strategies that focus on enduring features that are important for biodiversity, yet less likely to be altered by climate change. The analogy is that it is important to conserve the 'stage' (the physical conditions that contribute to high levels of biodiversity) for whatever 'actors' (for example, species and populations) find those areas suitable in the future. In Northeastern United States, biodiversity is fairly well correlated with the diversity of geophysical settings (for example, geology, latitude, elevation) (Anderson and Ferree, 2010). Therefore, Anderson and Ferree (2010) advocate for designing reserves based on the location and diversity of geophysical settings rather than species distributions and numbers. Similarly, Beier and Brost (2010) promote the design of reserves and linkages based on the diversity and interspersion of land facets, or "recurring landscape units with uniform topographic and soil attributes." Since land facets and geophysical settings may not serve as surrogates for all species, and conservation goals will continue to focus on specific species rather than biodiversity in the aggregate, these 'conserving the stage' approaches are generally recommended as a complement to species-and ecosystem-specific conservation planning approaches (Beier and Brost, 2010; Groves and others, In Press).

Conservation of biodiversity vs. ecological functions and services

As climate change alters species' habitats, it will be increasingly challenging to achieve conservation goals for some highly vulnerable species. Therefore, a common suggestion is that conservation goals may need to shift from those that focus on preserving current patterns of species numbers and compositions at particular locations towards goals focused on maintaining processes, both ecological and evolutionary (Pressey and others, 2007; AFWA, 2009; Prober and Dunlop, 2011). While planners have increasingly advocated for inclusion of ecological processes and functions in conservation planning (for example, Leroux and others, 2007; Manning and others, 2009), explicitly defining conservation goals focused on these processes is a significant challenge (Groves and others, In Press). Indeed, understanding the explicit role of biodiversity components in ecosystem functioning has been an increasingly active area of research over the last 15 years (Loreau, 2010).

The relationship between biodiversity and ecosystem services is still ambiguous, stemming from a lack of consistency in how biodiversity is included within assessment frameworks (Mace and others, 2011). The treatment of biodiversity can range from being considered interchangeable with ecosystem services (that is, managing for one also manages for the other), to biodiversity being considered as one of many benefits derived from ecosystems. One potential solution to this conundrum is the redefining of biodiversity into the multiple roles it can play within an ecosystem services rubric (Mace and others, 2012): as a regulator of ecosystem processes (for example, nutrient cycling), as a final ecosystem service (for example, genetic diversity conferring resilience to climate change), and as a good (for example, flagship and umbrella species). It is therefore important for adaptation strategies to include management endpoints that relate to the multiple components of biodiversity.

Policy Considerations

In addition to the relatively direct consequences of climate change for species and other components of biodiversity described throughout this chapter, climate change will also impact issues of direct importance to humans such as water availability, food security, built infrastructures, human health, and economies. How humans respond to those impacts could have negative and, potentially, positive impacts on biodiversity (Turner and others, 2010; Bradley and others, 2012). It will be critical to guard against maladaptive human responses to climate change that may exacerbate existing threats to biodiversity (for example, armoring of coastlines or the installation of levees along flood plains to abate inundation risks), especially where natural systems may provide a more cost-effective long-term solution than an engineered solution (Kamali and others, 2010; Pérez and others, 2010). For example, protecting and augmenting coastal wetlands could buffer inland communities against sea level rise and increased storm surge, while also accruing other ecological benefits and services. The development of renewable energy resources in an effort to mitigate greenhouse gas emissions may also have unintended consequences for biodiversity. Wind and solar energy development can have large terrestrial footprints, and may potentially cause marked increases in habitat loss and fragmentation (McDonald and others, 2009; Kiesecker and others, 2011; Lovich and Ennen, 2011). Yet, there are opportunities to minimize such adverse effects. Development on lands that are already disturbed, and that have a relatively lower conservation value could keep development pressure off of more ecologically intact natural lands (Kiesecker and others, 2011; Cameron and others, 2012).

2.5.2. Techniques and Approaches for Understanding the Impacts of Climate Change and Human Actions on Biodiversity

Understanding the impacts of a changing climate on biodiversity requires coincident monitoring of climate and biodiversity. The integration of *in situ* and satellite climate observation systems, the development of Internet-accessible databases of climate information, advances in global and regional climate models, and the evolution of an international policy mechanism focused on climate change have together resulted in a global framework that provides climate information and projections to policy makers, researchers, and the general public (IPCC, 2007). A similar global framework is currently being developed for biodiversity (Scholes and others 2008; Andrefouet and others, 2008; Horning and others, 2010; Perrings and others, 2011; Scholes and others, 2012). Technical advances in observations and modeling along with increasing computing capacity are significantly improving our ability to observe and understand biodiversity patterns and processes across a range of spatial scales and levels of organization. However, greater coordination among observations, databases, modeling, and emerging policy mechanisms will be necessary to provide biodiversity information and projections commensurate with those now available for climate. Classic approaches, such as field comparisons of organismal performance and habitat responses across latitude and altitude as well as laboratory experiments with warming temperatures, continue to play a critical role in global change ecology (Ibáñez and others, written communication 2012), and have increasing power when combined with emerging techniques.

Observations

Airborne and satellite remote sensing of the Earth provide observations of ecosystems, and increasingly at the level of species. For example, multispectral sensors and more advanced cross-spectrum imaging spectrometers (for example, hyperspectral sensors) provide data on ecosystem composition and function, while active lidar and radar systems provide information on the three-dimensional structure of ecosystems. Airborne studies that combine these different types of sensors on the same aircraft are enabling significant strides in our ability to detect species and plant functional types remotely, to understand how they are changing over time, and why (Asner and others, 2008; Vierling and others, 2011; Swatantran and others, 2012). The eventual deployment of these airborne sensors on satellite platforms will enable global observations of key elements of biodiversity. Remote sensing imagery, both airborne and satellite, requires *in situ* data to validate and elaborate upon the patterns of biodiversity detected remotely. *In situ* sensors (for example, camera traps, bioacoustic recorders, and animal tracking devices), provide vital organismal level information as stand-alone datasets and may also be used in conjunction with airborne and satellite remote sensing (Boelman and others, 2007; Pennisi, 2011).

Progress in nanotechnology has enabled tracking of some of the smallest organisms such as the planktonic crustacean, *Daphnia* (Lard and others, 2010). For these technologies to inform large-scale processes such as geographic range shifts of species, tracking technologies must work over meters to kilometers, but such capabilities are limited to date. An alternative to physical tracking is measuring historic movements recorded in the genetic code of populations. Emerging high-throughput sequencing technologies (for example, 454, Illumina) have recently increased the amount of sequence information that can be captured per unit time, significantly reduced sequencing costs (Metzker 2010), and increased our abilities to make fine scale measurements of organism movements. High-volume sequencing has also brought functional studies of genetic diversity within reach of field research on non-model organisms. Such research has begun to reveal the distribution and genetic basis of organism's responses to climate change and rapid advances in the field of ecological genetics are expected in the near future (**Box 2.5**).

The integration of different types of observations at different spatial scales (for example, combining satellite pixels, species presence/absence and abundance data sets, and genetic sequencing information) offers the potential for significant advances in our understanding of biodiversity (Palumbi and others, 2009). In addition, developing biophysical techniques to translate metrics such as air temperature into organismal relevant metrics (for example, body temperature) will improve our ability to predict future species' range shifts (Helmuth and others, 2010).

Box 2.5. Recent Advances in the Genetics and Evolution of Climate Responses

When paired with field experiments and laboratory simulations, recent breakthroughs in molecular biology, particularly high-throughput genetic sequencing, are transforming our understanding of biology and climate change (Stapley and others, 2010; Ekblom and Galindo, 2011). Whole-genome sequencing is increasingly affordable so that genomes are now known for several species of ecological significance (for example, Weinstock and others, 2006, Nene and others, 2007), and these genomes provide a backbone for future genetic research and phenotypic assays. Whole-genome sequences are useful for studying genome evolution under varying environmental conditions and evolutionary change due to climate change. For example, common garden experiments in mouse-ear cress (*Arabidopsis thaliana*), combined with genome-wide genetic markers, identified locally adapted genotypes that performed best in their local climate (Fournier-Level and others, 2011). This suggests that populations within a species may deal with—and respond to—climate change in different ways. The expressed genome (all transcribed genes) also has been sequenced for several species to understand ecological responses including climatic sensitivity (for example, Meyer and others, 2009; O'Neil and others, 2010). Short-read sequencing technologies can reveal single nucleotide polymorphisms (SNPs) for mapping traits in the genome, revealing the genetic basis of phenotypes, and identifying sequences under rapid evolution, including traits and genes related to climatic tolerance (for example, Hancock and others, 2011; Ellison and others, 2011). SNPs also can be used to identify fine-scale genetic structure that reflects underlying genetic differences among populations and to quantify gene flow (for example, Hohenlohe and others, 2010), an indirect measure of dispersal capability that can indicate future capacity for geographic movement in a species (Allendorf and others, 2010). High-throughput technologies also have a key role to play in revealing novel genetic diversity and its response to changing conditions, such as in microbial communities with a preponderance of species that cannot be grown in culture (for example, Hill and others, 2002; Tringe and others, 2005). The importance of genetic variation in determining species' ecological and evolutionary responses to climate change is just beginning to be revealed, but genetics and genomics are likely to play a large role in climate change biology in the coming years.

Data Networks

In order to make connections between climate and biodiversity observations, frameworks that bring together time series of biodiversity information with coincident climate time series across spatial and temporal scales are necessary (Jones and others, 2010; McMahon and others, 2011). This integration of biodiversity and climate observations is urgently necessary whether one is trying to assess vulnerability, gage adaptation strategies, or attribute changes in biodiversity to climate or other forcing factors (Dawson and others, 2011; Pau and others, 2011). Assembling temporally rich datasets requires the construction of databases and data networks that organize, make accessible, and archive observations. Furthermore, these networks should catalyze the development of standardized protocols for collecting and analyzing data and foster the recognition of key metrics and indices of biodiversity condition for use by researchers and decision makers. A growing number of facilities serve various types of biodiversity information, and are expanding their geographic coverage (**Table 2.3**).

Citizen science offers a growing and potentially vital source of observations in a world still constrained by a lack of biodiversity time series data. Mobilizing citizen scientists not only provides additional and much needed observations, but also increases awareness of and hopefully appreciation for, biodiversity amongst the general public. If citizen science is to flourish, it needs readily accessible networks in which to deposit and share data. The U.S. National Phenology Network provides an example of such a network. It is a focal point for the public to track and share information about changes in phenology across the United States. It also enables links to organizations making phenological observations around the globe.

Genetic researchers are now required by most publishing outlets to deposit sequence information in databases that are open-access to other researchers (for example, GenBank; Benson and others, 2011). This sharing should enable comparison across species and reveal general phenomena across all branches of life. Genetic databases of annotated genomes are also helping researchers identify genes involved in climatic tolerance for non-model organisms (for example, Basic Local Alignment Search Tool (BLAST)). In addition, the Group on Earth Observations Biodiversity Observation Network (GEO BON) has worked to coordinate the growing number of regionally and taxonomically focused biodiversity networks around the world, and support the convergence of approaches for managing and making biodiversity data accessible (Scholes and others, 2008; Amaral-Zettler and others, 2011; Scholes and others, 2012).

Models

Models allow linkage of observations across spatial and temporal scales, provide understanding of the processes driving changes in biodiversity patterns seen in the observations, and enable testing of assumptions as to cause and effect. Ultimately, improved understanding of the responses of genes, species, and ecosystems to climate change rests upon combining models and observations from the realms of ecology, evolutionary biology, and climatology (Pau and others, 2011). Nonetheless, significant challenges remain in how best to combine models across different time horizons and spatial resolutions to make accurate projections for particular species and ecosystems in regions of interest. While model interoperability is necessary, it requires methods for tracking uncertainty across a diversity of links among systems, models, and organizations.

Impacts of Climate Change on Biodiversity, Ecosystems, and Ecosystem Services |
Technical Input to the 2013 National Climate Assessment

Chapter 2
Biodiversity

Table 2.3. Examples of facilities and networks that organize and archive observations of biodiversity on national and global scales for public use. Note this list is not comprehensive.

Program Name	Acronym	Program description
Global Biodiversity Information Facility	GBIF	Internet-based, globally-distributed network of databases focusing on making species information from museum collections and field observations available to the public
International Union for Conservation of Nature	IUCN	Maintains Red List of Threatened Species which is an evaluation of the conservation status of global plant and animal species
World Database on Protected Areas	WDPA	Global, spatially-explicit dataset of terrestrial and marine protected areas
UN Environment Programme World Conservation Monitoring Centre	UNEP-WCMC	Works with the IUCN and collaborating nongovernmental organizations to support the WDPA and a number of other products, for example, biodiversity atlases
Long Term Ecological Research Network	LTER	National Science Foundation (NSF) funded collaboration of scientists and students investigating ecological processes at 26 research sites
National Ecological Observatory Network	NEON	NSF funded continental-scale ecological observation platform that will detect and forecast ecological change and related impacts.
Breeding Bird Survey	BBS	USGS sponsored long-term ecological data sets on bird populations in the United States
Earth Observing System Data and Information System	EOSDIS	NASA sponsored online database of satellite-derived information, includes biodiversity- relevant information, for example, land cover condition
Comprehensive Large Array-Data Stewardship System	CLASS	NOAA sponsored online database of satellite-derived information that includes biodiversity-relevant information, for example, sea surface conditions

Policy Framework

Understanding the impacts of climate change on biodiversity requires knowledge of how biodiversity is responding to changes in climate across local, regional and global scales. There are numerous Federal, State, and other efforts within the United States that monitor elements of biodiversity. However, a lack of coordination among these efforts means that there is no national biodiversity monitoring program within the United States that addresses the wide-reaching impacts of climate or for that matter any other driver of change, on biodiversity. This limits our ability both to track the impacts of climate change on biodiversity and understand how elements of biodiversity are adapting (or not) to change. A recent report by the President's

Council of Advisors on Science and Technology (PCAST, 2011) states that U.S. Federal agencies currently spend more than $10 billion a year on conserving biodiversity and protecting or restoring ecosystem services in the United States. We know too little about the results of these investments.

In 2010, the UN General Assembly agreed to establish the Intergovernmental Platform on Biodiversity and Ecosystem Services (IPBES). Akin to an Intergovernmental Panel on Climate Change (IPCC) for biodiversity, IPBES seeks to establish an international assessment regime for biodiversity and ecosystem services between the scientific community and policymakers (Perrings and others, 2011). The PCAST 2011 report recommends that the United States contribution to IPBES take the form of a quadrennial ecosystems service trends assessment that would provide an integrated and comprehensive overview of the condition of the country's ecosystems, with projected trends in ecosystem change. It would also be closely coordinated with National Climate Assessments, currently mandated by U.S. law. Thus, IPBES provides an international policy framework for domestic action linking its assessments of climate to those for biodiversity and ecosystem services. However, conducting a proper assessment will necessitate a review of current biodiversity observation systems and data networks in order to identify where gaps exist, what needs to be done to fill them, and likely result in the identification of indicators or key biodiversity variables for monitoring systems to provide to the greater community.

2.6. SYNTHESIS OF IMPACTS ON BIODIVERSITY

The findings of this report demonstrate that climate change is having, and will continue to have, widespread and varied impacts across all components of biodiversity. The wealth of information gained from recent studies reinforces the main conclusions of the 2009 National Climate Assessment, and provides a more comprehensive understanding of the complex ways that biodiversity is responding and adapting to climate change. New technologies and approaches have largely been responsible for increasing our abilities to detect and evaluate biological and evolutionary responses to climate change, and have enabled new insights into past impacts on modern biogeography (Hoffmann and Sgro, 2011; Sandel and others, 2011). Here we return to our Key Findings to summarize our knowledge of the current and future impacts of climate change on biodiversity, key vulnerabilities and risks, and potential strategies that may be implemented to reduce risk.

Climate change is causing many species to shift their geographical ranges, distributions, and phenologies at faster rates than were previously thought; however, these rates are not uniform across species. In the Northern Hemisphere, springtime temperatures are advancing by an average of 2.08 days/decade in the oceans and by 1.46 days/decade on land; most, but not all, marine and terrestrial populations are advancing their springtime phenologies to track these warming patterns. The velocity of range shifts for marine taxa exceeds those reported for terrestrial organisms, leading to numerous local extinctions in sub-polar regions, the tropics, and semi-enclosed seas. Together with invasions from warmer latitudes, these extinctions are expected to result in species turnover of greater than 60 percent in the world oceans. New evidence suggests that terrestrial organisms are moving up in elevation at rates 2 – 3 times greater than was previously estimated (Burrows and others, 2011; Chen and others, 2011).

Impacts of Climate Change on Biodiversity, Ecosystems, and Ecosystem Services |
Technical Input to the 2013 National Climate Assessment

Chapter 2
Biodiversity

However, geographical range and distribution shifts are not consistent among species and populations, and some are not shifting at all. Species and populations that are unable to shift their geographic distributions or have narrow environmental tolerances are at an increased risk of extinction.

Increasing evidence suggests that range shifts and novel climates will result in new community assemblages, new associations among species, and promote interactions that have not existed in the past. Shifts in the seasonal and spatial distributions of flora and fauna within marine, aquatic, and terrestrial environments would result in trophic mismatches, asynchronies, and altered population dynamics. New species assemblages would substantially alter the structure, function, and flow of energy through ecosystems. Biological interactions are complex, difficult to predict, and have resulted in counterintuitive outcomes.

Differences in how organisms respond to climate change determine which species or populations will benefit (winners), and which will decline and possibly go extinct (losers) in response to climate change. There is increasing evidence of population declines and localized extinctions that can be directly attributed to climate change. This is in part because there are both biotic (for example, genetic) and abiotic (for example, habitat) limits to the degree to which organisms and systems can cope with climate change. Environmental and ecological shifts caused by climate change may be favorable to some elements of biodiversity thereby promoting range and population growth. Species turnover is projected to be greatest at high latitudes and at high altitudes as organisms move poleward, up in elevation, and decline due to loss of suitable habitat. The cumulative effect of climate change is projected to result in a net loss of global biodiversity.

The potential for biodiversity to respond to climate change over short (plasticity) and long (evolutionary) time scales is enhanced by increased genetic diversity; however, the rate of climate change may outpace species' and population's capacity to adjust to environmental change. Climate induced range shifts and population declines are expected to increase the prevalence of population bottlenecks, and reduce genetic diversity within and among species. Long-lived species are particularly vulnerable to climate changes because they experience longer generation times, lower population turnover rates, and slower rates of evolution. The potential for biodiversity to cope with the impacts of climate change can be maximized by maintaining high genetic diversity among and within species and population, conserving environmental heterogeneity, and reducing barriers to dispersal.

Identifying highly vulnerable species and understanding why they are vulnerable are critical to developing climate change adaptation strategies and reducing biodiversity loss in the coming decades. Biodiversity's exposure, sensitivity, and adaptive capacity to climate change is very likely to be non-uniform across the United States, thus different organisms and ecosystems face greater risk of loss than others. Ecological specialists, species that live at high altitudes and latitudes, and species that live at or near their thermal limits are particularly vulnerable to climate change. Climate-induced changes in species' abundance, can lead to local and global extinctions that have consequences for ecosystem function and services. Human responses to climate change have the potential to exacerbate impacts on biodiversity; therefore, mangers need to integrate risk-based analyses and adaptation principles into their decision making process.

Impacts of Climate Change on Biodiversity, Ecosystems, and Ecosystem Services |
Technical Input to the 2013 National Climate Assessment

Chapter 2
Biodiversity

Existing environmental regulations currently lack criteria for categorizing the degree of species imperilment posed by climate change, and how those considerations factor into listing or delisting species once they are recognized under governmental protection. Vulnerability Assessments and other decision support tools will be critical to identify species most at risk to climate change, and to develop adaptation strategies that reduce extinction potential; however many of these frameworks are still being tested.

As *species shift in space and time in response to climate change, effective management and conservation decisions require consideration of uncertain future projections as well as historic conditions.* Human responses to climate change can have unintended impacts on biodiversity. Therefore, risk-based framing, scenario development, and engagement of stakeholders will be essential in enhancing our ability to respond to the impacts of climate change. Furthermore, greater coordination among observations, databases, modeling, and policy mechanisms will increase our ability to detect, track, project, and understand climate induced changes in biodiversity.

Broader and more coordinated monitoring efforts across Federal and State agencies are necessary to support biodiversity research, management, assessment, and policy. Evaluating status, trends, and gaps in national and global biodiversity will require integrated research and monitoring efforts as species and ecosystem boundaries shift due to climate change. Existing monitoring networks could be improved by integrating biodiversity and climate observations, data networks, models, and policy frameworks to detect and attribute the impacts of climate change on biodiversity.

2.7. CRITICAL GAPS IN KNOWLEDGE, RESEARCH, AND DATA NEEDS

The findings in this report are supported by a broad spectrum of high-quality and recent research, and highlight the challenges resource managers now face in adapting to the impacts of climate change. We conclude by identifying critical gaps and the activities needed to support ongoing climate change research initiatives and best management practices.

We are just beginning to develop an understanding of the underlying mechanisms that influence biological responses to climate change. There is abundant evidence and agreement that the degree to which organisms will tolerate new conditions imposed by climate change will vary across species and populations, but we cannot yet predict the extent to which phenotypic plasticity, evolutionary adaptation, and non-genetic parental effects will allow species to adjust. Basic information on species and population traits ranging from physiology to behavior, life history characteristics, current distributions, dispersal abilities, and ecological relationships is needed to understand why some species and populations are able to adjust to the impacts of climate change (while others decline), and will be critical for building better models to forecast future biological responses and vulnerabilities (McMahan and others, 2011).

As climate change continues to cause species to shift their geographical ranges, distributions, and phenologies there will likely be disruptions to community interactions. A growing body of research is focused on trophic mismatches, yet few studies have been able to document the ramifications of asynchronous species interactions. This is presumably because

Impacts of Climate Change on Biodiversity, Ecosystems, and Ecosystem Services |
Technical Input to the 2013 National Climate Assessment

Chapter 2
Biodiversity

many of the effects are only apparent over longer time periods than the studies were able to evaluate. There is high agreement among researchers that trophic mismatches will likely have negative implications for ecosystem processes, ecosystem services, and our capacity for climate change adaptation (Miller-Rushing and others, 2010; Thackeray and others, 2010; Yang and Rudolf, 2010). However, additional experimental and field research is needed to improve our abilities to detect, attribute, and predict changes in these relationships as well as the emergence of novel interactions and species assemblages.

Projecting climate change impacts on biodiversity involves many uncertainties (Pereira and others, 2010; Bellard and others, 2012) stemming from variability in climate projections (particularly precipitation patterns), uncertainties in future emissions, and assumptions and uncertainties in the models used to project species responses and extinctions (He and Hubbell, 2011). Some of these uncertainties are inevitable given that we are trying to predict the future; nonetheless, techniques and modeling approaches are becoming more sophisticated and able to evaluate myriad influences such as biotic interactions and dispersal abilities that were previously deficient. Projections are also complicated by uncertainty about where and how human responses to climate change are likely to impact biodiversity. Sustainable energy development and infrastructure, changes in agricultural practices, human migrations, and changes in water extraction and storage practices in response to climate change are all very likely to have impacts on biodiversity. Predicting where these mitigation and adaptation responses will occur, and how they will impact biodiversity will be a critical step in developing credible future climate change impact scenarios. Although many tools for forecasting climate change impacts on ecosystem services exist (Kareiva and others, 2011), fewer methods for anticipating how people will respond to those impacts have been developed or incorporated into projected impacts on biodiversity.

Our collective understanding of biodiversity and its importance to ecosystems and the services they provide is advancing across all scales of biological organization. While longstanding threats remain such as habitat conversion and loss, the impacts of climate change on biodiversity are evident and are likely to become increasingly significant in the future. Establishing and implementing climate adaptation planning will be critical to the success of resource management under uncertain future conditions. Vulnerability assessments are one tool that will assist adaptation planning, however the rigor and application of these frameworks are still in the process of being developed and require further testing (Angert and others, 2011). Efforts that incorporate adaptive management principles into their practices should be closely monitored with an aim to inform and improve the effectiveness of future adaptation planning and response. Policies for protecting and managing biodiversity will also need to incorporate new flexibility to allow actions to be taken under uncertainty. Lastly, improved observation capabilities, more sophisticated data infrastructures and modeling platforms, as well as coordinated, landscape-level monitoring approaches will continue to be essential in improving climate change research.

Impacts of Climate Change on Biodiversity, Ecosystems, and Ecosystem Services |
Technical Input to the 2013 National Climate Assessment

Chapter 2
Biodiversity

2.8. LITERATURE CITED

AFWA (Association of Fish and Wildlife Agencies). 2009. Voluntary guidance for States to incorporate climate change into State Wildlife Action Plans and other management plans. A collaboration of the AFWA climate change and teaming With Wildlife Committees. Association of Fish and Wildlife Agencies.

Aldridge G, Inouye DW, Forrest JRK, Barr WA, and Miller-Rushing AJ. 2011. Emergence of a mid-season period of low floral resources in a montane meadow ecosystem associated with climate change. *Journal of Ecology* **99**: 905-913.

Allendorf FW, Hohenlohe PA, and Luikart G. 2010. Genomics and the future of conservation genetics. *Nature Reviews Genetics* **11**: 697-709.

Alo CA, and Wang G. 2008. Potential future changes of the terrestrial ecosystem based on climate projections by eight general circulation models. *Journal of Geophysical Research-Biogeosciences* **113**: G01004.

Altermatt F. 2010. Tell me what you eat and I'll tell you when you fly: diet can predict phenological changes in response to climate change. *Ecology Letters* **13**: 1475-1484.

Amaral-Zettler L, Duffy JE, Fautin D, Paulay G, Rynearson T, Sosik H, and Stachowicz J. 2011. Attaining an operational marine biodiversity observation network (BON) synthesis report. Available at: http://www.oceanleadership.org/tag/observing-network/

Anderegg WRL, Berry JA, Smith DD, Sperry JS, Anderegg LDL, and Field CB. 2012. The roles of hydraulic and carbon stress in a widespread climate-induced forest die-off. *Proceedings of the National Academy of Sciences* **109**: 233-237.

Anderson MG, and Ferree CE. 2010. Conserving the stage: Climate change and the geophysical underpinnings of species diversity. *PLoS ONE* **5**: e11554.

Andrefouet S, Costello MJ, Faith DP. 2008. The GEO Biodiversity Observation Network: Concept Document. Group on Earth Observations, Geneva, Switzerland. 43 p. Available at: http://nora.nerc.ac.uk/5620/

Angert AL, Crozier LG, Rissler LJ, Gilman SE, Tewksbury JJ, and Chunco AJ. 2011. Do species' traits predict recent shifts at expanding range edges? *Ecology Letters* **14**: 677-689.

Araujo MB, and New M. 2007. Ensemble forecasting of species distributions. *Trends in Ecology and Evolution* **22**: 42-47.

Ashcroft MB. 2010. Identifying refugia from climate change. *Journal of Biogeography* **37**: 1407-1413.

Asner GP, Hughes RF, Vitousek PM, Knapp DE, Kennedy-Bowdoin T, Boardman J, Martin RE, Eastwood M, and Green RO. 2008. Invasive plants transform the three-dimensional structure of rain forests. *Proceedings of the National Academy of Sciences* **105**: 4519-4523.

Ault TR, Macalady AK, Pederson GT, Betancourt JL, and Schwartz MD. 2011. Northern Hemisphere modes of variability and the timing of spring in western North America. *Journal of Climate* **24**: 4003-4014.

Bagne KE, Friggens MM, and Finch DM. 2011. A system for assessing vulnerability of species (SAVS) to climate change. Department of Agriculture, Forest Service, Rocky Mountain Research Station, Fort Collins, CO. Available at: http://www.treesearch.fs.fed.us/pubs/37850

Barnosky AD, Matzke N, Tomiya S, Wogan GOU, Swartz B, Quental TB, Marshall C, McGuire JL, Lindsey EL, Maguire KC, Mersey B, and Ferrer EA. 2011. Has the Earth's sixth mass extinction already arrived? *Nature* **471**: 51-57.

Bartomeus I, Ascher JS, Wagner D, Danforth BN, Colla S, Kornbluth S, and Winfree R. 2011. Climate-associated phenological advances in bee pollinators and bee-pollinated plants. *Proceedings of the National Academy of Sciences* **108**: 20645-20649.

Battin J, Wiley MW, Ruckelshaus MH, Palmer RN, Korb E, Bartz KK, and Imaki H. 2007. Projected impacts of climate change on salmon habitat restoration. *Proceedings of the National Academy of Sciences* **104**: 6720–6725.

Beaugrand G, Edwards M, and Legendre L. 2010. Marine biodiversity, ecosystem functioning, and carbon cycles. *Proceedings of the National Academy of Sciences* **107**: 10120-10124.

Beaumont LJ, Pitman A, Perkins S, Zimmermann NE, Yoccoz NG, and Thuiller W. 2011a. Impacts of climate change on the world's most exceptional ecoregions. *Proceedings of the National Academy of Sciences* **108**: 2306-2311.

Beier P, and Brost B. 2010. Use of Land Facets to Plan for climate change: Conserving the arenas, not the actors. *Conservation Biology* **24**: 701-710.

Bellard C, Bertelsmeier C, Leadley P, Thuiller W, and Courchamp F. 2012. Impacts of climate change on the future of biodiversity. *Ecology Letters* **15**: 365–377.

Benson DA, Karsch-Mizrachi I, Lipman DJ, Ostell J, and Sayers EW. 2011. GenBank. *Nucleic Acids Research* **39**: D32-D37.

Bergengren JC, Waliser DE, and Yung YL. 2011. Ecological sensitivity: a biospheric view of climate change. *Climatic Change* **107**: 433-457.

Bertness M, Garrity S, and Levings S. 1981. Predation pressure and gastropod foraging: a tropical-temperate comparison. *Evolution* **35**: 995–1007.

Best AS, Johst K, Muenkemueller T, and Travis JMJ. 2007. Which species will succesfully track climate change? The influence of intraspecific competition and density dependent dispersal on range shifting dynamics. *Oikos* **116**: 1531-1539.

Bindoff NL, Willebrand J, Artale V, Cazenave A, Gregory J, Gulev S, Hanawa K, Le Quéré C, Levitus S, Nojiri Y, Shum CK, Talley LD, and Unnikrishnan A. 2007. Observations: oceanic climate change and sea level. *In* Solomon S, Qin D, Manning M, Chen Z, Marquis M, Averyt KB, Tignor M, and Miller HL (Eds), Climate Change 2007: The Physical Science Basis: Contribution of Working Group I to the Fourth Assessment Report of the Intergovernmental Panel on Climate Change. Cambridge University Press, Cambridge, United Kingdom and New York NY. 385-432 p.

Boelman NT, Asner GP, Hart PJ, and Martin RE. 2007. Multi-trophic invasion resistance in Hawaii: Bioacoustics, field surveys, and airborne remote sensing. *Ecological Applications* **17**: 2137-2144.

Bolser RC, and Hay ME. 1996. Are tropical plants better defended? Palatability and defenses of temperate vs tropical seaweeds. *Ecology* **77**: 2269-2286.

Botero CA, and Rubenstein DR. 2012. Fluctuating environments, sexual selection and the evolution of flexible mate choice in birds. *PLoS ONE* **7**: e32311.

Both C, van Asch M, Bijlsma RG, van den Burg AB, and Visser ME. 2009. Climate change and unequal phenological changes across four trophic levels: constraints or adaptations? *Journal of Animal Ecology* **78**: 73-83.

Botkin DB, Saxe H, Araujo MB, Betts R, Bradshaw RHW, Cedhagen T, Chesson P, Dawson TP, Etterson JR, Faith DP, Ferrier S, Guisan A, Hansen AS, Hilbert DW, Loehle C, Margules C, New M, Sobel MJ, and Stockwell DRB. 2007. Forecasting the effects of global warming on biodiversity. *BioScience* **57**: 227-236.

Bouchet P. 2006. The magnitude of marine biodiversity. *In* Duarte C (Ed), The exploration of marine biodiversity: scientific and technological challenges. Fundación BBVA. 31 - 62 p.

Bradley BA, Estes LD, Hole DG, Holness S, Oppenheimer M, Turner WR, Beukes H, Schulze RE, Tadross MA, and Wilcove DS. 2012. Predicting how adaptation to climate change could affect ecological conservation: secondary impacts of shifting agricultural suitability. *Diversity and Distributions* **18**: 425-437.

Brander K. 2010. Impacts of climate change on fisheries. *Journal of Marine Systems* **79**: 389-402.

Brook BW, Sodhi NS, and Bradshaw CJA. 2008. Synergies among extinction drivers under global change. *Trends in Ecology and Evolution* **23**: 453-460.

Bruno JF, Selig ER, Casey KS, Page CA, Willis BL, Harvell CD, Sweatman H, and Melendy AM. 2007. Thermal stress and coral cover as drivers of coral disease outbreaks. *PLoS ONE Biology* **5**: 1220-1227.

Buckley LB, Urban MC, Angilletta MJ, Crozier LG, Rissler LJ, and Sears MW. 2010. Can mechanism inform species' distribution models? *Ecology Letters* **13**: 1041-1054.

Buisson L, and Grenouillet G. 2009. Contrasted impacts of climate change on stream fish assemblages along an environmental gradient. *Diversity and Distributions* **15**: 613-626.

Burns CE, Johnston KM, and Schmitz OJ. 2003. Global climate change and mammalian species diversity in U.S. national parks. *Proceedings of the National Academy of Sciences* **100**: 11474-11477

Burrows MT, Schoeman DS, Buckley LB, Moore P, Poloczanska ES, Brander KM, Brown C, Bruno JF, Duarte CM, Halpern BS, Holding J, Kappel CV, Kiessling W, O'Connor MI, Pandolfi JM, Parmesan C, Schwing FB, Sydeman WJ, and Richardson AJ. 2011. The pace of shifting climate in marine and terrestrial ecosystems. *Science* **334**: 652-655.

Butchart SHM, Walpole M, Collen B, van Strien A, Scharlemann JPW, Almond REA, Baillie JEM, Bomhard B, Brown C, Bruno J, Carpenter KE, Carr GM, Chanson J, Chenery AM, Csirke J, Davidson NC, Dentener F, Foster M, Galli A, Galloway JN, Genovesi P, D Gregory R, Hockings M, Kapos V, Lamarque J-F, Leverington F, Loh J, McGeoch MA, McRae L, Minasyan A, Hernández Morcillo M, Oldfield TEE, Pauly D, Quader S, Revenga C, Sauer JR, Skolnik B, Spear D, Stanwell-Smith D, Stuart SN, Symes A, Tierney M, Tyrrell TD, Vié J-C, and Watson R. 2010. Global biodiversity: indicators of recent declines. *Science* **328**: 1164-1168.

Byrnes JE, Reynolds PL, and Stachowicz JJ. 2007. Invasions and extinctions reshape coastal marine food webs. *PLoS ONE* **2**:e295.

Bürger R, and Lynch M. 1995. Evolution and extinction in a changing environment: a quantitative-genetic analysis. *Evolution* **49**: 151-163.

Cameron DR, Cohen B, and Morrison S. 2012. An approach to enhance the conservation-compatibility of solar energy development. *PLoS ONE* 7(6): e38437. doi:10.1371/journal.pone.0038437.

Carpenter KE, Abrar M, Aeby G, Aronson RB, Banks S, Bruckner A, Chiriboga A, Cortes J, Delbeek JC, DeVantier L, Edgar GJ, Edwards AJ, Fenner D, Guzman HM, Hoeksema

BW, Hodgson G, Johan O, Licuanan WY, Livingstone SR, Lovell ER, Moore JA, Obura DO, Ochavillo D, Polidoro BA, Precht WF, Quibilan MC, Reboton C, Richards ZT, Rogers AD, Sanciangco J, Sheppard A, Sheppard C, Smith J, Stuart S, Turak E, Veron JEN, Wallace C, Weil E, and Wood E. 2008. One-third of reef-building corals face elevated extinction risk from climate change and local impacts. *Science* **321**: 560-563.

Carroll C. 2007. Interacting effects of climate change, landscape conversion, and harvest on carnivore populations at the range margin: marten and lynx in the Northern Appalachians. *Conservation Biology* **21**: 1092-1104.

Chapin FS, Walker BH, Hobbs RJ, Hooper DU, Lawton JH, Sala OE, and Tilman D. 1997. Biotic control over the functioning of ecosystems. *Science* **277**: 500-504.

Chen IC, Hill JK, Ohlemuller R, Roy DB, and Thomas CD. 2011. Rapid range shifts of species associated with high levels of climate warming. *Science* **333**: 1024-1026.

Chester CC, Hilty JA, and Trombulak SC. In Press. Climate change science, impacts and opportunities. *In* Hilty J.A, Chester CC, and Cross MS (Eds), Climate and Conservation: Landscape and Seascape Science, Planning and Action. Island Press, Washington, DC.

Cheung WWL, Lam VWY, Sarmiento JL, Kearney K, Watson R, and Pauly D. 2009. Projecting global marine biodiversity impacts under climate change scenarios. *Fish and Fisheries* **10**: 235-251.

Clavel J, Julliard R, and Devictor V. 2011. Worldwide decline of specialist species: toward a global functional homogenization? *Frontiers in Ecology and the Environment* **9**: 222-228.

Clavero M, Villero D, and Brotons L. 2011. Climate change or land use dynamics: Do we know what climate change indicators indicate? *PLoS ONE* **6**: e18581.

Collette BB, Carpenter KE, Polidoro BA, Juan-Jorda MJ, Boustany A, Die DJ, Elfes C, Fox W, Graves J, Harrison LR, McManus R, Minte-Vera CV, Nelson R, Restrepo V, Schratwieser J, Sun CL, Amorim A, Brick Peres M, Canales C, Cardenas G, Chang SK, Chiang WC, de Oliveira Leite N, Jr., Harwell H, Lessa R, Fredou FL, Oxenford HA, Serra R, Shao KT, Sumaila R, Wang SP, Watson R, and Yanez E. 2011. High value and long life-double jeopardy for tunas and billfishes. *Science* **333**: 291-292.

Collie JS, Wood AD, and Jeffries HP. 2008. Long-term shifts in the species composition of a coastal fish community. *Canadian Journal of Fisheries and Aquatic Sciences* **65**: 1352-1365.

Craine JM, Towne EG, Joern A, and Hamilton RG. 2009. Consequences of climate variability for the performance of bison in tallgrass prairie. *Global Change Biology* **15**: 772-779.

Crausbay SD, and Hotchkiss SC. 2010. Strong relationships between vegetation and two perpendicular climate gradients high on a tropical mountain in Hawai'i. *Journal of Biogeography* **37**: 1160-1174.

Crimmins TM, Crimmins MA, and Bertelsen CD. 2009. Flowering range changes across an elevation gradient in response to warming summer temperatures. *Global Change Biology* **15**: 1141-1152.

Crimmins TM, Crimmins MA, and Bertelsen CD. 2010. Complex responses to climate drivers in onset of spring flowering across a semi-arid elevation gradient. *Journal of Ecology* **98**: 1042-1051.

Crimmins TM, Crimmins MA, and Bertelsen CD. 2011a. Onset of summer flowering in a 'Sky Island' is driven by monsoon moisture. *New Phytologist* **191**: 468-479.

Impacts of Climate Change on Biodiversity, Ecosystems, and Ecosystem Services | Technical Input to the 2013 National Climate Assessment

Chapter 2
Biodiversity

Crimmins SM, Dobrowski SZ, Greenberg JA, Abatzoglou JT, and Mynsberge AR. 2011b. Changes in Climatic Water Balance Drive Downhill Shifts in Plant Species' Optimum Elevations. *Science* **331**: 324-327.

Cross MS, McCarthy PD, Garfin G, Gori D, and Enquist CAF. Accepted. Accelerating climate change adaptation for natural resources in southwestern United States. *Conservation Biology*.

Daufresne M, Lengfellner K, and Sommer U. 2009. Global warming benefits the small in aquatic ecosystems. *Proceedings of the National Academy of Sciences* **106**: 12788-12793.

Dawson TP. 2011. Beyond predictions: Biodiversity conservation in a changing climate (vol 332, pg 53, 2011). *Science* **332**: 664-664.

deBuys W. 2008. Welcome to the anthropocene. *Rangelands* **30**: 31-35.

Deutsch CA, Tewksbury JJ, Huey RB, Sheldon KS, Ghalambor CK, Haak DC, and Martin PR. 2008. Impacts of climate warming on terrestrial ectotherms across latitude. *Proceedings of the National Academy of Sciences* **105**: 6668-6672.

Devictor V, Julliard R, Clavel J, Jiguet F, Lee A, and Couvet D. 2008. Functional biotic homogenization of bird communities in disturbed landscapes. *Global Ecology and Biogeography* **17**: 252-261.

Diamond SE, Frame AM, Martin RA, and Buckley LB. 2011. Species' traits predict phenological responses to climate change in butterflies. *Ecology* **92**: 1005-1012.

Diaz S, Symstad AJ, Chapin FS, Wardle DA, and Huenneke LF. 2003. Functional diversity revealed by removal experiments. *Trends in Ecology and Evolution* **18**: 140-146.

Dijkstra JA, Boudreau J, and Dionne M. 2011. Species-specific mediation of temperature and community interactions by multiple foundation species. *Oikos* **121**(5): 646-654.

Dirzo R, and Raven PH. 2003. Global state of biodiversity and loss. *Annual Review of Environment and Resources* **28**: 137-167.

Doak DF, and Morris WF. 2010. Demographic compensation and tipping points in climate-induced range shifts. *Nature* **467**: 959-962.

Domenici P, Allan B, McCormick MI, and Munday PL. 2012. Elevated carbon dioxide affects behavioural lateralization in a coral reef fish. *Biology Letters* **8**: 78-81.

Donelson JM, Munday PL, McCormick MI, and Pitcher CR. 2011. Rapid transgenerational acclimation of a tropical reef fish to climate change. *Nature Climate Change* **2**: 30-32.

Doney SC, Ruckelshaus M, Duffy JE, Barry JP, Chan F, English CA, Galindo HM, Grebmeier JM, Hollowed AB, Knowlton N, Polovina J, Rabalais NN, Sydeman WJ, and Talley LD. 2012. Climate change impacts on marine ecosystems. *Annual Reviews in Marine Science* **4**: 4.1–4.27.

Dubois NS, DeWan A, Boshoven JL, and Parsons DC. 2011. Using species-level vulnerability assessments to inform conservation planning under climate change. Defenders of Wildlife.

Dunnell KL, and Travers SE. 2011. Shifts in the flowering phenology of the Northern Great Plains: Patterns over 100 years. *American Journal of Botany* **98**: 935-945.

Durak T. 2010. Long-term trends in vegetation changes of managed versus unmanaged Eastern Carpathian beech forests. *Forest Ecology and Management* **260**: 1333-1344.

Eakin CM, Morgan JA, Heron SF, Smith TB, Liu G, Alvarez-Filip L, Baca B, Bartels E, Bastidas C, Bouchon C, Brandt M, Bruckner AW, Bunkley-Williams L, Cameron A, Causey BD, Chiappone M, Christensen TRL, Crabbe MJC, Day O, de la Guardia E,

Impacts of Climate Change on Biodiversity, Ecosystems, and Ecosystem Services | Technical Input to the 2013 National Climate Assessment

Chapter 2
Biodiversity

Diaz-Pulido G, DiResta D, Gil-Agudelo DL, Gilliam DS, Ginsburg RN, Gore S, Guzman HM, Hendee JC, Hernandez-Delgado EA, Husain E, Jeffrey CFG, Jones RJ, Jordan-Dahlgren E, Kaufman LS, Kline DI, Kramer PA, Lang JC, Lirman D, Mallela J, Manfrino C, Marechal J-P, Marks K, Mihaly J, Miller WJ, Mueller EM, Muller EM, Orozco Toro CA, Oxenford HA, Ponce-Taylor D, Quinn N, Ritchie KB, Rodriguez S, Rodriguez Ramirez A, Romano S, Samhouri JF, Sanchez JA, Schmahl GP, Shank BV, Skirving WJ, Steiner SCC, Villamizar E, Walsh SM, Walter C, Weil E, Williams EH, Roberson KW, and Yusuf Y. 2010. Caribbean corals in crisis: Record thermal stress, bleaching, and mortality in 2005. *PLoS ONE* **5**: e13969.

Early R, and Sax DF. 2011. Analysis of climate paths reveals potential limitations on species range shifts. *Ecology Letters* **14**: 1125-1133.

Eastmann LM, Morelli TL, Rowe KC, Conroy CJ, and Moritz C. 2012. Size increase in high elevation ground squirrels over the last century. *Global Change Biology* **18**(5): 1499-1508.

Edwards CJ, Suchard MA, Lemey P, Welch JJ, Barnes I, Fulton TL, Barnett R, O'Connell TC, Coxon P, Monaghan N, Valdiosera CE, Lorenzen ED, Willerslev E, Baryshnikov GF, Rambaut A, Thomas MG, Bradley DG, and Shapiro B. 2011. Ancient hybridization and an Irish origin for the modern polar bear matriline. *Current Biology* **21**: 1251-1258.

Ehrhardt NM, and Deleveaux VKW. 2007. The Bahamas' Nassau grouper (*Epinephelus striatus*) fishery - two assessment methods applied to a data-deficient coastal population. *Fisheries Research* **87**: 17-27.

Ekblom R, and Galindo J. 2011. Applications of next generation sequencing in molecular ecology of non-model organisms. *Heredity* **107**: 1-15.

Ellison CE, Hall C, Kowbel D, Welch J, Brem RB, Glass NL, and Taylor JW. 2011. Population genomics and local adaptation in wild isolates of a model microbial eukaryote. *Proceedings of the National Academy of Sciences* **108**: 2831-2836.

Erwin T. 1982. Tropical forests: their richness in Coleoptera and other arthropod species. *Coleopterists Bulletin* **36**: 74-75.

Erwin TL. 1991. How Many Species are there - Revisited. *Conservation Biology* **5**: 330-333.

ESA (Endangered Species Act). 1973. 16 U.S.C. 1531-1544, 87 Stat. 884 -- Public Law 93-205.

Fabry VJ, Seibel BA, Feely RA, and Orr JC. 2008. Impacts of ocean acidification on marine fauna and ecosystem processes. *ICES Journal of Marine Science* **65**: 414-432.

Fautin D, Dalton P, Incze LS, Leong JAC, Pautzke C, Rosenberg A, Sandifer P, Sedberry G, Tunnell JW, Abbott I, Brainard RE, Brodeur M, Eldredge LG, Feldman M, Moretzsohn F, Vroom PS, Wainstein M, and Wolff N. 2010. An overview of marine biodiversity in United States waters. *PLoS ONE* **5**: e11914.

Feder ME. 2010. Physiology and global climate change. *Annual Review of Physiology* **72**: 123-125.

Feder ME, and Hofmann GE. 1999. Heat-shock proteins, molecular chaperones, and the stress response: Evolutionary and ecological physiology. *Annual Review of Physiology* **61**: 243-282.

Feely RA, Doney SC, and Cooley SR. 2009. Ocean acidification: Present conditions and future changes in a high-CO_2 world. *Oceanography* **22**: 36-47.

Impacts of Climate Change on Biodiversity, Ecosystems, and Ecosystem Services |
Technical Input to the 2013 National Climate Assessment

Chapter 2
Biodiversity

Ferrari MCO, Elvidge CK, Jackson CD, Chivers DP, and Brown GE. 2010. The responses of prey fish to temporal variation in predation risk: sensory habituation or risk assessment? *Behavioral Ecology* **21**: 532-536.

Ferrari MCO, Manassa RP, Dixson DL, Munday PL, McCormick MI, Meekan MG, Sih A, and Chivers DP. 2012. Effects of Ocean Acidification on Learning in Coral Reef Fishes. *PLoS ONE* **7**: e31478.

Flather CH, Wilson KR, Dean DJ, and McComb WC. 1997. Identifying gaps in conservation networks: Of indicators and uncertainty in geographic-based analyses. *Ecological Applications* **7**: 531-542.

Flaxman M, and Vargas-Moreno JC. 2011. Considering Climate Change in State Wildlife Action Planning: A Spatial Resilience Planning Approach. Department of Urban Studies and Planning, Massachusetts Institute of Technology, Cambridge, MA.

Fodrie FJ, Heck KL, Powers SP, Graham WM and Robinson KL. 2010. Climate-related, decadal-scale assemblage changes of seagrass-associated fishes in the northern Gulf of Mexico. *Global Change Biology* **16**: 48-59.

Ford SE, and Smolowitz R. 2007. Infection dynamics of an oyster parasite in its newly expanded range. *Marine Biology* **151**: 119-133.

Forrest JRK, and Thomson JD. 2011. An examination of synchrony between insect emergence and flowering in Rocky Mountain meadows. *Ecological Monographs* **81**: 469-491.

Forster J, Hirst AG, and Woodward G. 2011. Growth and Development Rates Have Different Thermal Responses. *American Naturalist* **178**: 668-678.

Fournier-Level A, Korte A, Cooper MD, Nordborg M, Schmitt J, and Wilczek. 2011. A Map of Local Adaptation in *Arabidopsis thaliana. Science* **334** (6052): 86-89.

Franks SJ, Sim S, and Weis AE. 2007. Rapid evolution of flowering time by an annual plant in response to a climate fluctuation. *Proceedings of the National Academy of Sciences* **104**: 1278-1282.

Franks SJ, and Weis AE. 2008. A change in climate causes rapid evolution of multiple life-history traits and their interactions in an annual plant. *Journal of Evolutionary Biology* **21**: 1321-1334.

Fuller A, Dawson T, Helmuth B, Hetem RS, Mitchell D, and Maloney SK. 2010. Physiological mechanisms in coping with climate change. *Physiological and Biochemical Zoology* **83**: 713-720.

FWC (Florida Fish and Wildlife Conservation Commission). 2011. Chapter 4: Florida adapting to climate change in Florida's Wildlife Legacy Initiative: Florida's State Wildlife Action Plan. Tallahassee, FL.

Game ET, Groves C, Andersen M, Cross M, Enquist C, Ferdaña Z, Girvetz E, Gondor A, Hall K, Higgins J, Marshall R, Popper K, Schill S, and Shafer SL. 2010. Incorporating climate change adaptation into regional conservation assessments. The Nature Conservancy, Arlington, Virginia.

Gardner JL, Peters A, Kearney MR, Joseph L, and Heinsohn R. 2011. Declining body size: a third universal response to warming? *Trends in Ecology and Evolution* **26**: 285-291.

Garroway CJ, Bowman J, Cascaden TJ, Holloway GL, Mahan CG, Malcolm JR, Steele MA, Turner G, and Wilson PJ. 2010. Climate change induced hybridization in flying squirrels. *Global Change Biology* **16**: 113-121.

Geyer J, Kiefer I, Kreft S, Chavez V, Salafsky N, Jeltsch F, and Ibisch PL. 2011. Classification of Climate-Change-induced stresses on biological diversity. *Conservation Biology* **25**: 708-715.

Gibbs AG. 2011. Thermodynamics of cuticular transpiration. *Journal of Insect Physiology* **57**: 1066-1069.

Glick P, Stein BA, and Edelson N. 2011. Scanning the Conservation Horizon: A Guide to Climate Change Vulnerability Assessment. National Wildlife Federation, Washington, DC.

Goldstein JB. 2010. Will climate change help or harm species listing? *Sustainable Development Law and Policy* **43**: 57.

Gonzalez P, Neilson RP, Lenihan JM, and Drapek RJ. 2010. Global patterns in the vulnerability of ecosystems to vegetation shifts due to climate change. *Global Ecology and Biogeography* **19**: 755-768.

Groves CR, Game ET, Anderson MG, Cross M, Enquist C, Ferdana Z, Girvetz E, Gondor A, Hall KR, Higgins J, Marshall R, Popper K, Schill S, and Shafer SL. In Press. Incorporating climate change into systematic conservation planning. *Biodiversity and Conservation.*

Halofsky JE, Peterson DL, O'Halloran KA, and Hawkins-Hoffman C. 2011. Adapting to climate change at Olympic National Forest and Olympic National Park. U.S. Department of Agriculture, Forest Service, Pacific Northwest Research Station, Portland, OR.

Hancock AM, Brachi B, Faure N, Horton MW, Jarymowycz LB, Sperone FG, Toomajian C, Roux F, and Bergelson J. 2011. Adaptation to climate across the *Arabidopsis thaliana* genome. *Science* **333**: 83-86.

Hare JA, Alexander MA, Fogarty MJ, Williams EH, and Scott JD. 2010. Forecasting the dynamics of a coastal fishery species using a coupled climate-population model. *Ecological Applications* **20**: 452-464.

Harley CDG. 2011. Climate change, keystone predation, and biodiversity loss. *Science* **334**: 1124-1127.

Harte J, and Kitzes J. 2011. The use and misuse of species area relationships in predicting climate driven extinction *In* Hannah L (Ed), Saving a million species: Extinction risk from climate change. Island Press, Washington, DC.

Harvell CD, Mitchell CE, Ward JR, Altizer S, Dobson AP, Ostfeld RS, and Samuel MD. 2002. Ecology - Climate warming and disease risks for terrestrial and marine biota. *Science* **296**: 2158-2162.

He F, and Hubbell SP. 2011. Species-area relationships always overestimate extinction rates from habitat loss. *Nature* **473**: 368-371.

Heidelberg KB, Gilbert JA, and Joint I. 2010. Marine genomics: at the interface of marine microbial ecology and biodiscovery. *Microbial Biotechnology* **3**: 531-543.

Heinz Center (The H. John Heinz III Center for Science, Economics and the Environment). 2002. The State of the Nation's Ecosystems: Measuring the Lands, Waters, and Living Resources of the United States: Cambridge University Press.

Heller NE, and Zavaleta ES. 2009. Biodiversity management in the face of climate change: A review of 22 years of recommendations. *Biological Conservation* **142**: 14-32.

Hellmann JJ, and Pineda-Krch M. 2007. Constraints and reinforcement on adaptation under climate change: Selection of genetically correlated traits. *Biological Conservation* **137**: 599-609.

Helmuth B, Broitman BR, Yamane L, Gilman SE, Mach K, Mislan KAS, and Denny MW. 2010. Organismal climatology: analyzing environmental variability at scales relevant to physiological stress. *Journal of Experimental Biology* **213**: 995-1003.

Hill JE, Seipp RP, Betts M, Hawkins L, Van Kessel AG, Crosby WL, and Hemmingsen SM. 2002. Extensive profiling of a complex microbial community by high-throughput sequencing. *Applied and Environmental Microbiology* **68**: 3055-3066.

Hobbs RJ, and Cramer VA. 2008. Restoration ecology: Interventionist approaches for restoring and maintaining ecosystem function in the face of rapid environmental change. *Annual Review of Environment and Resources* **33**: 39–61.

Hobbs RJ, Higgs E, and Harris JA. 2009. Novel ecosystems: implications for conservation and restoration. *Trends in Ecology and Evolution* **24**: 599-605.

Hoffmann AA, and Sgrò CM. 2011. Climate change and evolutionary adaptation. *Nature* **470**: 479-485.

Hofmann GE, Barry JP, Edmunds PJ, Gates RD, Hutchins DA, Klinger T, and Sewell MA. 2010. The effect of ocean acidification on calcifying organisms in marine ecosystems: An organism-to-ecosystem perspective. *Annual Review of Ecology, Evolution, and Systematics* **41**: 127-147.

Hofmann GE, and Todgham AE. 2010. Living in the now: Physiological mechanisms to tolerate a rapidly changing environment. *Annual Review of Physiology* **72**: 127-145.

Hohenlohe PA, Bassham S, Etter PD, Stiffler N, Johnson EA, and Cresko WA. 2010. Population genomics of parallel adaptation in threespine stickleback using sequenced RAD tags. *PLos Genetics* **6** (2): e1000862-e1000862.

Hooper DU, Chapin FS, Ewel JJ, Hector A, Inchausti P, Lavorel S, Lawton JH, Lodge DM, Loreau M, Naeem S, Schmid B, Setala H, Symstad AJ, Vandermeer J, and Wardle DA. 2005. Effects of biodiversity on ecosystem functioning: A consensus of current knowledge. *Ecological Monographs* **75**: 3-35.

Horning N, Robinson JA, Sterling EJ, Turner W, and Spector S. 2010. Remote Sensing for Ecology and Conservation. Oxford: Oxford University Press.

Hunter ML, Jacobson GL, and Webb T. 1988. Paleoecology and the coarse-filter approach to maintaining biological diversity. *Conservation Biology* **2**: 375-385.

Ibáñez I, Clark JS, and Dietze MC. 2008. Evaluating the sources of potential migrant species: Implications under climate change. *Ecological Applications* **18**: 1664-1678.

Inouye DW. 2008. Effects of climate change on phenology, frost damage, and floral abundance of montane wildflowers. *Ecology* **89**: 353-362.

IPCC (Intergovernmental Panel on Climate Change). 2007. Climate Change 2007: The Physical Science Basis. Contribution of Working Group I to the Fourth Assessment Report of the Intergovernmental Panel on Climate Change, Cambridge , UK New York, NY.

IUCN (International Union for Conservation of Nature). 2011. IUCN Red List of Threatened Species. IUCN (International Union for Conservation of Nature). 2010. Building resilience to climate change: Ecosystem-based adaptation and lessons from the field. IUCN, Gland, Switzerland.

Iverson LR, Schwartz MW, and Prasad AM. 2004. Potential colonization of new available tree species habitat under climate change an analysis for five eastern U.S. species. *Landscape Ecology* **19**: 787-799.

Iverson LR, Prasad AM, Matthews SN, and Peters M. 2008. Estimating potential habitat for 134 eastern U.S. tree species under six climate scenarios. *Forest Ecology and Management* **254**: 390-406.

Jetz W, Wilcove DS, and Dobson AP. 2007. Projected impacts of climate and land-use change on the global diversity of birds. *PLos Biology* **5**: 1211-1219.

Jones KB, Bogena H, Vereecken H, and Weltzin JF. 2010. Design and Importance of multi-tiered ecological monitoring networks. *In* Müller F, Baessler C, Schubert H, and Klotz S (Eds), Long-Term Ecological Research. Springer, Netherlands. 355 - 374 p.

Julius SH, and West J (Eds). 2008. Preliminary review of adaptation options for climate sensitive ecosystems and resources. U.S. Environmental Protection Agency, Washington, DC.

Kamali B, Hashim R, and Akib S. 2010. Efficiency of an integrated habitat stabilisation approach to coastal erosion management. *International Journal of the Physical Sciences* **5**: 1401-1405.

Kareiva P, Daily G, Ricketts T, Tallis H, and Polasky S. 2011. Natural Capital, Theory and Practice of Mapping Ecosystem Services: Oxford University Press. 432 p.

Kaschner K, Tittensor DP, Ready J, Gerrodette T, and Worm B. 2011. Current and future patterns of global marine mammal biodiversity. *PLoS ONE* **6**.

Kearney M, and Porter W. 2009. Mechanistic niche modelling: combining physiological and spatial data to predict species' ranges. *Ecology Letters* **12**: 334-350.

Keith DA, Akcakaya HR, Thuiller W, Midgley GF, Pearson RG, Phillips SJ, Regan HM, Araujo MB, and Rebelo TG. 2008. Predicting extinction risks under climate change: coupling stochastic population models with dynamic bioclimatic habitat models. *Biology Letters* **4**: 560-563.

Kelly MW, Sanford E, and Grosberg RK. 2012. Limited potential for adaptation to climate change in a broadly distributed marine crustacean. *Proceedings of the Royal Society B-Biological Sciences* **279**: 349-356.

Kiesecker JM, Copeland HE, McKenney BA, Pocewicz A, and Doherty KE. 2011. Energy by design: Making mitigation work for conservation and development. *In* Naugle DE (Ed), Energy Development and Wildlife Conservation in Western North America. Island Press, Washington, D.C. 159 - 182 p.

Kimball S, Angert AL, Huxman TE, and Venable DL. 2010. Contemporary climate change in the Sonoran Desert favors cold-adapted species. *Global Change Biology* **16**(5): 1555-1565.

Kroeker K, Kordas R, Crim R, and Singh G. 2009. Meta-analysis reveals negative yet variable effects of ocean acidification on marine organisms. *Ecological Letters* **13**: 1419–1434.

Lard M, Backman J, Yakovleva M, Danielsson B, and Hansson L-A. 2010. Tracking the small with the smallest - Using nanotechnology in tracking zooplankton. *PLoS ONE* **5**.

Lawler JJ, Shafer SL, White D, Kareiva P, Maurer EP, Blaustein AR, and Bartlein PJ. 2009. Projected climate-induced faunal change in the Western Hemisphere. *Ecology* **90**: 588-597.

Lawler JJ, Shafer SL, and Blaustein AR. 2010. Projected Climate Impacts for the Amphibians of the Western Hemisphere. *Conservation Biology* **24**: 38-50.

Impacts of Climate Change on Biodiversity, Ecosystems, and Ecosystem Services |
Technical Input to the 2013 National Climate Assessment

Chapter 2
Biodiversity

Lawton RJ, Messmer V, Pratchett MS, and Bay LK. 2011. High gene flow across large geographic scales reduces extinction risk for a highly specialised coral feeding butterflyfish. *Molecular Ecology* **20**: 3584-3598.

Leadley P, Pereira HM, Alkemade R, Fernandez-Manjarrés JF, Proença V, Scharlemann JPW, Walpole MJ. 2010. Biodiversity scenarios: projections of 21st century change in biodiversity and associated ecosystem services. Convention on Biological Diversity, Montreal, Canada.

Le Lann C, Wardziak T, van Baaren J, and van Alphen JJM. 2011. Thermal plasticity of metabolic rates linked to life-history traits and foraging behaviour in a parasitic wasp. *Functional Ecology* **25**: 641-651.

Leroux SJ, Schmiegelow FKA, Cumming SG, Lessard RB, and Nagy J. 2007. Accounting for system dynamics in reserve design. *Ecological Applications* **17**: 1954-1966.

Levitus S, Antonov JI, Boyer TP, Locarnini RA, Garcia HE, and Mishonov AV. 2009. Global ocean heat content 1955-2008 in light of recently revealed instrumentation problems. *Geophysical Research Letters* **36**: L07608.

Loarie SR, Duffy PB, Hamilton H, Asner GP, Field CB, and Ackerly DD. 2009. The velocity of climate change. *Nature* **462**: 1052-1057.

Loreau M. 2010. Linking biodiversity and ecosystems: toward a unifying ecological theory. *Philosophical Transactions of the Royal Society B-Biological Sciences* **365**: 49-60.

Lovich JE, and Ennen JR. 2011. Wildlife conservation and solar energy development in the Desert Southwest, United States. *BioScience* **61**: 982-992.

Loya Y, Sakai K, Yamazato K, Nakano Y, Sambali H, and van Woesik R. 2001. Coral bleaching: the winners and the losers. *Ecology Letters* **4**: 122-131.

Lynch M, and Lande R. 1993. Evolution and extinction in response to environmental change. *In* Kareiva P, Kingsolver J, and Huey R (Eds), Biotic Interactions and Global Change. Sinauer Associates. 234-250 p.

Lyons KG, and Schwartz MW. 2001. Rare species loss alters ecosystem function - invasion resistance. *Ecology Letters* **4**: 358-365.

Lyons KG, Brigham CA, Traut BH, and Schwartz MW. 2005. Rare species and ecosystem functioning. *Conservation Biology* **19**: 1019-1024.

MA (Millennium Ecosystem Assessment). 2005. Ecosystems and Human Well-being: Biodiversity Synthesis. World Resources Institute, Washington, D.C.

Mace GM, Norris K, and Fitter AH. 2012. Biodiversity and ecosystem services: a multilayered relationship. *Trends in Ecology and Evolution* **27**: 19-26.

Maclean IMD, and Wilson RJ. 2011. Recent ecological responses to climate change support predictions of high extinction risk. *Proceedings of the National Academy of Sciences* **108**: 12337-12342.

MacMynowski DP, and Root TL. 2007. Climate and the complexity of migratory phenology: sexes, migratory distance, and arrival distributions. *International Journal of Biometeorology* **51**: 361-373.

MacMynowski DP, Root TL, Ballard G, and Geupel GR. 2007. Changes in spring arrival of Nearctic-Neotropical migrants attributed to multiscalar climate. *Global Change Biology* **13**: 2239-2251.

Malcolm JR, Liu C, Neilson RP, Hansen L, and Hannah LEE. 2006. Global warming and extinctions of endemic species from biodiversity hotspots. *Conservation Biology* **20**: 538-548.

Manning AD, Fischer J, Felton A, Newell B, Steffen W, and Lindenmayer DB. 2009. Landscape fluidity - a unifying perspective for understanding and adapting to global change. *Journal of Biogeography* **36**: 193-199.

Mantyka-Pringle CS, Martin TG, and Rhodes JR. 2011. Interactions between climate and habitat loss effects on biodiversity: a systematic review and meta-analysis. *Global Change Biology* **18**: 1239-1252.

Marsico TD, Hellmann JJ, and Romero-Severson J. 2009. Patterns of seed dispersal and pollen flow in *Quercus garryana* (Fagaceae) following post-glacial climatic changes. *Journal of Biogeography* **36**: 929-941.

Massot M, Clobert J, and Ferriere R. 2008. Climate warming, dispersal inhibition and extinction risk. *Global Change Biology* **14**: 461-469.

Master L, Faber-Langendoen D, Bittman R, Hammerson GA, Heidel B, Nichols J, Ramsay L, and Tomaino A. 2009. NatureServe Conservation Status Assessments: Factors for Assessing Extinction Risk. NatureServe, Arlington, VA.

Matthews SN, O'Connor RJ, Iverson LR, and Prasad AM. 2004. Atlas of climate change effects in 150 bird species of the eastern United States. USDA Forest Service, Northeastern Research Station, Newtown Square, PA.

May RM. 2011. Why should we be concerned about loss of biodiversity. *Comptes Rendus Biologies* **334**: 346-350.

May RM. 1988. How many species are there on earth. *Science* **241**: 1441-1449.

McClenachan L, Cooper AB, Carpenter KE, and Dulvy NK. 2012. Extinction risk and bottlenecks in the conservation of charismatic marine species. *Conservation Letters* **5**: 73-80.

McDonald RI, Fargione J, Kiesecker J, Miller WM, and Powell J. 2009. Energy sprawl or energy efficiency: Climate Policy impacts on natural habitat for the United States of America. *PLoS ONE* **4**: e6802.

McKelvey KS, Copeland JP, Schwartz MK, Littell JS, Aubry KB, Squires JR, Parks SA, Elsner MM, and Mauger GS. 2011. Climate change predicted to shift wolverine distributions, connectivity, and dispersal corridors. *Ecological Applications* **21**: 2882-2897.

McMahon SM, Harrison SP, Armbruster WS, Bartlein PJ, Beale CM, Edwards ME, Kattge J, Midgley G, Morin X, and Prentice IC. 2011. Improving assessment and modelling of climate change impacts on global terrestrial biodiversity. *Trends in Ecology and Evolution* **26**: 249-259.

Metzker ML. 2010. Sequencing technologies - the next generation. *Nature Reviews Genetics* **11**: 31-46.

Meyer E, Aglyamova GV, Wang S, Buchanan-Carter J, Abrego D, Colbourne JK, Willis BL, and Matz MV. 2009. Sequencing and de novo analysis of a coral larval transcriptome using 454 GSFlx. *BMC Genomics* **10**: 219.

Millar CI, Stephenson NL, and Stephens SL. 2007. Climate change and forests of the future: Managing in the face of uncertainty. *Ecological Applications* **17**: 2145-2151.

Miller-Rushing AJ, Lloyd-Evans TL, Primack RB, and Satzinger P. 2008. Bird migration times, climate change, and changing population sizes. *Global Change Biology* **14**: 1959-1972.

Miller-Rushing AJ, Hoye TT, Inouye DW, and Post E. 2010. The effects of phenological mismatches on demography. *Philosophical Transactions of the Royal Society B-Biological Sciences* **365**: 3177-3186.

Montoya JM, and Raffaelli D. 2010. Climate change, biotic interactions and ecosystem services. *Philosophical Transactions of the Royal Society B-Biological Sciences* **365**: 2013-2018.

Monzón J, Moyer-Horner L, and Palamar MB. 2011. Climate change and species range dynamics in protected areas. *BioScience* **61**: 752.

Moore KA, and Jarvis JC. 2008. Environmental factors affecting recent summertime eelgrass diebacks in the lower Chesapeake Bay: Implications for long-term persistence. *Journal of Coastal Research*: 135-147.

Mora C, Tittensor DP, Adl S, Simpson AGB, and Worm B. 2011. How many species are there on earth and in the ocean? *PLoS ONE Biology* **9** (8): e1001127-e1001127.

Moritz C, Patton JL, Conroy CJ, Parra JL, White GC, and Beissinger SR. 2008. Impact of a century of climate change of small-mammal communities in Yosemite National Park, USA. *Science* **322**: 261-264.

Munday PL, Dixson DL, Donelson JM, Jones GP, Pratchett MS, Devitsina GV, and Doving KB. 2009. Ocean acidification impairs olfactory discrimination and homing ability of a marine fish. *Proceedings of the National Academy of Sciences* **106**: 1848-1852.

Myers N. 1988. Threatened biotas: "Hot Spots" in tropical forests. *The Environmentalist* **8**: 187-208.

Myers N. 1991. Extinction hot spots. *Science* **254**: 919-919.

Myers N, Mittermeier RA, Mittermeier CG, da Fonseca GAB, and Kent J. 2000. Biodiversity hotspots for conservation priorities. *Nature* **403**: 853-858.

Naeem S. 2009. Ecology: Gini in the bottle. *Nature* **458**: 579-580.

National Audubon S. 2009. Birds and Climate Change: Ecological Disruption in Motion. National Audubon Society, New York.

NatureServe. 2011. NatureServe's Central Databases. Available at: http://www.natureserve.org/

Nei M, Maruyama T, and Chakraborty R. 1975. The bottleneck effect and genetic variability of populations. *Evolution* **29**: 1-10.

Nene V, Wortman JR, Lawson D, Haas B, Kodira C, Tu Z, Loftus B, Xi Z, Megy K, Grabherr M, Ren Q, Zdobnov EM, Lobo NF, Campbell KS, Brown SE, Bonaldo MF, Zhu J, Sinkins SP, Hogenkamp DG, Amedeo P, Arensburger P, Atkinson PW, Bidwell S, Biedler J, Birney E, Bruggner RV, Costas J, Coy MR, Crabtree J, Crawford M, deBruyn B, DeCaprio D, Eiglmeier K, Eisenstadt E, El-Dorry H, Gelbart WM, Gomes SL, Hammond M, Hannick LI, Hogan JR, Holmes MH, Jaffe D, Johnston JS, Kennedy RC, Koo H, Kravitz S, Kriventseva EV, Kulp D, LaButti K, Lee E, Li S, Lovin DD, Mao C, Mauceli E, Menck CFM, Miller JR, Montgomery P, Mori A, Nascimento AL, Naveira HF, Nusbaum C, O'Leary S, Orvis J, Pertea M, Quesneville H, Reidenbach KR, Rogers Y-H, Roth CW, Schneider JR, Schatz M, Shumway M, Stanke M, Stinson EO, Tubio JMC, VanZee JP, Verjovski-Almeida S, Werner D, White O, Wyder S, Zeng Q, Zhao Q, Zhao Y, Hill CA, Raikhel AS, Soares MB, Knudson DL, Lee NH, Galagan J, Salzberg SL, Paulsen IT, Dimopoulos G, Collins FH, Birren B, Fraser-Liggett CM, and Severson DW. 2007. Genome sequence of *Aedes aegypti*, a major arbovirus vector. *Science* **316**: 1718-1723.

Nowicki JP, Miller GM, and Munday PL. 2012. Interactive effects of elevated temperature and CO_2 on foraging behavior of juvenile coral reef fish. *Journal of Experimental Marine Biology and Ecology* **412**: 46-51.

NOAA (National Oceanographic and Atmospheric Administration). 2010. NOAA Proposes Listing Ringed and Bearded Seals as Threatened Under Endangered Species Act: Available at: http://www.noaanews.noaa.gov/stories2010/20101203_sealsesa.html

NRC (National Research Council). 2010. America's Climate Choices: Adapting to the Impacts of Climate Change. National Academies Press, Washington, DC.

Nye, J. A., J. S. Link, J. A. Hare, and W. J. Overholtz. 2009. Changing spatial distribution of fish stocks in relation to climate and population size on the Northeast United States continental shelf. *Marine Ecology-Progress Series* **393**:111-129

O'Connor MI, Bruno JF, Gaines SD, Halpern BS, Lester SE, Kinlan BP, and Weiss JM. 2007. Temperature control of larval dispersal and the implications for marine ecology, evolution, and conservation. *Proceedings of the National Academy of Sciences* **104**: 1266-1271.

O'Connor MI, Piehler MF, Leech DM, Anton A, and Bruno JF. 2009. Warming and resource availability shift food web structure and metabolism. *PLoS ONE Biology* **7**.

O'Neil ST, Dzurisin JDK, Carmichael RD, Lobo NF, Emrich SJ, and Hellmann JJ. 2010. Population-level transcriptome sequencing of nonmodel organisms *Erynnis propertius* and *Papilio zelicaon. BMC Genomics* **11**.

Olsson M, Wapstra E, Schwartz T, Madsen T, Ujvari B, and Uller T. 2011. In hot pursuit: fluctuating mating system and sexual selection in sand lizards. *Evolution* **65**: 574-583.

Ozgul A, Childs DZ, Oli MK, Armitage KB, Blumstein DT, Olson LE, Tuljapurkar S, and Coulson T. 2010. Coupled dynamics of body mass and population growth in response to environmental change. *Nature* **466**: 482-U485.

Palumbi SR, Sandifer PA, Allan JD, Beck MW, Fautin DG, Fogarty MJ, Halpern BS, Incze LS, Leong J-A, Norse E, Stachowicz JJ, and Wall DH. 2009. Managing for ocean biodiversity to sustain marine ecosystem services. *Frontiers in Ecology and the Environment* **7**: 204-211.

Parmesan C, and Yohe G. 2003. A globally coherent fingerprint of climate change impacts across natural systems. *Nature* **421**: 37-42.

Parmesan C. 2006. Ecological and evolutionary responses to recent climate change. *Annual Review of Ecology Evolution and Systematics* **37**: 637-669.

Parmesan C, Duarte C, Poloczanska E, Richardson AJ, and Singer MC. 2011. Commentary: Overstretching attribution. *Nature Climate Change* **1**: 2-4.

Parmesan C. 2007. Influences of species, latitudes and methodologies on estimates of phenological response to global warming. *Global Change Biology* **13**: 1860-1872.

Pau S, Wolkovich EM, Cook BI, Davies TJ, Kraft NJB, Bolmgren K, Betancourt JL, and Cleland EE. 2011. Predicting phenology by integrating ecology, evolution and climate science. *Global Change Biology* **17**: 3633-3643.

PCAST (President's Council of Advisors on Science and Technology). 2011. Sustaining environmental capital: protecting society and the economy. 145 p. Available at: http://www.whitehouse.gov/administration/eop/ostp/pcast/docsreports

Pearson RG. 2006. Climate change and the migration capacity of species. *Trends in Ecology and Evolution* **21**: 111-113.

Pearson RG, and Dawson TP. 2005. Long-distance plant dispersal and habitat fragmentation: identifying conservation targets for spatial landscape planning under climate change. *Biological Conservation* **123**: 389-401.

Peery MZ, Gutiérrez RJ, Kirby R, LeDee OE, and LaHaye W. 2012. Climate change and spotted owls: potentially contrasting responses in the Southwestern United States. *Global Change Biology* **18**: 865-880.

Pelini SL, Keppel JA, Kelley AE, and Hellmann JJ. 2010. Adaptation to host plants may prevent rapid insect responses to climate change. *Global Change Biology* **16**: 2923-2929.

Pennings SC, and Silliman BR. 2005. Linking biogeography and community ecology: Latitudinal variation in plant-herbivore interaction strength. *Ecology* **86**: 2310-2319.

Pennisi E. 2011. Animal Ecology Global Tracking of small animals gains momentum. *Science* **334**: 1042-1042.

Pereira HM, Leadley PW, Proenca V, Alkemade R, Scharlemann JPW, Fernandez-Manjarres JF, Araujo MB, Balvanera P, Biggs R, Cheung WWL, Chini L, Cooper HD, Gilman EL, Guenette S, Hurtt GC, Huntington HP, Mace GM, Oberdorff T, Revenga C, Rodrigues P, Scholes RJ, Sumaila UR, and Walpole M. 2010. Scenarios for global biodiversity in the 21st Century. *Science* **330**: 1496-1501.

Pérez AA, Fernández BH, and Roberto CG (eds). 2010. Building resilience to climate change: Ecosystem-based adaptation and lessons from the field. IUCN, Gland, Switzerland. Available at: http://www.iucn.org/what/tpas/climate/resources/publications/?uPubsID=4185

Perrings C, Duraiappah A, Larigauderie A, and Mooney H. 2011. The biodiversity and ecosystem services science-policy interface. *Science* **331**: 1139-1140.

Petchey OL, McPhearson PT, Casey TM, and Morin PJ. 1999. Environmental warming alters food-web structure and ecosystem function. *Nature* **402**: 69-72.

Peterson DL, Millar CI, Joyce LA, Furniss MJ, Halofsky JE, Neilson RP, and Morelli TL. 2011. Responding to climate change in national forests: a guidebook for developing adaptation options. U.S. Department of Agriculture, Forest Service, Pacific Northwest Research Station, Portland, OR.

Philippart CJM, van Aken HM, Beukema JJ, Bos OG, Cadee GC, and Dekker R. 2003. Climate-related changes in recruitment of the bivalve *Macoma balthica*. *Limnology and Oceanography* **48**: 2171-2185.

Pimm S. 2008. Biodiversity: Climate change or habitat loss — Which will kill more species. *Current Biology* **18**: R117-R119.

Pimm S, Russell G, Gittleman J, and Brooks T. 1995. The future of biodiversity. *Science* **269**: 347.

Pöertner HO, and Farrell AP. 2008. Ecology physiology and climate change. *Science* **322**: 690-692.

Poiani KA, Goldman RL, Hobson J, Hoekstra JM, and Nelson KS. 2011. Redesigning biodiversity conservation projects for climate change: examples from the field. *Biodiversity and Conservation* **20**: 185-201.

Post E, Pedersen C, Wilmers CC, and Forchhammer MC. 2008. Warming, plant phenology and the spatial dimension of trophic mismatch for large herbivores. *Proceedings of the Royal Society B-Biological Sciences* **275**: 2005-2013.

Impacts of Climate Change on Biodiversity, Ecosystems, and Ecosystem Services | Chapter 2
Technical Input to the 2013 National Climate Assessment
Biodiversity

Pressey RL, Cabeza M, Watts ME, Cowling RM, and Wilson KA. 2007. Conservation planning in a changing world. *Trends in Ecology and Evolution* **22**: 583-592.

Price TD, Qvarnstrom A, and Irwin DE. 2003. The role of phenotypic plasticity in driving genetic evolution. *Proceedings of the Royal Society of London Series B-Biological Sciences* **270**: 1433-1440.

Prober S, and Dunlop M. 2011. Climate change: A cause for new biodiversity conservation objectives but let's not throw the baby out with the bathwater. *Ecological Management and Restoration* **12**: 2-3.

Purvis A, Jones KE, and Mace GM. 2000. Extinction. *Bioessays* **22**: 1123-1133.

Pérez AA, Fernández BH (Eds). 2010. Building resilience to climate change: Ecosystem-based adaptation and lessons from the field. IUCN, Gland, Switzerland.

Reed TE, Schindler DE, and Waples RS. 2011. Interacting Effects of Phenotypic Plasticity and Evolution on Population Persistence in a Changing Climate. *Conservation Biology* **25**: 56–63.

Rehfeldt GE, Crookston NL, Sáenz-Romero CM, and Campbell EM. 2012. North American vegetation model for land-use planning in a changing climate: a solution to large classification problems. *Ecological Applications* **22**: 119-141.

Roberts CM, and Hawkins JP. 1999. Extinction risk in the sea. *Trends in Ecology and Evolution* **14**: 241-246.

Rode KD, Amstrup SC, and Regehr EV. 2010. Reduced body size and cub recruitment in polar bears associated with sea ice decline. *Ecological Applications* **20**: 768-782.

Rogers BM, Neilson RP, Drapek R, Lenihan JM, Wells JR, Bachelet D, and Law BE. 2011. Impacts of climate change on fire regimes and carbon stocks of the U.S. Pacific Northwest. *Journal Of Geophysical Research* **116**: G03037, doi:03010.01029/02011JG001695.

Root TL, Price JT, Hall KR, Schneider SH, Rosenzweig C, and Pounds JA. 2003. Fingerprints of global warming on wild animals and plants. *Nature* **421**: 57-60.

Rosenzweig C, Karoly D, Vicarelli M, Neofotis P, Wu Q, Casassa G, Menzel A, Root TL, Estrella N, Seguin B, Tryjanowski P, Liu C, Rawlins S, and Imeson A. 2008. Attributing physical and biological impacts to anthropogenic climate change. *Nature* **453**: 353-U320.

Rowland EL, Davison JE, and Graumlich LJ. 2011. Approaches to evaluating climate change impacts on species: A guide to initiating the adaptation planning process. *Environmental Management* **47**: 322-337.

Rubidge EM, Patton JL, Lim M, Burton AC, Brashares JS, and Moritz C. 2012. Climate-induced range contraction drives genetic erosion in an alpine mammal. *Nature Climate Change* **2**: 285-288.

Sala E, and Knowlton N. 2006. Global marine biodiversity trends. *Annual Review of Environment and Resources* **31**: 93-122.

Sala OE, Chapin FS, III, Armesto JJ, Berlow E, Bloomfield J, Dirzo R, Huber-Sanwald E, Huenneke LF, Jackson RB, Kinzig A, Leemans R, Lodge DM, Mooney HA, Oesterheld M, Poff NL, Sykes MT, Walker BH, Walker M, and Wall DH. 2000. Global biodiversity scenarios for the year 2100. *Science* **287**: 1770-1774.

Sandel B, Arge L, Dalsgaard B, Davies RG, Gaston KJ, Sutherland WJ, and Svenning JC. 2011.
    The influence of Late Quaternary climate-change velocity on species endemism. *Science*
    334: 660-664.

Sanford E. 1999. Regulation of keystone predation by small changes in ocean temperature.
    Science **283**: 2095-2097.

Schloss CA, Nuñez TA, and Lawler JJ. 2012. Dispersal will limit ability of mammals to track
    climate change in the Western Hemisphere. *Proceedings of the National Academy of
    Science* **109** (22): 8606-8611.

Schmidt JH, Lindberg MS, Johnson DS, Conant B, and King J. 2009. Evidence of Alaskan
    Trumpeter Swan Population Growth Using Bayesian Hierarchical Models. *Journal of
    Wildlife Management* **73**: 720-727.

Schmidt JH, Lindberg MS, Johnson DS, and Verbyla DL. 2011. Season length influences
    breeding range dynamics of trumpeter swans *Cygnus buccinator*. *Wildlife Biology* **17**:
    364-372.

Scholes RJ, Mace GM, Turner W, Geller GN, Juergens N, Larigauderie A, Muchoney D,
    Walther BA, and Mooney HA. 2008. Ecology - Toward a global biodiversity observing
    system. *Science* **321**: 1044-1045.

Scholes RJ, M. Walters, E. Turak, H. Saarenmaa, C.H.R. Heip, E. O′ Tuama, D.P. Faith, H.A.
    Mooney, S. Ferrier, R.H.G. Jongman, I.J. Harrison, T. Yahara, H.M. Pereira, A.
    Larigauderie, and Geller G. 2012. Building a global observing system for biodiversity.
    Current Opinion in Environmental Sustainability **4**: 139-146.

Schwartz MD, Ahas R, and Aasa A. 2006. Onset of spring starting earlier across the Northern
    Hemisphere. *Global Change Biology* **12**: 343-351.

Schwartz M, Hellmann J, McLachlan J, Sax D, Borevitz J, Brennan J, Camacho A, Ceballos G,
    Rappaport Clark J, Doremus H, Early R, Etterson J, Fielder D, Gill J, Gonzalez P, Green
    N, Hannh L, Jamieson D, Javeline D, Minteer B, Odenbaugh J, Polasky S, Richardson D,
    Root T, and Safford H. 2012. Managed relocation: integrating the scientific, regulatory
    and ethical challenges. *BioScience* **62**: 732-743.

Scott JM, Goble DD, Haines AM, Wiens JA, and Neel MC. 2010. Conservation-reliant species
    and the future of conservation. *Conservation Letters* **3**: 91-97.

Seehausen O, Takimoto G, Roy D, and Jokela J. 2008. Speciation reversal and biodiversity
    dynamics with hybridization in changing environments. *Molecular Ecology* **17**: 30-44.

Sekercioglu CH, Schneider SH, Fay JP, and Loarie SR. 2008. Climate change, elevational range
    shifts, and bird extinctions. *Conservation Biology* **22**: 140-150.

Sheridan JA, and Bickford D. 2011. Shrinking body size as an ecological response to climate
    change. *Nature Climate Change* **1**: 401-406.

Simenstad C, Reed D, and Ford M. 2006. When is restoration not? Incorporating landscape-scale
    processes to restore self-sustaining ecosystems in coastal wetland restoration. *Ecological
    Engineering* **26**: 27-39.

Sinervo B, Méndez-de-la-Cruz, F, Miles DB, Heulin B, Bastiaans E, Villagrán-Santa Cruz M,
    Lara-Resendiz R, Martínez-Méndez N, Calderón-Espinosa ML, Meza-Lázaro RN,
    Gadsden H, Avila LJ, Morando M, De la Riva IJ, Sepulveda PV, Duarte Rocha CF,
    Ibargüengoytía N, Puntriano CA, Massot M, Lepetz V, Oksanen TA, Chapple DG, Bauer
    AM, Branch WR, Clobert J, and Sites JW. 2010. Erosion of lizard diversity by climate
    change and altered thermal niches. *Science* **328**: 1354-1354.

Singer MC, and Parmesan C. 2010. Phenological asynchrony between herbivorous insects and their hosts: signal of climate change or pre-existing adaptive strategy? *Philosophical Transactions of the Royal Society B-Biological Sciences* **365**: 3161-3176.

Sitch S, Huntingford C, Gedney N, Levy PE, Lomas M, Piao SL, Betts R, Ciais P, Cox P, Friedlingstein P, Jones CD, Prentice IC, and Woodward FI. 2008. Evaluation of the terrestrial carbon cycle, future plant geography and climate-carbon cycle feedbacks using five Dynamic Global Vegetation Models (DGVMs). *Global Change Biology* **14**: 2015-2039.

Smith CR, Grange, LJ, Honi, DL, Naudts, L, Huber B, Guidi L, and Domack E. 2012. A large population of king crabs in Palmer Deep on the west Antarctic Peninsula shelf and potential invasive impacts. *Proceedings of the Royal Society B. Biological Sciences* **279** (1730): 1017-1026.

Sørensen JG, Vermeulen CJ, Flik G, and Loeschcke V. 2009. Stress specific correlated responses in fat content, Hsp70 and dopamine levels in *Drosophila melanogaster* selected for resistance to environmental stress. *Journal of Insect Physiology* **55**: 700-706.

Sperry JH, Blouin-Demers G, Carfagno GLF, and Weatherhead PJ. 2010. Latitudinal variation in seasonal activity and mortality in ratsnakes (*Elaphe obsoleta*). *Ecology* **91**: 1860-1866.

Stapley J, Reger J, Feulner PGD, Smadja C, Galindo J, Ekblom R, Bennison C, Ball AD, Beckerman AP, and Slate J. 2010. Adaptation genomics: the next generation. *Trends in Ecology and Evolution* **25**: 705-712.

Steigenga MJ, and Fischer K. 2007. Ovarian dynamics, egg size, and egg number in relation to temperature and mating status in a butterfly. *Entomologia Experimentalis Et Applicata* **125**: 195-203.

Stein BA, Kutner LS, and Adam JS. 2000. Precious heritage: The status of biodiversity in the United States. New York, NY: Oxford University Press.

Steinacher M, Joos F, Froelicher TL, Plattner GK, and Doney SC. 2009. Imminent ocean acidification in the Arctic projected with the NCAR global coupled carbon cycle-climate model. *Biogeosciences* **6**: 515-533.

Storlazzi CD, Elias E, Field ME, and Presto MK. 2011. Numerical modeling of the impact of sea-level rise on fringing coral reef hydrodynamics and sediment transport. *Coral Reefs* **30**: 83-96.

Stralberg D, Jongsomjit D, Howell CA, Snyder MA, Alexander JD, Wiens JA, and Root TL. 2009. Re-shuffling of species with climate disruption: A no-analog future for California birds? *PLoS ONE* **4**: e6825.

Suttle KB, Thomsen MA, and Power ME. 2007. Species interactions reverse grassland responses to changing climate. *Science* **315**: 640-642.

Swanson DL, and Palmer JS. 2009. Spring migration phenology of birds in the Northern Prairie region is correlated with local climate change. *Journal of Field Ornithology* **80**: 351-363.

Swatantran A, Dubayah R, Goetz S, Hofton M, Betts MG, Sun M, Simard M, and Holmes R. 2012. Mapping migratory bird prevalence using remote sensing data fusion. *PLoS ONE* **7**: e28922.

Taylor SG. 2008. Climate warming causes phenological shift in pink salmon, *Oncorhynchus gorbuscha*, behavior at Auke Creek, Alaska. *Global Change Biology* **14**: 229-235.

Thackeray SJ, Sparks TH, Frederiksen M, Burthe S, Bacon PJ, Bell JR, Botham MS, Brereton TM, Bright PW, Carvalho L, Clutton-Brock T, Dawson A, Edwards M, Elliott JM,

Harrington R, Johns D, Jones ID, Jones JT, Leech DI, Roy DB, Scott WA, Smith M, Smithers RJ, Winfield IJ, and Wanless S. 2010. Trophic level asynchrony in rates of phenological change for marine, freshwater and terrestrial environments. *Global Change Biology* **16**: 3304-3313.

Thomas CD, Cameron A, Green RE, Bakkenes M, Beaumont LJ, Collingham YC, Erasmus BFN, de Siqueira MF, Grainger A, Hannah L, Hughes L, Huntley B, van Jaarsveld AS, Midgley GF, Miles L, Ortega-Huerta MA, Peterson AT, Phillips OL, and Williams SE. 2004. Extinction risk from climate change. *Nature* **427**: 145-148.

Tittensor DP, Mora C, Jetz W, Lotze HK, Ricard D, Vanden Berghe E, and Worm B. 2010. Global patterns and predictors of marine biodiversity across taxa. *Nature* **466**: 1098-1101.

Todd BD, Scott DE, Pechmann JHK, and Gibbons JW. 2011. Climate change correlates with rapid delays and advancements in reproductive timing in an amphibian community. *Proceedings of the Royal Society B-Biological Sciences* **278**: 2191-2197.

Tomanek L. 2010. Variation in the heat shock response and its implication for predicting the effect of global climate change on species' biogeographical distribution ranges and metabolic costs. *Journal of Experimental Biology* **213**: 971-979.

Tringe SG, von Mering C, Kobayashi A, Salamov AA, Chen K, Chang HW, Podar M, Short JM, Mathur EJ, Detter JC, Bork P, Hugenholtz P, and Rubin EM. 2005. Comparative metagenomics of microbial communities. *Science* **308**: 554-557.

Tuomainen U, and Candolin U. 2011. Behavioural responses to human-induced environmental change. *Biological Reviews* **86**: 640–657.

Turner WR, Bradley BA, Estes LD, Hole DG, Oppenheimer M, and Wilcove DS. 2010. Climate change: helping nature survive the human response. *Conservation Letters* **3**: 304-312.

Urban MC, Tewksbury JJ, and Sheldon KS. 2012. On a collision course: competition and dispersal differences create no-analogue communities and cause extinctions during climate change. *Proceedings of the Royal Society B-Biological Sciences*: 1471-2954.

USEPA (U.S. Environmental Protection Agency). 2009. A framework for categorizing the relative vulnerability of threatened and endangered species to climate change. National Center for Environmental Assessment, Washington, D.C.

USGCRP (U. S. Global Change Research Program). 2009. Global Climate Change Impacts in the United States: Cambridge University Press.

USFWS (U.S. Fish and Wildlife Service). 2008. Determination of Threatened Status for the Polar Bear.

USFWS (U.S. Fish and Wildlife Service). 2012. Endangered Species Program. www.fws.gov/endangered/. USFWS (U.S. Fish and Wildlife Service). 2010. Endangered and Threatened Wildlife and Plants: 12-month Finding on a Petition to List the American Pika as Threatened or Endangered. Federal Register.

Van Buskirk J, Mulvihill RS, and Leberman RC. 2009. Variable shifts in spring and autumn migration phenology in North American songbirds associated with climate change. *Global Change Biology* **15**: 760-771.

Van Buskirk J, Mulvihill RS, and Leberman RC. 2010. Declining body sizes in North American birds associated with climate change. *Oikos* **119**: 1047-1055.

van Mantgem PJ, Stephenson NL, Byrne JC, Daniels LD, Franklin JF, Fule PZ, Harmon ME, Larson AJ, Smith JM, Taylor AH, and Veblen TT. 2009. Widespread increase of tree mortality rates in the western United States. *Science* **323**: 521-524.

van Woesik R, Shiroma K, and Koksal S. 2010. Phenotypic variance predicts symbiont population densities in corals: A modeling approach. *PLoS ONE* **5**.

Vierling KT, Baessler C, Brandl R, Vierling LA, Weiss I, and Mueller J. 2011. Spinning a laser web: predicting spider distributions using LiDAR. *Ecological Applications* **21**: 577-588.

Visser ME, and Both C. 2005. Shifts in phenology due to global climate change: the need for a yardstick. *Proceedings of the Royal Society B-Biological Sciences* **272**: 2561-2569.

Vitousek PM, Mooney HA, Lubchenco J, and Melillo JM. 1997. Human domination of Earth's ecosystems. *Science* **278**: 21-21.

Voigt W, Perner J, Davis AJ, Eggers T, Schumacher J, Bahrmann R, Fabian B, Heinrich W, Kohler G, Lichter D, Marstaller R, and Sander FW. 2003. Trophic levels are differentially sensitive to climate. *Ecology* **84**: 2444-2453.

Vonlanthen P, Bittner D, Hudson AG, Young KA, Muller R, Lundsgaard-Hansen B, Roy D, Di Piazza S, Largiader CR, and Seehausen O. 2012. Eutrophication causes speciation reversal in whitefish adaptive radiations. *Nature* **482**: 357-NIL_1500.

Wallace BP, DiMatteo AD, Bolten AB, Chaloupka MY, Hutchinson BJ, Abreu-Grobois FA, Mortimer JA, Seminoff JA, Amorocho D, Bjorndal KA, Bourjea J, Bowen BW, Duenas RB, Casale P, Choudhury BC, Costa A, Dutton PH, Fallabrino A, Finkbeiner EM, Girard A, Girondot M, Hamann M, Hurley BJ, Lopez-Mendilaharsu M, Marcovaldi MA, Musick JA, Nel R, Pilcher NJ, Troeng S, Witherington B, and Mast RB. 2011. Global conservation priorities for marine turtles. *PLoS ONE* **6**(9): e24510.

Walls SC. 2009. The role of climate in the dynamics of a hybrid zone in Appalachian salamanders. *Global Change Biology* **15**: 1903-1910.

Walther GR, Post E, Convey P, Menzel A, Parmesan C, Beebee TJC, Fromentin JM, Hoegh-Guldberg O, and Bairlein F. 2002. Ecological responses to recent climate change. *Nature* **416**: 389-395.

Walther GR, Roques A, Hulme PE, Sykes MT, Pysek P, Kuhn I, Zobel M, Bacher S, Botta-Dukat Z, Bugmann H, Czucz B, Dauber J, Hickler T, Jarosik V, Kenis M, Klotz S, Minchin D, Moora M, Nentwig W, Ott J, Panov VE, Reineking B, Robinet C, Semenchenko V, Solarz W, Thuiller W, Vila M, Vohland K, and Settele J. 2009. Alien species in a warmer world: risks and opportunities. *Trends in Ecology and Evolution* **24**: 686-693.

Walther G-R. 2010. Community and ecosystem responses to recent climate change. *Philosophical Transactions of the Royal Society B-Biological Sciences* **365**: 2019-2024.

Weeks D, Malone P, and Welling L. 2011. Climate change scenario planning: A tool for managing parks into uncertain futures. *ParkScience* **28**: 26-33.

Weinstock GM, Robinson GE, Gibbs RA, Worley KC, Evans JD, Maleszka R, Robertson HM, Weaver DB, Beye M, Bork P, Elsik CG, Hartfelder K, Hunt GJ, Zdobnov EM, Amdam GV, Bitondi MMG, Collins AM, Cristino AS, Lattorff HMG, Lobo CH, Moritz RFA, Nunes FMF, Page RE, Jr., Simoes ZLP, Wheeler D, Carninci P, Fukuda S, Hayashizaki Y, Kai C, Kawai J, Sakazume N, Sasaki D, Tagami M, Albert S, Baggerman G, Beggs KT, Bloch G, Cazzamali G, Cohen M, Drapeau MD, Eisenhardt D, Emore C, Ewing MA, Fahrbach SE, Foret S, Grimmelikhuijzen CJP, Hauser F, Hummon AB, Huybrechts J,

Impacts of Climate Change on Biodiversity, Ecosystems, and Ecosystem Services |
Technical Input to the 2013 National Climate Assessment

Chapter 2
Biodiversity

Jones AK, Kadowaki T, Kaplan N, Kucharski R, Leboulle G, Linial M, Littleton JT, Mercer AR, Richmond TA, Rodriguez-Zas SL, Rubin EB, Sattelle DB, Schlipalius D, Schoofs L, Shemesh Y, Sweedler JV, Velarde R, Verleyen P, Vierstraete E, Williamson MR, Ament SA, Brown SJ, Corona M, Dearden PK, Dunn WA, Elekonich MM, Fujiyuki T, Gattermeier I, Gempe T, Hasselmann M, Kadowaki T, Kage E, Kamikouchi A, Kubo T, Kucharski R, Kunieda T, Lorenzen MD, Milshina NV, Morioka M, Ohashi K, Overbeek R, Ross CA, Schioett M, Shippy T, Takeuchi H, Toth AL, Willis JH, Wilson MJ, Gordon KHJ, Letunic I, Hackett K, Peterson J, Felsenfeld A, Guyer M, Solignac M, Agarwala R, Cornuet JM, Monnerot M, Mougel F, Reese JT, Vautrin D, Gillespie JJ, Cannone JJ, Gutell RR, Johnston JS, Eisen MB, Iyer VN, Iyer V, Kosarev P, Mackey AJ, Solovyev V, Souvorov A, Aronstein KA, Bilikova K, Chen YP, Clark AG, Decanini LI, Gelbart WM, Hetru C, Hultmark D, Imler J-L, Jiang H, Kanost M, Kimura K, Lazzaro BP, Lopez DL, Simuth J, Thompson GJ, Zou Z, De Jong P, Sodergren E, Csuroes M, Milosavljevic A, Osoegawa K, Richards S, Shu C-L, Duret L, Elhaik E, Graur D, Anzola JM, Campbell KS, Childs KL, Collinge D, Crosby MA, Dickens CM, Grametes LS, Grozinger CM, Jones PL, Jorda M, Ling X, Matthews BB, Miller J, Mizzen C, Peinado MA, Reid JG, Russo SM, Schroeder AJ, St Pierre SE, Wang Y, Zhou P, Jiang H, Kitts P, Ruef B, Venkatraman A, Zhang L, Aquino-Perez G, Whitfield CW, Behura SK, Berlocher SH, Sheppard WS, Smith DR, Suarez AV, Tsutsui ND, Wei X, Wheeler D, Havlak P, Li B, Liu Y, Sodergren E, Jolivet A, Lee S, Nazareth LV, Pu L-L, Thorn R, Stolc V, Newman T, Samanta M, Tongprasit WA, Claudianos C, Berenbaum MR, Biswas S, de Graaf DC, Feyereisen R, Johnson RM, Oakeshott JG, Ranson H, Schuler MA, Muzny D, Chacko J, Davis C, Dinh H, Gill R, Hernandez J, Hines S, Hume J, Jackson L, Kovar C, Lewis L, Miner G, Morgan M, Nguyen N, Okwuonu G, Paul H, Santibanez J, Savery G, Svatek A, Villasana D, Wright R, and Honeybee Genome Sequencing C. 2006. Insights into social insects from the genome of the honeybee *Apis mellifera. Nature* **443**: 931-949.

Wenger SJ, Isaak DJ, Luce CH, Neville HM, Fausch KD, Dunham JB, Dauwalter DC, Young MK, Elsner MM, Rieman BE, Hamlet AF, and Williams JE. 2011. Flow regime, temperature, and biotic interactions drive differential declines of trout species under climate change. *Proceedings of the National Academy of Sciences* **108**: 14175-14180.

Wiebe KL, and Gerstmar H. 2010. Influence of spring temperatures and individual traits on reproductive timing and success in a migratory woodpecker. *Auk* **127**: 917-925.

Wiens JA, Seavy NE, and Jongsomjit D. 2011. Protected areas in climate space: What will the future bring? *Biological Conservation* **144**: 2119-2125.

Wilcove DS, Rothstein D, Dubow J, Phillips A, and Losos E. 1998. Quantifying threats to imperiled species in the United States. *BioScience* **48**: 607-615.

Wilcove DS, and Master LL. 2005. How many endangered species are there in the United States? *Frontiers in Ecology and the Environment* **3**: 414-420.

Williams CM, Marshall KE, MacMillan HA, Dzurisin JDK, Hellmann JJ, and Sinclair BJ. 2012. Thermal variability increases the impact of autumnal warming and drives metabolic depression in an overwintering butterfly. *PLoS ONE* **7**(3): e34470.

Williams EH. 1989. Endangered Aquatic Animals. *American Scientist* **77**: 318-318.

Williams JW, and Jackson ST. 2007. Novel climates, no-analog communities, and ecological surprises. *Frontiers in Ecology and the Environment* **5**: 475-482.

Williams JW, Jackson ST, and Kutzbach JE. 2007. Projected distributions of novel and disappearing climates by 2100 AD. *Proceedings of the National Academy of Sciences* **104**: 5738–5742.

Williams SE, Shoo LP, Isaac JL, Hoffmann AA, and Langham G. 2008. Towards an integrated framework for assessing the vulnerability of species to climate change. *PLoS ONE Biology* **6**: 2621-2626.

Willis CG, Ruhfel B, Primack RB, Miller-Rushing AJ, and Davis CC. 2008. Phylogenetic patterns of species loss in Thoreau's woods are driven by climate change. *Proceedings of the National Academy of Sciences* **105**: 17029-17033.

Wilson EO. 1988. Biodiversity. Smithsonian Institution, Washington D.C.

Winder M, and Schindler DE. 2004. Climate change uncouples trophic interactions in an aquatic ecosystem. *Ecology* **85**: 2100-2106.

Wood AJM, Collie JS, and Hare JA. 2009. A comparison between warm-water fish assemblages of Narragansett Bay and those of Long Island Sound waters. *Fishery Bulletin* **107**: 89-100.

Yang D-S, Conroy CJ, and Moritz C. 2011. Contrasting responses of *Peromyscus* mice of Yosemite National Park to recent climate change. *Global Change Biology* **17**: 2559-2566.

Yang LH, and Rudolf VHW. 2010. Phenology, ontogeny and the effects of climate change on the timing of species interactions. *Ecology Letters* **13**: 1-10.

Yom-Tov Y, and Yom-Tov J. 2005. Global warming, Bergmann's rule and body size in the masked shrew *Sorex cinereus* Kerr in Alaska. *Journal of Animal Ecology* **74**: 803-808.

Yom-Tov Y, Yom-Tov S, and Jarrell G. 2008. Recent increase in body size of the American marten Martes americana in Alaska. *Biological Journal of the Linnean Society* **93**: 701-707.

Young B, Byers E, Gravuer K, Hall K, Hammerson G, and Redder A. 2009. Guidelines for using the NatureServe climate change vulnerability index, Release 1.0.

Chapter 3. Impacts of Climate Change on Ecosystem Structure and Functioning

Convening Lead Authors: Nancy B. Grimm and F. Stuart Chapin III
Lead Authors: Britta Bierwagen, Patrick Gonzalez, Peter M. Groffman, Yiqi Luo, Forrest Melton, Knute Nadelhoffer, Amber Pairis, Peter Raymond, Josh Schimel, Craig E. Williamson
Contributing Author: Michael J. Bernstein

Key findings

- Changes in terrestrial plant species ranges attributable to climate change are shifting the location and extent of biomes, altering ecosystem structure and functioning.

- Changes in precipitation regimes and extremes (more intense storms, altered seasonality, increased drought), coupled with warming or independently, can cause ecosystem transitions (state change).

- Forests have responded to climate change, with faster growth in some humid areas and slower growth in some drier areas. Longer growing seasons and warmer winters are enhancing pest outbreaks, leading to tree mortality and to more severe and extensive fires.

- Changes in winter (for example, soil freezing, snow cover) have big and surprising effects, in terms of carbon sequestration, decomposition, and carbon export, which influence agricultural and forest production.

- Intensification of the hydrologic cycle increases movement of nutrients and pollutants to downstream ecosystems, restructuring processes, biota, and habitats.

- Both lakes and oceans are experiencing warmer air temperatures and elevated organic inputs, leading to greater thermal stratification and lower water clarity, which can increase "dead zones," harmful algal blooms, human and other parasites, and alter nutrient recycling and biological productivity.

- Feedbacks from altered ecosystem functioning to climate change via greenhouse gas emissions are potentially important but remain unclear.

- Federal and State agencies are integrating climate change research into resource management plans and adaptation actions to address impacts of climate change.

3.1. INTRODUCTION

Climate exerts a fundamental control over the distribution, structure, and functioning of ecosystems through direct effects and through interactions mediated by other controlling factors, such as geological substrate, the distribution and activity of species (including humans), and patterns of disturbance (Jenny, 1941; Vitousek, 2004; **Figure 3.1**). Climate has changed more rapidly during the 20th century than at any time since the peak of the last ice ages (Loarie and others, 2009) and is projected to change even more rapidly in the next 50-100 years. In this chapter, we synthesize evidence for the effects of recent climate change on ecosystems and explore further changes that might be anticipated if climate change continues as projected. We identify the changes that may have the greatest likelihood and consequence for ecosystems and society.

Ecosystems are places with a defined boundary that contain biotic and abiotic components that interact with each other and with adjacent ecosystems. In this chapter, we focus

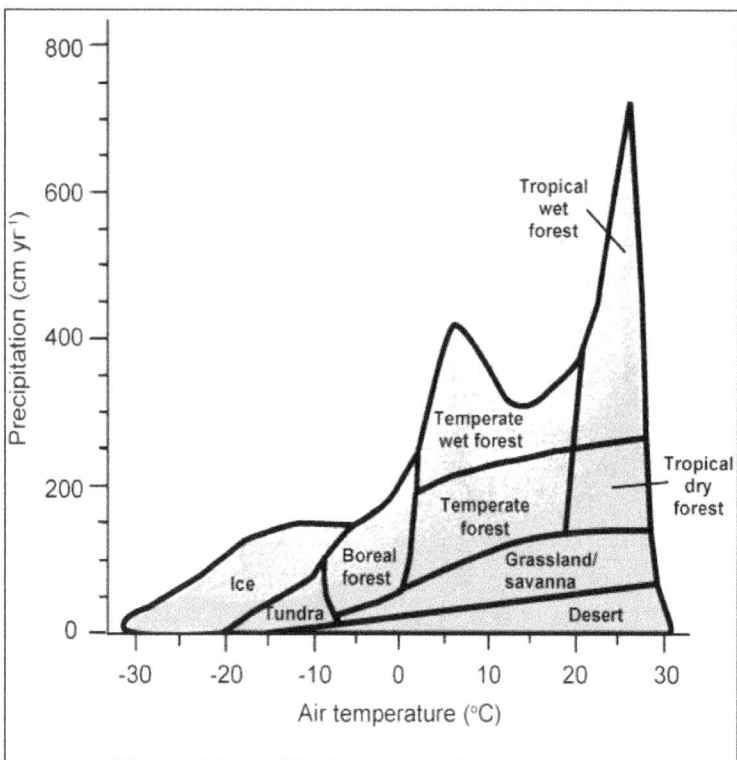

Figure 3.1. Distribution of the world's biomes (general categories of ecosystems) with respect to mean annual temperature and total annual precipitation (Chapin and others, 2011; used with permission). Gray dots show the temperature-precipitation regime of all terrestrial locations (excluding Antarctica) at 18.5 km resolution (data from New and others 2002).

specifically on structural elements of ecosystems such as biomass and stored soil carbon, and on functions such as the fluxes of energy and matter. Ecosystems include people, although the human role varies substantially among ecosystems, from a limited role in tundra, to an intense management role in cropped systems, to a completely dominant role in cities. Research focused on social-ecological systems has embraced an interdisciplinary perspective that includes people as integral component of feedbacks that control the dynamics of ecosystems (Chapin and others, 2010a; Collins and others, 2011; Liu and others, 2007; Matson, 2009; Turner and others, 2003a,b). For example, urban ecosystems house over half of the world's population yet depend on rural ecosystems for food, water, and waste processing; they are both drivers of global environmental change and potentially adaptive human response to it (Grimm and others, 2008).

3.1.1. Ecosystem impacts in context

During the last 50 years, human activities have changed ecosystems more rapidly and extensively than at any comparable period of human history (Steffen and others, 2004; Foley and others, 2005; MA, 2005). Human activities have directly modified 80 percent of Earth's ice-free land surface, in large part to support agriculture (Ellis and Ramankutty, 2008; Bringezu and others, 2012). People have indirectly affected all terrestrial, freshwater, and marine ecosystems on Earth through changes in climate and atmospheric chemistry (Kareiva and others, 2007) and have created new ecosystems with novel species composition and ecosystem properties (Hobbs and others, 2006). These multiple sources of environmental change make the task of elucidating climate-specific impacts challenging. Not only do multiple anthropogenic stressors influence ecosystems (Heathwaite, 2010; Strayer, 2010), but these stressors also interact in ways that can be synergistic, antagonistic, or additive and can feed back to exacerbate the effects of global climate change.

People experience climate change impacts on ecosystem structure and functioning through changes in ecosystem services (see *Chapter 4*). People depend on ecosystems for supplying harvestable resources (that is, food, fuel, fiber, water), regulating the movement of materials and disturbances among ecosystems (for example, retaining or transforming pollutants, modulating climate, moderating flooding), and recreational, cultural, and aesthetic values (consider, for example, the exposure to nature afforded by New York's Central Park or the Grand Canyon National Park). Biodiversity is a structural attribute of ecosystems that is a concern in its own right (see *Chapter 2*), but one that also has consequences for ecosystem functioning and the services we derive from an ecosystem. Particularly important are changes in dominant and keystone species (that is, species that have an influence on ecosystem functioning disproportional to their abundance; Power and others, 1996) or in species interactions that alter energy and material fluxes and, through these, change ecosystem services (Díaz and others, 2008).

3.1.2. A conceptual framework for ecosystem change

Ecosystems change gradually over long time periods (that is, decades to centuries) in response to changes in *average* climatic conditions (for example, average temperature and precipitation) (Iverson and Prasad, 2001). However, *seasonal shifts* and *extreme conditions* will exert the greatest impacts in the coming years to decades, because it is the droughts, frosts, and winter thaws that directly kill organisms or change their competitive balance. Furthermore, disturbances associated with extreme climate conditions (floods, wildfires, hurricanes) strongly influence the dynamics of ecosystems over decades to centuries (Peters and others, 2011a). Climatic changes that alter *disturbance regimes*, i.e., the spatial and temporal patterns of different disturbance types, can radically alter ecosystem properties. For example, warming and drying of wet regions can increase the frequency of fire, to which many species in these regions may be poorly adapted (Bond and Keeley, 2005).

A conceptual framework for understanding impacts of climate change on ecosystem structure and functioning therefore requires consideration of: 1) average climatic conditions that control ecosystem distribution; 2) shifts in seasonality; 3) individual disturbances (extreme events); and 4) disturbance regimes. Classical disturbance and succession theories, which argue that ecosystems return toward pre-disturbance states after a disturbance, are inadequate in a changing climate because pre-disturbance conditions may no longer exist and because ecosystems may have multiple stable states (Scheffer and others, 2001; Peters and others, 2011b). In other words, as the factors driving ecosystem dynamics change, tipping points may be breached, leading to changes in structure, functioning, and feedbacks, or to complete transformation (see **Box 3.1**).

To evaluate the risks of climate change impacts on ecosystems, we considered the structural and functional changes that may occur (or may have occurred) within a given ecosystem (for example, changes in species composition and fluxes of carbon and nutrients), the connections among ecosystems that are generally associated with regulating services (for example, spread of flood waters, fire or disease; downstream delivery of nutrients, pollutants, and sediments; feedbacks to the climate system), and the consequences of these changes for society. *Risk* is the product of likelihood and consequence (Yohe, 2008, NRC, 2010). *Vulnerability* of ecosystems to climate change is a function of the character, magnitude, and rate of climate change to which a system is exposed (*exposure*), as well as the system's *sensitivity* and its *capacity to adapt* to change (IPCC, 2007). When the *resilience* of an ecosystem or a

Impacts of Climate Change on Biodiversity, Ecosystems, and Ecosystem Services |
Technical Input to the 2013 National Climate Assessment

Chapter 3
Ecosystems

**Box 3.1. A Resilience-Based Framework for Considering
Climate Change Impacts on Ecosystem Structure and Functioning**

Resilience is the capacity of a system to absorb disturbance and reorganize while undergoing change so as to retain the same essential function, structure, identity, and feedbacks (Walker and others, 2004). When a system crosses a threshold, it may change to a new state, a phenomenon also called a "regime shift." These concepts are illustrated in (A) and (B) below for the classic case of lakes that exhibit regime shifts from clear to turbid lakes when phosphorus inputs exceed a threshold (Carpenter, 2003; used with permission). Before crossing the threshold (for example, threshold 1 in (B)), the ecosystem varies within a range of water-column phosphorus concentrations (an indicator variable for a host of ecosystem structural and functional attributes that comprise its "state" or "regime"); however, once it crosses that threshold it enters a new regime (turbid) that can be difficult to reverse. The diagram in (C) illustrates that early management can postpone the crossing of thresholds and consequent changes in state; however, management may not be sufficient ultimately to prevent regime shifts. In (D), the classical idea of succession is illustrated by change in ecosystem state variables following disturbance (from Chapin and others, 2011; used with permission).

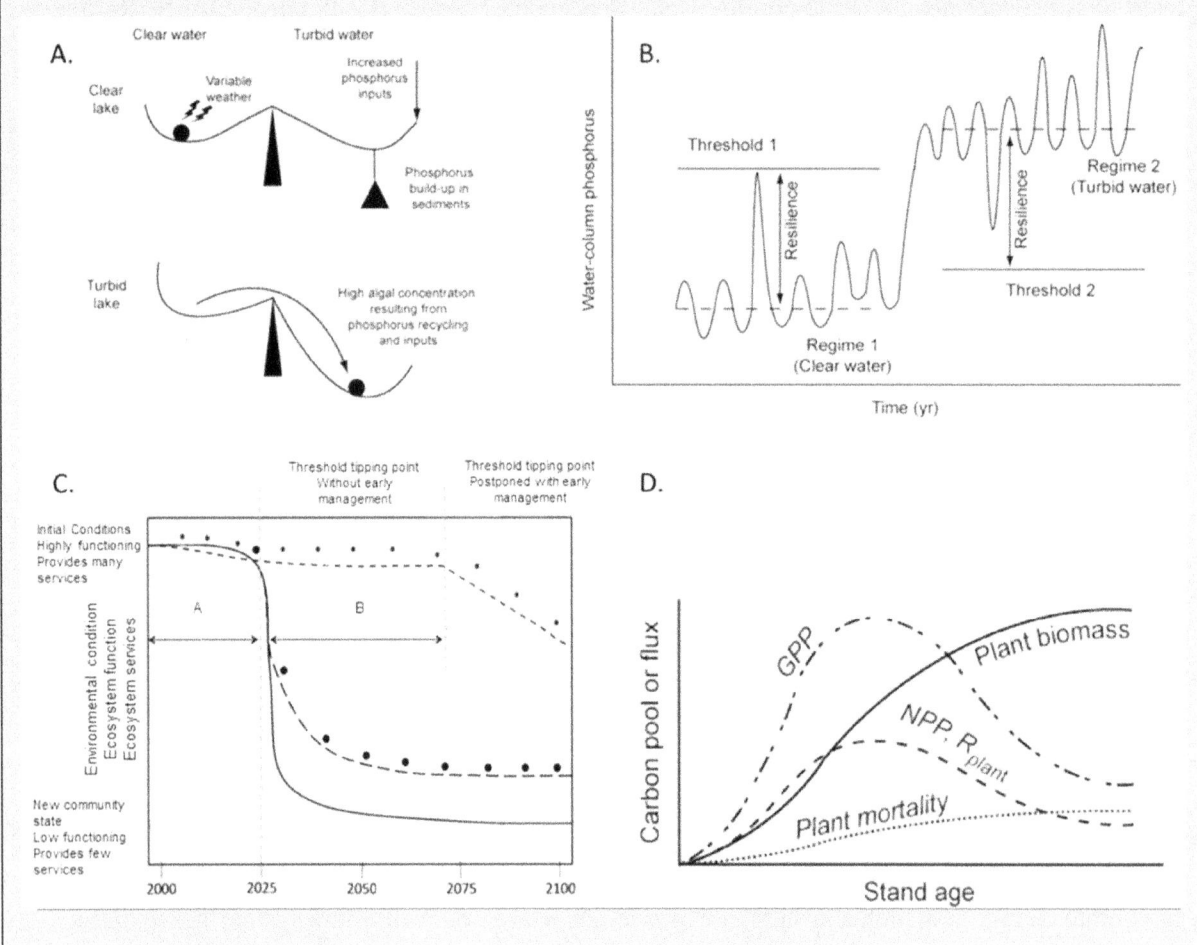

Impacts of Climate Change on Biodiversity, Ecosystems, and Ecosystem Services |
Technical Input to the 2013 National Climate Assessment

Chapter 3
Ecosystems

social-ecological system—defined as the capacity of a system to absorb disturbance and reorganize so as to retain essential function, structure, identity, and feedbacks (Walker and others, 2004)—is exceeded, transformation to a new state is likely (Scheffer and others, 2001; Walker and others, 2004; Chapin and others, 2010b). Climate change vulnerability assessments provide a qualitative and, in some cases, a quantitative way to determine which species, habitats, and ecosystems will respond and why they will respond to climate change (Metzger and others, 2008; Mawdsley, 2011; Diaz and others, 2007). However, measuring vulnerability consistently across systems and across spatial and temporal scales is challenging (Preston and others, 2011). Despite these challenges, assessing vulnerability is a crucial step for evaluating the risks of climate change.

3.2. HISTORICAL CHANGES, CURRENT STATUS, AND PROJECTED CHANGES IN ECOSYSTEM STRUCTURE AND FUNCTIONING

At the time of the last National Climate Assessment (2009), vulnerable ecosystems in Arctic sea ice, arid regions, and coastal zones were already affected by, respectively, warming and melt; complex interactions of increased dryness, fire, invasive species, and erosion; and multiple stresses of sea-level rise, storms, warming ocean temperature, and ocean acidification. Shifts in species ranges were well documented, as was increasing asynchrony between seasonal transitions and phenologies. However, impacts on *ecosystem functioning* were not always clear: for example, the balance between primary production (which takes carbon dioxide from the atmosphere) and ecosystem respiration (which releases it), each governed by a host of factors including temperature, water availability and seasonality, showed regional and local variation.

This report, organized by the key findings listed above, will summarize the impacts of climate change on several aspects of ecosystem structure and functioning. For each key finding we identify the most important climate driver controlling the ecosystem property and, where possible, the proportion of change that can be attributed to that driver and whether a change in the mean, extremes, or regime (temporal pattern) of the driver exerts the largest impact. We ask the following questions: What impacts have been observed, and what are the consequences?

Ecosystems exist in a context of large regions with a heterogeneous mix of different ecosystem types that, in turn, are embedded in a continent with strong climatic gradients; ecosystems are connected within and across these spatial scales (Peters and others, 2004). Scaling and interactions among ecosystem processes are thus important considerations; impacts of climate change that alter some aspect of ecosystem functioning in one place may be seen in a corresponding change at a distant place or at a larger scale (for example, effect of global-scale warming that is amplified at high latitudes, causing loss of sea ice, which in turn magnifies rates of warming at the global scale). Much of this "teleconnection" is mediated by the vectors of connectivity: wind, water, organisms, and people (Peters and others, 2004, 2008). A classic example is the connectivity within a watershed, wherein water transports materials from upland terrestrial ecosystems via a drainage network to downstream, recipient ecosystems. Watersheds exist at a range of scales and at the very largest scales may integrate disparate climate effects across half the continent.

3.2.1. Biome Shifts

Context. Biomes are major ecosystem types characterized by a dominant plant life form (Woodward and others, 2004). We define a biome shift as one in which at least one species of a dominant life form changes its distribution (for example, the movement of trees into a treeless tundra or grassland biome). A biome shift, as we have defined it, does not imply that all species of a biome synchronously shift their geographic distribution. A change in the location or extent of a biome can alter vegetation structure and ecosystem processes, which vary widely among biomes, affecting plant and animal habitats and the provision of ecosystem services.

Key Finding: **Changes in terrestrial plant species ranges attributable to climate change are shifting the location and extent of biomes, altering ecosystem structure and functioning.**

Climate. Spatial and temporal patterns of temperature and precipitation, modified by disturbances such as wildfires, determine biome location and extent. The velocity of 20th century climate change (Loarie and others, 2009) is elevated for a substantial part of the world, including the United States (Burrows and others, 2011). The 1960-2009 climate change velocities of 20 km y^{-1} in parts of Alaska, California, the Midwest, and the Southwest (Burrows and others, 2011) exceed an average 20th century range shift of ~0.6 km y^{-1} observed for terrestrial plant and animal taxa in Europe and the Americas (Chen and others, 2011). Climate change-velocities for the period from the last glacial maximum 21,000 years ago to today are less than 0.002 km y^{-1} (Sandel and others, 2011).

Projected climate velocities under Intergovernmental Panel on Climate Change (IPCC) emissions scenario A1B, which assumes substantial reductions in fossil fuel emissions (IPCC, 2007), exceed 1 km y^{-1} for most of the area between the Rocky Mountains and the Appalachians, Florida, and the mid-Atlantic region (Loarie and others, 2009). Biomes with the highest projected climate change velocities globally are flooded grasslands and savannas, mangroves, and deserts; biomes with the lowest velocities globally are montane grasslands and shrublands, temperate conifer forests, and tropical and subtropical conifer forests (Loarie and others, 2009).

Historical Biome Shifts. Field research has detected elevational and latitudinal shifts of plant species attributable to climate change that have shifted the location and extent of biomes around the world (Gonzalez and others 2010) and within the United States (Suarez and others, 1999; Lloyd and Fastie, 2003; Millar and others, 2004; Dial and others, 2007; Beckage and others, 2008). Using field data to examine biome location for greater than 30-year periods, researchers have detected latitudinal and elevational shifts of boreal forest into Alaskan tundra (Suarez and others, 1999; Lloyd and Fastie, 2003; Dial and others, 2007; Wilmking and others, 2004); elevational shifts of boreal and subalpine forest into the tundra and alpine biome in the Sierra Nevada, California (Millar and others, 2004); and an elevational shift of temperate broadleaf forest into boreal conifer forest in the Green Mountains, Vermont (Beckage and others, 2008). Climate predominated over local human factors in causing these changes.

Other research has found evidence of biome shifts consistent with, but not definitively attributed to climate change. These include an elevational shift of temperate shrubland into temperate conifer forest in Bandelier National Monument, New Mexico (Allen and Breshears, 1998), a latitudinal shift of boreal forest into tundra across Alaska (Beck and others, 2011), and upslope shifts of temperate mixed forest into temperate conifer forest in Southern California (Kelly and Goulden, 2008).

Biome shifts have altered ecosystem functioning. Growth of individual trees has tended to increase in forest stands already present in poleward and upslope ecosystems and in non-forest areas with newly developing shrub and tree cover (Suarez and others, 1999; Lloyd and Fastie, 2003; Millar and others, 2004; Dial and others, 2007; Beckage and others, 2008). Due to drought

Impacts of Climate Change on Biodiversity, Ecosystems, and Ecosystem Services | Chapter 3
Technical Input to the 2013 National Climate Assessment
Ecosystems

stress in Alaska, net primary productivity (NPP) declined at the trailing edge of shifting boreal forest (Beck and others, 2011).

Because field data from forests show that NPP and net ecosystem productivity (NEP) generally decrease with latitude (Pregitzer and Euskirchen, 2004), poleward biome shifts may be a contributing factor to a modeled 20th century increase in NPP at the continental scale (Piao and others, 2009) and to observed, remotely sensed increases from 1982 to 2009 across the United States, except for the Southwest (Nemani and others, 2003; Zhao and Running, 2010). Biome shifts often involve changes in disturbance regime that alter conditions for seedling establishment and therefore the suite of species that colonizes after disturbance (Johnstone and others, 2010; Turner, 2010).

Vulnerability to Future Biome Shifts. Projections of potential future vegetation indicate substantial vulnerability of ecosystems in the United States to continued biome shifts. Five dynamic global vegetation models (DGVMs) and one equilibrium model—performed for a range of the IPCC (2007) general circulation model (GCM) runs of emissions scenarios B1, A1B, A2, and A1FI—project biome changes on about 5 to about 20 percent of the land area of the United States from about 1990 to 2100 (Alo and Wang, 2008; Bergengren and others, 2011; Gonzalez and others, 2010; Sitch and others, 2008). An analysis that combined 1901–2002 historical climate changes and 1990–2100 DGVM projections indicates that one-seventh to one-third of North America may be vulnerable to biome shifts due to climate change (Gonzalez and others, 2010; **Figure 3.2**).

Projections generally agree on extensive poleward shifts of vegetation, although spatial distributions differ among GCMs, emissions scenarios, and vegetation models. Biomes and areas that show the highest vulnerability include tundra in Alaska, alpine in the Sierra Nevada and Rocky Mountains, temperate conifer forests in Southern California and the sky islands (isolated mountains and ranges in the Basin and Range province) of the Southwest, and temperate mixed forest around the Great Lakes.

Compensatory changes in demographic rates may buffer southern populations of tundra plants against warming, slowing northward range shifts (Doak and Morris, 2010). High-severity fires in boreal conifer stands in Alaska alter conditions for regrowth favorable to temperate broadleaf forest (Johnstone and others, 2010), a biome shift that affected 39 percent of areas burned in 2004 (Barrett and others, 2011). At the ecotone between the boreal forest and the Great Plains, projected increases in drought and forest disturbances (such as fire, windthrow, and pests) could shift the grassland–boreal forest ecotone farther north (Frelich and Reich, 2010). Wind dispersal abilities of individual species may limit shifts of temperate deciduous forest (Nathan and others, 2011).

Biome shifts that increase tree cover would tend to increase standing biomass and carbon, NPP, radiation use efficiency, canopy closure, and leaf area, while decreasing grass:tree and root:shoot ratios (Alo and Wang, 2008; Euskirchen and others, 2009; Garbulsky and others, 2010). In contrast, regional tree dieback in the Southwest (Breshears and others, 2005) could potentially convert temperate woodlands into temperate grasslands and produce opposite trends. For Alaska, models project increased NPP in all plant functional types in the tundra and boreal conifer forest biomes under IPCC emissions scenarios B1, B2, and A2, although increases in heterotrophic respiration may result in decreases or no net change in NEP (McGuire and others, 2010; Euskirchen and others, 2009). At high latitudes, increases in vegetation cover and aboveground biomass may reduce albedo, increase regional heat absorption, and initiate a positive feedback to climate warming (Chapin and others, 2005; Bala and others, 2007); reduced

Impacts of Climate Change on Biodiversity, Ecosystems, and Ecosystem Services |
Technical Input to the 2013 National Climate Assessment

Chapter 3
Ecosystems

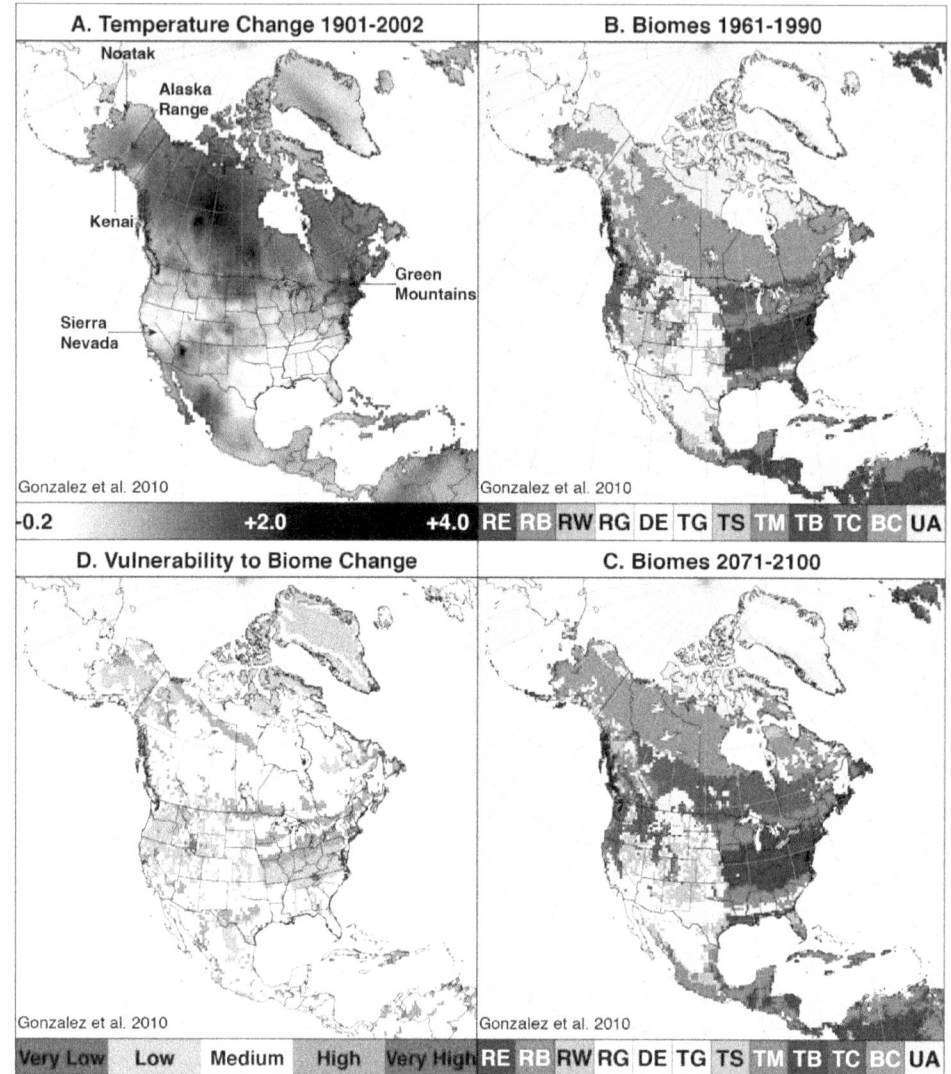

Figure 3.2. Biome shifts. (A) Observed linear temperature trend 1901-2002 (°C century⁻¹) and field sites of detected shifts: Alaska Range (Lloyd and Fastie, 2003), Green Mountains, Vermont (Beckage and others 2008), Kenai Mountains, Alaska (Dial and others, 2007), Noatak National Preserve, Alaska (Suarez and others, 1999), Sierra Nevada, California (Millar and others, 2004). (B) Potential vegetation under observed 1961–90 climate. Biomes: tropical evergreen broadleaf forest (RE), tropical deciduous broadleaf forest (RD), tropical woodland (RW), tropical grassland (RG), desert (DE), temperate grassland (TG), temperate shrubland (TS), temperate mixed forest (TM), temperate broadleaf forest (TB), temperate conifer forest (TC), boreal conifer forest (BC), tundra and alpine (UA). (C) Potential vegetation under projected 2071–2100 climate where any of nine GCM– emissions scenario combinations project change. (D) Vulnerability of ecosystems to biome shifts based on historical climate and projected vegetation. Vulnerability classes: very low (confidence less than 0.05), low (0.05 ≤ confidence less than 0.2), medium (0.2 ≤ confidence less than 0.8), high (0.8 ≤ confidence less than 0.95), very high (confidence ≥ 0.95). Data from Gonzalez and others (2010); used with permission.

snow cover due to a shorter snow season may cause even greater reductions in albedo (Euskirchen and others, 2009; 2010).

Climate change is projected to strongly influence disturbance regime. The MC1 DGVM (Dynamic Global Vegetation Model) projects wildfire changes on half of the area of North America by 2100 under IPCC emissions scenarios B1, A1B, and A2, with equal areas of increase and decrease, and high confidence of potential biome shifts on 35–40 percent of the continent (Gonzalez and others, 2010). The LPJ DGVM (Scholze and others, 2006) and an equilibrium model of fire (Krawchuk and others, 2009) project even more extensive areas of increased wildfire, a change that is unlikely to be prevented by increased suppression effort (Westerling and others, 2011).

Conclusion. Climate-induced biome shifts are projected to occur in about 5–20 percent of the United States from 1990 to 2100. Areas of high to very high vulnerability to biome change are home to one fifth to one half of the human population of North America (Gonzalez and others, 2010). Potential biome shifts in natural ecosystems have prompted the United States Department of the Interior to initiate the development of specific adaptation measures in national parks (Baron and others, 2009; Gonzalez, 2011) and national wildlife refuges (Griffith and others, 2009). These include targeting fire management to areas of potential changes in fire regime and biome type in Sequoia National Park, and landscape-scale conservation of potential refuge areas outside of parks.

3.2.2. Ecosystem state transitions

Context. Changes in precipitation regimes and temperature are expected to cause changes in vegetation distributions (see section 3.2.1, Biome Shifts), but these shifts are not always gradual or predictable. In fact, ecosystems may undergo rapid transitions from one state to another, where states are characterized by a suite of variables, including vegetation type, spatial pattern, nutrient dynamics and pools, etc. Ecosystem state transitions can dramatically alter ecosystem function and service provision (Brock and Carpenter, 2006). For example, the Sahel changed from a tropical forest to grassland and then to desert within a few thousand years (Kröpelin, 2008). Rapid or abrupt transitions, such as desertification or collapse of coral reefs, may occur when a threshold is crossed (Scheffer and others, 2001; see also **Box 3.1**). Although dramatic changes in ecosystem state have potentially profound impacts on the Earth system, they are incompletely understood and do not appear in models. They also can be difficult to predict, although theoretical and empirical research indicates that changes in variance may precede the transition (Brock and Carpenter, 2006, Scheffer and others, 2009, Carpenter and others, 2011).

Key finding: **Changes in precipitation regimes and extremes (more intense storms, altered seasonality, increased drought), coupled with warming or independently, can cause ecosystem transitions (state change).**

Warming-induced changes in species composition have been broadly observed in grasslands (Sherry and others, 2011; Yang and others, 2011; Zavaleta and others, 2003) and tundra ecosystems (Chapin and others, 1995; Klein and others, 2004 and 2007; Walker and others, 2006). These changes have potential functional consequences, including changes in NPP, water and nutrient cycling, regulation of regional climate, and trophic interactions (Eviner and Chapin, 2003; Tilman and others, 1997; Zavaleta and others, 2006). Woody invasion of high-latitude, herb-dominated ecosystems has been reported by several investigators to result from warming (Harte and Shaw, 1995; Chapin and others, 1995; Sturm and others, 2001). In a meta-analysis of warming experiments involved in the International Tundra Experiment, Walker and others (2006) found that plant community responses to warming were rapid, with height and cover of deciduous shrubs and graminoids increasing while cover of mosses and lichens decreased. The increase in shrubs and graminoids could reduce the competitive performance of other members of the community and therefore change the competitive hierarchy within a community (Niu and Wan, 2008). In temperate grasslands, abundant C_3 and C_4 grasses, representing the two major photosynthetic pathway classes of plants, are predicted to respond differentially to climate change: C_4 grasses are predicted to become more abundant at the expense of C_3 grasses in the temperate grasslands within North and South America as temperatures increase (Epstein and others, 2002). Experimental warming was also reported to consistently increase species richness of shrubs and decrease grass species richness in a consecutive five-year measurement of a temperate steppe (Yang and others, 2011). However, in an unproductive grassland in northern England species composition did not respond to simulated warming (Grime and others, 2008), possibly because plants in infertile ecosystems are already stress-adapted, with nutrient retention and tissue-protection strategies that could promote resilience to climate warming.

In the Chihuahuan Desert of New Mexico, ecosystem transitions may be controlled by a combination of temperature and precipitation dynamics. Experimental research has shown that chronic drought reduces cover of native grasses, yet has limited impacts on creosotebush, a ubiquitous desert shrub. Thus, droughts may contribute to shrub encroachment throughout the region. The shrubs exert a positive feedback on climate by altering surface energy balance and promoting higher nighttime temperatures (D'Odorico and others, 2010). Continued regional warming may further promote shrub expansion, as freezing temperatures limit the range of creosotebush. Given the responsiveness of grasses to rainfall variation, a more variable rainfall regime (more extreme events) during the summer monsoon will likely increase grass production and soil respiration, and may promote resilience of the grassland state. Warming favors the Chihuahuan grass, black grama, so it may increase in abundance with climate warming, and if rainfall increases this can also help the system resist shrub encroachment.

Changes in precipitation regime are likely to have strong effects on arid and semi-arid ecosystems, and may have the potential to reverse historical regime shifts, such as the desertification of grasslands (that is, transition to dominance by woody shrubs; Peters and others, 2012). One climatic phenomenon that can accentuate high interannual variability in precipitation is the El Nino Southern Oscillation (ENSO; Holmgren and others, 2006). Productivity in arid lands may respond to very wet years with greatly increased productivity, which can propagate in food webs to higher trophic levels, increasing the productivity of herbivores and predators (see examples in Holmgren and others, 2006). The complex interactions of grazing, interannual precipitation variability, precipitation seasonality, fire, and pests can result in rapid ecosystem transitions, for example between stable states with high and low vegetation biomass (Holmgren and Scheffer, 2001), although these impacts cannot be solely attributed to climate change. However, the changes in regimes brought about by increasing frequency and magnitude of extreme events or any changes in climate phenomena that, like ENSO, have a strong impact on interannual variability will affect arid and semi-arid ecosystems. A decade of climate change experiments manipulating these variables has begun to reveal the role of precipitation seasonality, timing, variability, and magnitude (Jentsch and others, 2007).

Experimental manipulations have shown reductions in aboveground NPP from drought, drought coupled with heavy rain, frost, heat events, and drought + heat events; belowground

NPP was affected by experimental imposition of drought + heavy rain (Jentsch and others, 2007). Knapp and others (2008) proposed a general framework for alteration of the "packaging" of rainfall—heavier, less frequent rain events, or heavier events without a change in frequency—and its impact on soil water dynamics. They proposed that mesic systems would show reduced productivity because of greater water stress, whereas dryland systems would be less frequently stressed because the greater amplitude of soil-water fluctuations would give them intermittent respite from the baseline, "usually stressed" situation (**Figure 3.3**). They also included wetlands in their model, suggesting that thresholds of anoxia (caused by waterlogging) would be crossed less frequently, potentially leading to increased NPP. Experimental tests of this model have agreed with its predictions (Heisler-White and others, 2008; Thomey and others, 2011).

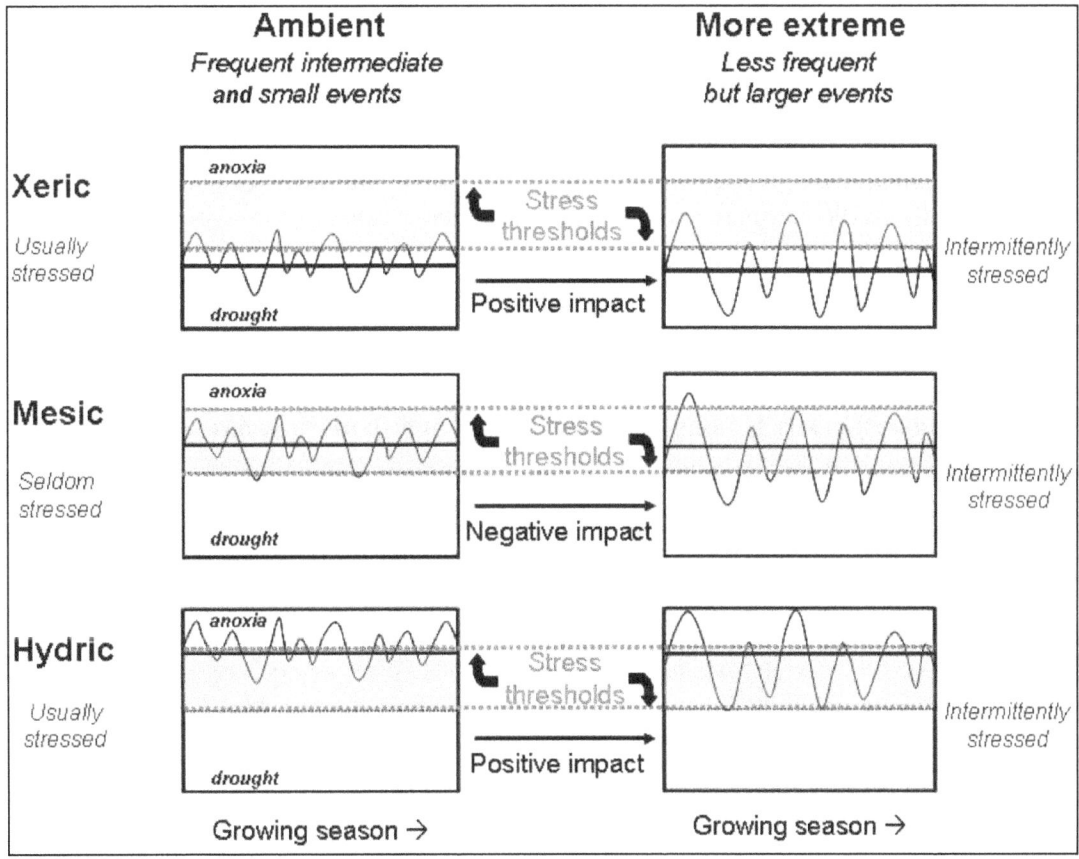

Figure 3.3. Hypothesized changes in soil moisture in xeric, mesic, and hydric systems with less frequent but larger precipitation events. Solid, heavy black lines represent mean soil moisture, solid, thin black lines represent the variability in soil moisture; and dotted red lines represent the stress thresholds above and below which ecological processes are limited by anoxia and drought, respectively. In this model, a shift to larger, less frequent events would have a positive impact in xeric and hydric soils, where drought and anoxia stresses (respectively) would become less chronic, but would have a negative impact in mesic soils where fluctuations in soil moisture would more frequently cross the stress thresholds. From Knapp and others (2008); used with permission.

Impacts of Climate Change on Biodiversity, Ecosystems, and Ecosystem Services |
Technical Input to the 2013 National Climate Assessment

Chapter 3
Ecosystems

Extreme events also are key to understanding how streams and riparian zones will respond to climate change. Although riparian forests will respond in some of the same ways as their upland counterparts (Perry and others, 2012), they are more strongly influenced by hydrologic variability, both flood and drought, and thus their future composition, extent, and functioning will depend upon how hydrologic regimes change (Poff and Zimmerman, 2010, Rood and others, 2008). Regional differences are expected in how hydrologic regimes change and thus the response of riparian ecosystems. For the Southwest, reduced baseflow and increased intermittency is likely to lead to conversion of native cottonwood-willow forest to exotic tamarisk or other non-native species that are more drought-tolerant (Rood and others, 2008, Stromberg and others, 2010). These ecosystem transitions fundamentally change the character of southwestern riparian ecosystems.

3.2.3. Forest growth, mortality, pests, and fire

Context. Projected biome shifts described in section 3.2.1 show major alterations in forest distribution, suggesting that changes in forest dynamics and distribution are critical consequences of climate change in the United States. These occur through changes in tree growth and mortality, and changes in carbon, water and energy fluxes, all of which respond directly to climatic drivers such as temperature and moisture, as well as responding indirectly to climate-induced changes in disturbance regimes.

Key Finding: **Forests have responded to climate change, with faster growth in some humid areas and slower growth in some drier areas. Longer growing seasons and warmer winters are enhancing pest outbreaks, leading to tree mortality and to more severe and extensive fires.**

Recent studies show net increases in forest productivity in the more humid eastern United States (McMahon and others, 2010; Cole and others, 2010), with further increases expected in response to projected climate changes (Ryan and others, 2008). However, in drier regions range expansions and invasions of insect pest and fungal pathogens are altering forest biomass and species composition (for example, Albani and others, 2010; Lamsal and others, 2011), trends that are projected to continue and reduce forest productivity in drier, western parts of the continent (Ryan and others, 2008). In particular, forest productivity and carbon uptake may decrease and tree death may increase in regions experiencing increased drought frequency, duration, or severity. Observations support these projections, with productivity responding to both changes in growth of individual trees and changes in pest/pathogen outbreaks and fire regime (see **Box 3.2** for a detailed case study). For example, reductions in the carbon sink potential, due to drought-related mortality and decreased productivity have occurred across a wide swath of western Canada, consistent with the projections of drier regions (Ma and others, 2012). Biotic disturbances due to pests and pathogens can dramatically change forest carbon uptake and storage, often decreasing both processes, but with unclear impacts over the long term (Hicke and others, 2012).

Increases in tree mortality have been reported at monitoring sites across the western United States; if trends continue, baseline tree mortality rates in western forests are projected to double every 17-29 years (van Mantgem and others, 2009). Such trends are consistent across elevation, tree size, forest type, and fire history. The observed patterns are attributed to warming of 0.3° to 0.4 °C per decade across these higher elevations, where forested ecosystems are generally found in the West. Regional warming in these ecosystems is associated with temperature-driven changes in hydrology, including declining snowpack, earlier spring snowmelt and runoff, and a lengthening of summer droughts (van Mantgem and others, 2009). Increases in fire frequency and severity from 1970 to 2003 in mid-elevation western conifer forests are consistent with spring and summer warming in that time period due to climate change (Westerling and others, 2006). Multivariate analysis of wildfire observations across the western United States from 1916 to 2003 indicate that climate was the dominant factor controlling burned area, even during periods of human fire suppression (Littell and others, 2009, Westerling and others, 2011).

Bark beetles have infested extensive areas of the western United States and Canada, killing extensive stands of temperate and boreal conifer forest across areas more extensive than any other outbreak in the last 125 years (Raffa and others, 2008). Climate change has been a major causal factor, with warmer temperatures allowing more beetles to survive winter, shorten their life cycle, and move to higher elevations and latitudes (Berg and others, 2006; Raffa and others, 2008; see also **Box 3.2**). Bark beetle outbreaks in the Greater Yellowstone Ecosystem are outside the historic range of variability (Logan and others, 2010). Continuation of warm winter conditions leaves western United States forests vulnerable to higher mortality from bark beetle infestations (Bentz and others, 2010).

Box 3.2. Climate Change and Ecosystem Disruption:
Whitebark Pine and Mountain Pine Beetles

The whitebark pine (*Pinus albicaulis)* has occupied its current range for approximately 8000 years, and was widespread across western North America at the end of the last ice age (MacDonald and others, 1989). The ongoing loss of this species from ecosystems across the northwestern U.S. and southwestern Canada will have significant impacts on services from these ecosystems (Logan and Powell, 2009), and provides an example of the ways in which changes in ecosystem composition and structure affect the functional properties of ecosystems, including changes in biogeochemistry, hydrology, and trophic dynamics. High mortality of whitebark pine across the northwestern U.S. has been attributed to severe outbreaks of mountain pine beetles driven by increases in winter temperatures and a reduced frequency of extreme low winter temperature (Logan and others, 2010, Hicke and others, 2006, Carroll and others, 2004, Logan and Powell, 2001). Susceptibility to mountain pine beetle outbreaks is enhanced by infestations of blister rust (*Cronartium ribicola*, pathogen), which was introduced and spread throughout the western U.S. in the first half of the 20th century (Tomback and Achuff, 2010).

Whitebark pine plays a key role in maintaining winter snow packs in high elevations through canopy interception and shading, and loss of this species leads to faster melt of the winter snowpack, increasing the flashiness of streams and reducing summer water availability at high elevations (Ellison and others, 2005). Mountain pine beetle outbreaks can also disrupt biogeochemical cycling, converting infested forests from a carbon sink to a carbon source (Kurz and others, 2008) and altering nitrogen cycling (Griffin and others, 2011). Whitebark pine nuts are an important food source for many species, including grizzly bears (*Ursus arctos horribilis*) which depend on them as a key food source in the fall prior to hibernation. Reductions in whitebark pine nuts have been tied to lower cub birth rates, lower over-winter survival rates, and increased conflicts between bears and humans. In 2010, a year with low whitebark cone production, there were 295 reported grizzly bear-human conflicts in the Greater Yellowstone Ecosystem, more than double the

---

Box 3.2, continued.

average number of conflicts from 1992-2009 (Gunther and others, 2010, Gunther and others, 2004). Aerial surveys conducted in 2009 found that 50 percent of whitebark pine stands in the Greater Yellowstone area exhibited high mortality of overstory trees (MacFarlane and others, 2009). In some areas of the Greater Yellowstone Ecosystem, mortality of cone bearing whitebark pines has exceeded 95 percent (Gibson and others, 2008), raising concerns about the long-term survival of grizzly bears and other species in the Greater Yellowstone Ecosystem that are ecologically dependent on whitebark pines. Projected increases in fire frequency and severity in the ecosystem are likely to further exacerbate the decline in whitebark pine. Recent modeling studies have concluded that based on projected changes in climate there is a high probability that the annual burned area in Greater Yellowstone will increase by 2020, and will exceed 100,000 ha annually after 2050, resulting in significant changes to species composition, hydrology, nutrient cycling, and carbon storage (Westerling and others, 2011).

While the threats to the whitebark pine ecosystem from climate change are severe, mitigation and adaptation strategies are being developed and implemented to attempt to prevent the loss of whitebark pine from the Greater Yellowstone Ecosystem. Ongoing mitigation actions include treatment of whitebark pines with carbaryl or verbenone to increase the resistance of individual trees to mountain pine beetles, pruning and thinning for fire protection, planting of whitebark pine seedlings, and establishment of whitebark pine orchards for production of blister rust resistant seedlings. These mitigation strategies are part of an adaptive management strategy which is being developed to ensure the preservation of this important ecosystem (GYCC, 2011).

Similar changes driven by increases in forests pests and pathogens associated with climate change and introduced species are also occurring in the other regions in the U.S., and include loss of the eastern hemlock in the northeastern U.S. (Dukes and others, 2009, Ellison and others, 2005), and the loss of spruce in Alaska (Berg and others, 2006).

Example images showing decline in whitebark pine ecosystems following infestation by mountain pine beetle (from Logan and others, 2010; image credit W.W. Macfarlane; used with permission). Left: looking north from the Bonneville Pass Trailhead, Absaroka Range, Shoshone National Forest, east-central Greater Yellowstone Ecosystem, USA. Right: Lazyman Hill, from the Gravelly Range Road, Beaverhead National Forest, northwest Greater Yellowstone Ecosystem, USA.

Impacts of Climate Change on Biodiversity, Ecosystems, and Ecosystem Services |
Technical Input to the 2013 National Climate Assessment

Chapter 3
Ecosystems

A key index of ecosystem productivity, the Normalized Difference Vegetation Index (NDVI, extracted from satellite sensor data), shows regional declines across boreal forests in Alaska and Canada; sustained negative trends have been detected from 1982-2006 on millions of acres of boreal forests (Goetz and others, 2005; Bunn and others, 2007; Verbyla, 2008; Parent and Verbyla, 2010; Beck and Goetz, 2011). Long-term satellite data records also show nearly 15 percent reductions in peak photosynthetic capacity over this same period for large areas in Alaska (Nemani and others, 2009). These satellite-sensor derived trends, corroborated by tree-ring studies (McGuire and others, 2010; Beck and others, 2011), have been attributed to higher evaporative demand and to temperature-induced drought stress associated with warming of more than 1.5 °C over the past 50 years in boreal Alaska (Verbyla, 2011; Stafford and others, 2000). In these regions, responses of vegetation growth to recent variability in spring and summer warming trends, including periods with cooler spring temperatures, have also been proposed as possible drivers of trends observed in the satellite data record (Wang and others, 2011). The NDVI changes observed in higher latitudes and elevations provide the best indication available of the likely response of water-limited, drought-vulnerable forested ecosystems as climate warming accelerates across the United States (Zhang and others, 2008).

Scientists continue to combine surface measurements, remote-sensing observations, and ecosystem models to accurately map and monitor biomass levels and carbon fluxes in United States and global ecosystems. Efforts such as the NASA Carbon Monitoring System, the USGS LandCarbon Assessment, Ameriflux, the North American Carbon Program, the Long-Term Ecological Research (LTER) network, the National Ecological Observatory Network (NEON), and the Department of Energy's Next-Generation Ecosystem Experiment are improving our ability to understand how climate change is affecting carbon fluxes from United States ecosystems. Collectively, these efforts address a number of key questions, such as the relationships between soil carbon, temperature, and decomposition rates, and the feedbacks between atmospheric CO_2 levels, forest productivity, disturbance rates, and nutrient and water availability. For example, net ecosystem exchange (NEE, a measure of carbon balance for relatively large areas based on eddy-correlation measures of CO_2 concentration) is continuously measured along an elevational gradient in New Mexico encompassing a range of climate and vegetation types. Inferring from the differential responses of NEE along the gradient to periods of warm, dry and cool, wet weather, Anderson-Teixeira and others (2011) concluded that transition to a warmer, drier climate (as projected by regional models) would shift NEE, so that much of the region would become a source of CO_2 because of a greater response of respiration than production. Over the next decade, efforts such as these are expected to significantly advance our ability to map and monitor biomass and carbon fluxes.

3.2.4. Changes in winter have surprising impacts

Context. While most research has focused on the effects of increases in temperature and changes in precipitation during the growing season, a recent development in climate change research has been a focus on changes during winter. This recent research suggests that climate change during winter is more marked than changes in summer and that these changes have strong effects on ecosystem structure and functioning. Seasonally snow-covered regions are especially susceptible to climate change, as small changes in temperature or precipitation may result in large changes in ecosystem structure and functioning. High-elevation and alpine ecosystems in the topographically diverse western United States are affected by changes in winter via the hydrologic connectivity of these upland regions to lowland regions.

Key Finding: **Changes in winter (for example, soil freezing, snow cover) have big and surprising effects, in terms of carbon sequestration, decomposition, and carbon export, which influence agricultural and forest production.**

At the Hubbard Brook Experimental Forest (HBEF) LTER site in New Hampshire, average annual air temperature has increased by 0.17 to 0.29 °C per decade over the last half-century, with more dramatic warming in winter than in summer (Campbell and others, 2007). These local trends in air temperature are characteristic of regional trends and are expected to continue into the future, with projected increases of 2.1 – 5.3 °C by 2100 (Burns and others, 2007; Hayhoe and others, 2007; Huntington and others, 2009). Precipitation has also increased by 3.5 to 6.7 cm per decade or 13 to 28 percent over 50 years (Campbell and others, 2007). Trends in precipitation at rain gages with the longest records are stronger due to the influence of a protracted drought in the mid-1960s. Winter precipitation has changed less than other seasons, which when combined with warmer winter air temperatures, have led to significant reductions in snowpack accumulation. Long-term snow measurements indicate that the maximum annual snowpack depth has declined by 4.8 cm per decade (1.4 cm snow water equivalent) and the number of days with snow cover has declined by 3.9 days per decade (Campbell and others, 2010). Annual and winter precipitation are projected to continue to increase by 7–14 percent and 12–30 percent, respectively, while summer precipitation is expected to show little change (Campbell and others, 2009).

Many of the key ecosystem effects of winter climate change are driven by changes in snow cover, which affects soil freezing and patterns of seasonal runoff (Brooks and others, 2011). Snow is important as an insulator of the soil; a lack of snow can produce the somewhat surprising phenomenon of colder/frozen soils in a warmer world. At the HBEF, manipulation experiments to simulate reductions in snow cover induced soil freezing, which led to increases in root mortality (Tierney and others, 2001; Cleavitt and others, 2008), decreases in decomposition (Christenson and others, 2010), marked increase in leaching losses of nitrogen, phosphorus and base cations (Fitzhugh and others, 2001, 2003), and increases in nitrous oxide flux (Groffman and others, 2006). Similar results have been observed in snow manipulation experiments in Canada (Boutin and Robitaille, 1995) and Colorado (Brooks and others, 1997; Williams and others, 1998). A recent review argues that the effects of winter climate change on the performance of temperate vegetation have been overlooked (Kreyling, 2010). These results suggest that winter climate change will increase the delivery of nutrients to receiving waters with negative effects on water quality.

There is considerable uncertainty about the nature, extent, and effects of winter climate change. First, it is not clear how the cooling effects of loss of snow insulation will play out against warming air temperatures and decreased albedo to ultimately determine the nature and extent of soil frost (Venalainen and others, 2001; Decker and others, 2003; Henry, 2008, Campbell and others, 2010). Second, while some snow manipulation experiments have produced marked increases in nutrient leaching losses and nitrous oxide flux, others have shown more muted effects (Austnes and Vestgarden, 2008; Hentschel and others, 2008; Hentschel and others, 2009). Much of the variation in response appears to be linked to dissolved organic carbon (DOC), which sometimes increases in response to soil frost, dampening the nitrogen response, but sometimes does not (Groffman and others, 2011; Haei and others, 2010).

Changes in winter conditions also influence patterns of runoff and provision of drinking water in water-supply watersheds. In the Colorado River basin, water shortages are expected as a consequence of changes in snowmelt timing (Barnett and Pierce, 2009). Recent modeling and observational studies in the Catskill Mountains in New York (the water supply for New York City) show that the combined effect of increased winter air temperatures, increased winter rain, and earlier snowmelt may result in more runoff during winter. This will cause reservoir storage levels, and water releases to increase during the winter and will cause reservoirs to refill earlier in the spring. An overall increase in precipitation will result in a reduction in number of days the system is under drought conditions, despite increased evapotranspiration later in the year (Matonse and others, 2011; Zion and others, 2011).

In alpine ecosystems, snowpack acts as a reservoir that supplies water to human populations at lower elevations. Alpine ecosystems are particularly vulnerable to climate change because warming is proceeding at a disproportionately rapid rate at high elevations (Bradley and others, 2004). Tree-ring reconstructions of historic snowpack point to a large decrease in snowpack in the late 20th century that may signal a shift in control of snowpack by precipitation to a control by temperature (Pederson and others, 2011). Acceleration of the annual melting of snowpack may reduce water availability later in the summer when it is most needed, particularly in more arid regions such as the western United States (Barnett and others, 2008). In the Colorado River basin, a recent modeling study suggests that substantially earlier peak snowmelt (for example, 2–3 weeks) results from the changes in radiative forcing brought about by dust deposition on snow, which changes albedo (Painter and others, 2010). The origin of this dust is disturbed soils of the Great Basin, and it has likely been affecting the duration of snow cover since ~1850 when American colonists began grazing, agricultural, and mining activities in the region (Neff and others, 2008).

Recent research in agricultural ecosystems suggests that winter climate change may reduce soil carbon levels and ecosystem carbon sequestration (Senthilkumar and others, 2009). At the Kellogg Biological Station LTER site, decreases in total soil carbon were observed in a wide range of agricultural management treatments and in never-tilled grassland between 1986 and 1988. Modeling analyses attributed the losses to higher rates of soil respiration during the dormant season, driven by increased winter temperatures.

3.2.5. Intensification of the hydrologic cycle

Context. Original studies on national streamflow trends from the 1990s documented changes in the amount of water transported in streams and rivers (Lins and Slack, 1999). These early studies demonstrated that baseflow and average streamflow have increased at a large number of streams, particularly in New England, mid-Atlantic, Midwest and south-central regions of the United States (Lins and Slack, 1999). A smaller but significant number of systems in the Pacific Northwest and Southeast demonstrated decreases in discharge. The arid Southwest is also projected to show declining discharge (Miller and others, 2011; Serrat-Capdevila and others, 2007), and reduced streamflow has been observed and attributed to climate change for three southwestern rivers (Barnett and others, 2008). The most important driver of these trends is changing precipitation (Groisman and others, 2004), although alterations in land cover and permafrost can also influence streamflow (Wang and Cai, 2010; Jones and Rinehart, 2010; Jones and others, 2012). Over the last 20 years, re-analysis of national streamflow data affirms these early findings (Groisman and others, 2004). In addition to alterations in baseflow and average flow, studies are reporting earlier snowmelt (Stewart and others, 2005; Steward, 2009; Hodgkins and Dudley, 2006) across northern states and an increase in very heavy precipitation events for New England (Euskirchen and others, 2007; Groisman and others, 2005; Kunkel and others, 2010).

These climate-induced changes in streamflow interact with human impacts, including dams, water withdrawals for human uses, and land use changes, to impact the structure and functioning of aquatic ecosystems (Poff and Zimmerman, 2010; Olden and Naiman, 2010). Water temperature and ecological flows are important metrics to understand current and future changes and to make informed management decisions, such as dam releases or timed withdrawals, in order to preserve aquatic ecosystem structure and functioning.

Ecological traits are a promising metric that may be useful in detecting and forecasting responses of ecosystem structure and functioning to changes in both water temperature and streamflow (Hamilton and others, 2010; Poff and others, 2010). For example, communities with greater numbers of cold-preference and flow-obligate stream benthic macroinvertebrates are likely to be more vulnerable to climate changes in the western United States (Poff and others, 2010).

Key Finding: **Intensification of the hydrologic cycle increases movement of nutrients and pollutants to downstream ecosystems, restructuring processes, biota, and habitats.**

The increase in water delivery and throughput can decrease the natural capacity of ecosystems to process biologically active elements such as nitrogen and will therefore exacerbate eutrophication problems. Several studies have demonstrated increased nitrogen losses from terrestrial ecosystems with increasing precipitation in the Northeast (Howarth and others, 2006) and California (Sobota and others, 2009). For the Mississippi drainage basin, which is the source of water to the Gulf of Mexico, where the nation's largest hypoxic zone occurs, studies have also demonstrated that increased precipitation leads to higher nitrogen deliveries (McIsaac and others, 2002; Justic and others, 2005). In addition, a re-arrangement of precipitation distribution on the landscape has led to changes in export (Raymond and others, 2008). It is now estimated that an additional 40 km^3 of water—equivalent to four Hudson Rivers—is originating from our nation's breadbasket each year, carrying with it the materials washed from those farmlands. Thus, nutrients and contaminants associated with agricultural systems will be more efficiently routed to inland and coastal waters with climate change.

Studies now demonstrate that even for rock-derived elements that become diluted by higher streamflows, fluxes from the terrestrial landscape increase in response to higher discharge (Godsey and others, 2009), and therefore fluxes are higher under increased precipitation. More high-resolution time series of more elements are needed to quantify the overall change in these fluxes; however, the flux of the main anion in the Mississippi—bicarbonate—has increased by about 45 percent over the past century (Raymond and others, 2008). Because major cations and anions are key to biological production in both soils and inland waters, it is likely that production will be reduced in terrestrial systems and increased in aquatic recipient systems.

In addition to altering material transport, changes in baseflow, mean flow, and flow extremes can affect the structure of inland waters, because discharge controls stream hydraulics, or the width, depth, and velocity of water (Leopold and others, 1964). The hydrologic disturbance regime of streams and rivers—seasonal and interannual patterns of very high (floods) and low (drying) streamflow—is characterized by timing, magnitude, type, and frequency of events that collectively affect stream biota and ecosystem functioning. New studies are linking changes in stream hydraulics to alterations of food-web structure (Sabo and others, 2010) and general biological integrity (Carlisle and others, 2011) in United States streams and rivers. For example, a period of extreme low flow in an Arizona desert stream resulted in complete loss of some macroinvertebrate species and ascendance of others, suggesting that a projected drier future for the Southwest portends major shifts in assemblages (Sponseller and others, 2010). Changes in the timing of spring melt due to warming can also affect species and habitat structure (Milner and others, 2011), which can lead to complex alterations of species assemblages (Perkins and others, 2010). The management of fish and invertebrates of inland waters in many regions of the United States must begin to account for alterations in habitat due to climate change (Hamilton and others, 2010).

The frequency and magnitude of extreme climate events are increasing in many regions of the country and these trends are projected to continue (Karl and others, 2009). Recent evidence suggests that human activities, including the production of greenhouse gases, are increasing the frequency and magnitude of extreme precipitation events (Allan, 2011; Min and others, 2011; Pall and others, 2011). Extreme precipitation events are often accompanied by extensive flooding, which can increase erosion and transport of dissolved and particulate materials including nutrients, contaminants, and pathogens. Over half of the outbreaks of waterborne diseases in the United States in the latter half of the 20th century were associated with extreme precipitation events (in the top 10 percent of rainfall events; (Curriero and others, 2001). While simple contaminant transport may be the primary vector of these increases in waterborne disease, changes in water transparency that reduce purification of surface waters by the ultraviolet (UV) in sunlight may also play a role (see section 3.2.5).

The increase in very heavy rainfall also has impacts on inland waters. In the most extreme cases, heavy rainfall can cause destructive losses to infrastructure and agriculture. Heavy rainfalls can also restructure stream geomorphology, altering habitat. The transport of sediment during these extreme events is massive, resulting in soil loss, disruption of reservoir water production, and filling of ponds and reservoirs (Inman and Jenkins, 1999). These extreme events can also extensively rework and redistribute sediments in river-dominated shelves (Allison and others, 2010) and coastal wetlands (Castaneda-Moya and others, 2010). Thus, management of important ecosystem services such as clean-water production, soil fertility, and wetland nutrient retention are very likely to be challenged by an intensification of heavy rains.

Although we have focused primarily on hydrologic drivers of change in stream, river and riparian ecosystems, thermal changes also are important. Stream and river temperatures in the United States have warmed by 1–3 °C over the period of record (Barnett and others, 2008; Kaushal and others, 2010), with higher rates of stream warming in part attributed to urbanization effects (Kaushal and others, 2010). Although few studies have examined the consequences at the ecosystem level, thermal regimes are important in controlling stream ecosystem functioning (Olden and Naiman, 2010); for example, through reductions in oxygen solubility and increased rates of microbial processes and organismal metabolism. Stream ecologists have documented numerous effects of an altered thermal regime at the population and community levels, on resident macroinvertebrate and fish populations, especially those that are sensitive to temperature (for example, Poff and others, 2010; Poff and Zimmerman, 2010; see also *Chapter 2*).

3.2.6. Ecosystem effects from physical changes in lakes and oceans

Context. In recent decades one of the most striking ecosystem scale responses to climate change at the global scale has been the loss of Arctic sea ice during the summer. A similar shortening of the duration of ice cover has occurred in inland waters. Ecosystem responses of the biota to climate are often stronger and more coherent than the actual physical climate factors that are forcing these changes (Hare and Mantua, 2000; Manca and DeMott, 2009). Several components of pelagic ecosystems in both oceans and lakes have shown dramatic shifts in terms of changes in primary productivity and zooplankton phenology and geographic distribution, as

Impacts of Climate Change on Biodiversity, Ecosystems, and Ecosystem Services |
Technical Input to the 2013 National Climate Assessment

Chapter 3
Ecosystems

well as changes at higher trophic levels (Richardson, 2008; Winder and others, 2009). While temperature was historically the component of climate change of primary concern in aquatic ecosystems, coastal and inland waters are being heavily influenced by increased precipitation that alters the flow of water and materials during extreme climate events ranging from drought to heavy precipitation. Signals of these changes are also archived in lake sediments, permitting interpretation of past climate events. As low points in a landscape, lakes are excellent indicators of climate change (Williamson and others, 2009a; Williamson and others, 2009b). Climate change threats to both inland and coastal oceans of primary concern to humans include expanding areas of low oxygen ("dead zones"), harmful algal blooms (HABs), and depressed fisheries production.

Key Finding: **Both lakes and oceans are experiencing warmer air temperatures and elevated organic inputs, leading to greater thermal stratification and lower water clarity, which can increase "dead zones," harmful algal blooms, human and other parasites, and alter nutrient recycling and biological productivity.**

The period of ice cover, water level, surface and deepwater temperatures, and a variety of physical and chemical variables respond to climate change and may signal impending consequences for aquatic ecosystems (Adrian and others, 2009) as well as human health. Warming of air temperatures over the past century has led to ice cover periods in lakes being an average of 12 days shorter per hundred years (Magnuson and others, 2000; Livingstone and others, 2010). The most notable climate-related reductions in ice cover are those observed in the Arctic Ocean (see **Box 3.3**), which is expected to have late-summer, ice-free periods by the middle of this century (Wang and Overland, 2009). These declines in summer Arctic sea ice are unprecedented in the past 1,450 years (Kinnard and others, 2011). Sea ice creates a critical habitat for organisms ranging from sea ice algae at the base of the food web, to many birds and mammals at higher trophic levels, including polar bears. Many coastal indigenous communities in Alaska and Canada depend nutritionally and culturally on these ice-associated foodwebs. In the Southern Hemisphere the population size of krill, a key crustacean food of whales and other marine vertebrates, is positively correlated with the extent of sea ice (Atkinson and others, 2004).

Large lakes of the world are warming at rates faster than the world's oceans (Verburg and Hecky, 2009) and at approximately twice the rate of the regional air-temperature warming (Moore and others, 2009; Schneider and others, 2009; Schneider and Hook, 2010). Why do warmer water temperatures matter? Warmer surface waters can stimulate harmful algal blooms, which may include toxic cyanobacteria that are favored at warmer temperatures (Paerl and Huisman, 2008). The cost of degradation of inland freshwaters by harmful algal blooms related to eutrophication (to which climate change is just one of many contributors) in the United States has been estimated at $2.2 billion (Dodds and others, 2009). Furthermore, warmer temperature stimulates biological processes, and therefore accelerates biogeochemical reactions. Accelerated metabolism can lead to changes in water quality, such as depletion of oxygen due to aerobic respiration. Oxygen depletion in aquatic ecosystems represents an important threshold, beyond which biogeochemical processes change fundamentally (that is, anaerobic metabolism predominates, and insoluble compounds and metals can dissolve and be released into the water column).

Warmer surface water can have the opposite effect in the open waters of large, deep lakes and the world's oceans, which are mostly distant from terrestrial runoff and nutrients. Stronger

Box 3.3. Ominous Signals from Ocean Ecosystems

Ocean ecosystems cover the majority of Earth's surface, and they are changing in fundamental ways in response to climate change. Two of the most ominous signals of this change are the rise in sea level and loss of Arctic sea ice, illustrated in the graphs below. Increases in sea level occur both because of expansion of the oceans due to warming and because of the melting of land-based ice. In addition to economic costs associated with adaptation to a higher sea level for the extensive coastal human populations and settlements, warmer ocean temperatures are causing severe bleaching of coral reefs, exacerbating acidification that inhibits skeletal formation in many invertebrates, and leading to decreases in ocean productivity. The anticipated disappearance of sea ice in the Arctic by the middle of this century will further increase heat absorption by the oceans, accelerating the rate of global warming. [Original figures; sea-level data from University of Colorado Sea Level Research Group (http://sealevel.colorado.edu/); sea-ice data from National Snow and Ice Data Center, Boulder, Colorado (ftp://sidads.colorado.edu/DATASETS/NOAA/G02135/Mar/N_03_area.txt)]

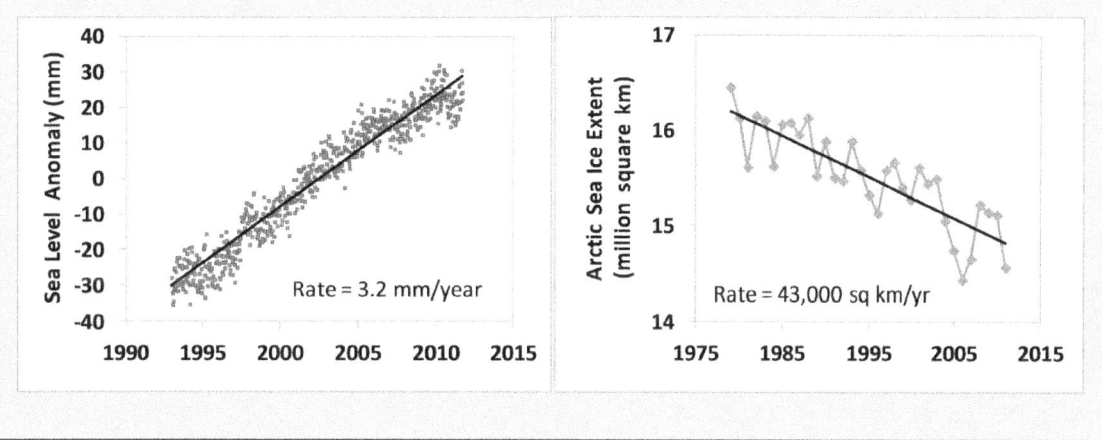

temperature and thus density differences between water layers lead to a separation of the nutrient-rich deeper waters from the surface waters where photosynthesis and primary productivity occur. This mechanism is thought to be largely responsible for recently observed decreases in primary productivity of up to one percent per year in eight out of 10 of the major ocean basins of the world (Boyce and others, 2010). These changes are similar to those observed previously in large tropical lakes (O'Reilly and others, 2003). Declines in primary productivity depress fisheries that sustain human populations.

One of the most striking changes in inland waters in recent decades is the approximate doubling of dissolved organic carbon (DOC; **Figure 3.4**), the tea-colored material that comes largely from decomposition of terrestrial plants and plays many important roles in aquatic ecosystems (Findlay, 2005; Evans and others, 2006). While the long-term trend is likely a "good news" story that results from decreases in acid deposition (Evans and others, 2006; Monteith and others, 2007), consequences of the increase in DOC may be deleterious to aquatic biota and ecosystem processes.

A primary driver of increased DOC in lakes and streams is increased precipitation (Pace and Cole, 2002; Raymond and Saiers, 2010; Zhang and others, 2010), which washes organic materials into lakes. Precipitation-driven increases in DOC concentration not only increase the cost of water treatment for municipal use (Haaland and others, 2010), but also may alter the ability of sunlight to inactivate parasites and pathogens in water, by absorbing ultraviolet radiation (UV) that would otherwise be an effective control. In an anthropocentric example, surveys have revealed that *Cryptosporidium*—a human pathogen potentially lethal to the elderly, babies, and people with compromised immune systems—is present in 55 percent of surface and 17 percent of drinking water supplies sampled in the United States (Rose and others, 1991). *Cryptosporidium* is inactivated by

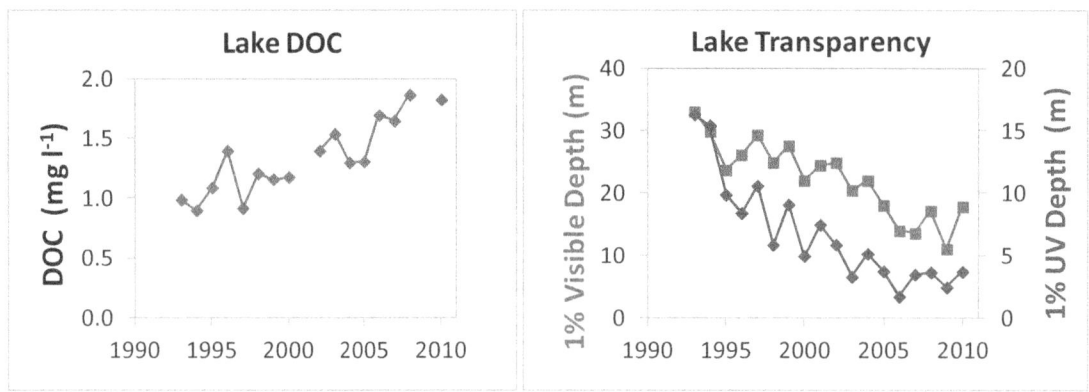

Figure 3.4. *An example of long-term trends in DOC concentration for a lake in northern Pennsylvania (left). As DOC concentration increases, lake transparency decreases (right), with consequences for primary productivity, the base of lake food webs. The 1 percent depths are estimated from surface mixed layer transparency and influenced by DOC quality as well as concentration. Data from Robert Moeller, Bruce Hargreaves, Don Morris, Jason Porter, and Craig Williamson with permission.*

UV levels contained in less than a day of sun exposure (Connelly and others, 2007; King and others, 2008). Similarly, fungal parasites that infect *Daphnia*, a keystone aquatic grazer and food source for fish, are sensitive to UV exposures as low as one percent of the amount of sunlight received at the surface of a lake on a clear day in June (Overholt and others, 2011): increasing DOC concentrations may thus reduce the ability of sunlight to inactivate these UV-sensitive parasites.

In cold, very clear lakes such as Lake Tahoe, UV may also play an important role in controlling invasions of warm-water fish species (Tucker and others, 2010). Increases in DOC that reduce UV transparency of lakes may also alter the spawning depth of fish such as the commercially valuable yellow perch (Huff and others, 2004).

In summary, physical changes in lakes and oceans are driven by warming and elevated organic inputs, which together lead to changes in ecosystem functioning (primary productivity, respiration, and anaerobic processes). These complex interactions between biogeochemical and physical processes can be difficult to untangle but their impacts are greater than we would expect from warming alone. No models currently include the changes in organic export driven by increased precipitation or their impacts on recipient systems.

3.2.7. Feedbacks from ecosystem functioning to climate

Context. One of the main ways in which ecosystems affect climate is through consumption and production of greenhouse gases. While CO_2 is the dominant trace gas in the climate system, CH_4 and N_2O follow in importance; both are present in much smaller concentrations than CO_2 but are much more potent greenhouse gases (IPCC, 2007). Methane is the second most important greenhouse gas controlling climate warming. Although methane is about 25 times as potent as carbon dioxide, the quantity of methane produced has been much less than that of carbon dioxide. Thus during the period from 1750-2000, the contribution of methane to greenhouse gas forcing has been a little over a third as important as carbon dioxide (Lacis and others, 2010).

Biological processes distributed among terrestrial ecosystems, including natural wetlands, rice agriculture, biomass burning, termites, and ruminant animal digestion account for roughly two-thirds of global CH_4 production (Dlugokencky and others, 2011). All of these sources are affected by both climate change and human land use decisions. The N_2O budget is dominated even more strongly by biological sources, notably from fertilized agriculture or systems receiving runoff from agriculture (IPCC, 2007).

Because all three major greenhouse gases, CO_2, CH_4, and N_2O, are sensitive to ecosystem perturbations, there are potentially strong feedbacks among climate change, land use decisions, and gas fluxes to the atmosphere. Because CO_2 is the common currency of biological energetics (consumed by photosynthesis and produced by respiration), its dynamics are relatively tractable; the potential "surprises" are driven by disturbance such as fire or land conversion. For CH_4 and N_2O, however, the specific magnitude and nature of the feedbacks to climate change are unclear and more difficult to predict, because both gases are produced by anaerobic microbial processes—CH_4 by Archaea and N_2O by denitrifying bacteria. CH_4 may also be rapidly consumed by aerobic bacteria (methanotrophs). Even modest shifts in soil aeration, which is controlled primarily by hydrology, might therefore substantially shift the amount of these gases released from ecosystems.

Key Finding: **Feedbacks from altered ecosystem functioning to climate change via greenhouse gas emissions are potentially important but remain unclear.**

One feedback to the climate system that is potentially large is carbon (C) release from thawing permafrost in the Arctic (Koven and others, 2011). Globally, permafrost soils contain approximately 1600 petagrams (Pg) C, which is roughly 50 percent of the total organic C reservoir and equivalent to twice the amount of CO_2 in the atmosphere (Schuur and others, 2011). As permafrost thaws, C becomes exposed to microbial attack and decomposition, with the potential to be released as either CO_2 or CH_4. The magnitude and nature of the feedback will depend on 1) how fast permafrost thaws, 2) the biodegradability of organic matter contained within the permafrost, and 3) whether it is released as CO_2 or CH_4. Questions 2 and 3 relate to soil hydrology. Less carbon is released when soils are anaerobic, but more of that C is released as CH_4 (Lee and others, 2012). In the permafrost zone of the United States, therefore, the nature of the feedbacks from thaw will be closely linked to changes in rainfall, runoff, and local patterns of inundation. If, for example, permafrost thaw on the Alaskan coastal plain causes the water table to drop, leaving a drier surface zone, CO_2 emissions will likely increase but CH_4 emissions might actually decrease.

Terrestrial ecosystems, including inland waters, are an important sink of human-produced greenhouse gases, absorbing on the order of 2.6 ± 1.7 Pg of carbon per year. Inland waters are an important element of the regulation of the balance of GHG between the atmosphere and continental land masses (Tranvik and others, 2009). Inland waters tend to be a source of GHG, outgassing both carbon dioxide and methane of largely terrestrial origin. A recent assessment of the methane emissions of 474 freshwater ecosystems estimates that methane emissions from inland waters are equal to 25 percent of the land-based greenhouse gas sink expressed in CO_2 equivalents (Bastviken and others, 2011), similar to the role of inland waters for CO_2 emissions (Kling and others,1991). Methane is produced when oxygen levels in aquatic ecosystems are depleted. These anoxic conditions are induced by increases in precipitation that increases runoff and the transport of nutrients and organic carbon to inland and coastal waters. These increases in nutrients and DOC reduce water transparency and further contribute to reductions in oxygen and the potential for methane emission from inland waters. Due to its relatively high energy content, methane can also contribute substantially to sediment-based food webs in inland waters, supporting certain fly species in particular that are highly tolerant of these low oxygen conditions (Jones and Grey, 2011).

Predicting N_2O emissions as a feedback from climate change is equally challenging. Producing N_2O requires coupling the obligate aerobic process of nitrification to the anaerobic process of denitrification; thus it is constrained to interfaces in either space or time: hot spots or hot moments (McClain and others, 2003; Harms and Grimm, 2008). At a fine spatial scale, denitrification can occur in anaerobic microsites in drained soils. In agricultural systems, the highest loss of applied N fertilizer as N_2O occurred in soils that remain at a water-filled pore space of 60–80 percent (Flechard and others, 2007), but at the landscape scale, riparian zones and freshwater wetlands are potential hotspots, when NO_3^-- rich waters reach saturated and C-rich soils and sediments (Ranalli and Macalady, 2010). These will be sensitive to changes in NO_3^- inputs and to land use decisions that affect the distribution of wet zones.

Temporally, N_2O fluxes spike following rainstorms that saturate the soil and following thaw. In agricultural soils of the temperate region, as much as two-thirds of annual N_2O emissions may occur at the spring thaw (Johnson and others, 2010). The dynamics of N_2O fluxes associated with freezing and thaw remain poorly quantified (Desjardins and others, 2010; Dietzel and others, 2011). Because CH_4 and N_2O dynamics are sensitive to small changes in threshold conditions of soil moisture and to the timing of specific weather events, it is almost certain that they will change with climate, but it remains difficult to predict either the sign or the magnitude of those changes.

3.2.8. Resource management in the context of climate change

Key Finding: **Federal and State agencies are integrating climate change research into resource management plans and adaptation actions to address impacts of climate change.**

Historical ecological impacts of climate change and a better understanding of future vulnerabilities have prompted natural resource management agencies to begin integrating climate change into resource management practices (Baron and others, 2009; California Natural Resources Agency, 2009; Gonzalez, 2011; Griffith and others 2009). Many State and Federal agencies are attempting to adapt natural resource management to climate change and also to reduce the greenhouse gas emissions that cause climate change. At the Federal level, Presidential Executive Order 13514 (October 5, 2009) directed Executive Branch agencies to develop adaptation approaches. Department of the Interior Secretarial Order 3289 (September 14, 2009) established Landscape Conservation Cooperatives (LCCs), which are science–management partnerships that include major land management agencies at the State and Federal levels along with other partners, and Climate Science Centers, which link the United States Geological Survey and universities to resource managers. By engaging Federal and State agencies, tribal and local governments, and non-governmental organizations across broad landscapes, LCCs present a unique opportunity for coordinated landscape-scale efforts that transcend political and jurisdictional boundaries. In addition, Federal, State, and tribal governments have joined together to develop the National Fish, Wildlife, and Plants Climate Adaptation strategy (http://www.wildlifeadaptationstrategy.gov/) to further develop adaptation measures based on ecological, not administrative, units. Further examples of adaptation planning and implementation efforts at multiple institutional scales are provided in *Chapter 6*.

While many State agencies have been in the process of adaptation planning, many have also been working to incorporate climate change research into existing activities. Sixteen States have existing or in-progress State adaptation plans that address multiple sectors, incorporating coastal, water resources, agriculture, forest and terrestrial ecosystems, bay and aquatic ecosystems, growth and land use, energy development, and public health constituencies. In addition, State fish and wildlife agencies are using their State Wildlife Action Plans (http://www.teaming.com/) as a platform from which to carry out climate change adaptation work. As of March 2011 all but twelve States were in the process of incorporating climate change into their State Wildlife Action Plan.

A key scientific basis for adaptation of natural resource management is the analysis of vulnerability of species and ecosystems to climate change (see **Box 3.4**). Vulnerability assessments are a key element to successful climate change adaptation as they reveal what systems, species, populations, and entities are most vulnerable to expected climatic changes, depending on three factors that define vulnerability: exposure, sensitivity, and adaptive capacity. With the increased recognition of the utility of vulnerability assessments, efforts to conduct these assessments are becoming more common. Robust vulnerability analyses combine historical and projected climate and ecological spatial data to identify vulnerable areas and potential refugia. Climate change vulnerability analyses in use by Federal agencies include analyses of the vulnerability of national wildlife refuges to habitat alteration (Magness and others, 2011), terrestrial ecosystems to biome shifts (Gonzalez and others 2010), coastal ecosystems to sea-level rise (Pendleton and others, 2010), plant and animal species to range shifts across the southwestern United States (Cole and others, 2011; Davison and others, 2012), and Pacific Northwest forests to vegetation shifts, changes in fire, and altered hydrology (Littell and others 2012). At the State level, numerous climate change vulnerability assessments are informing adaptation planning efforts, cross-sector coordination, and revision of State Wildlife Action Plans. For example, in California, assessments cover topics such as climate change vulnerability of California's at-risk birds, freshwater fish, rare plants, and a soil vulnerability index to identify drought-sensitive areas. California also has an assessment for water utility practices and public-health impacts (http://www.dfg.ca.gov/Climate_and_Energy/Vulnerability_Assessments/).

Using information from vulnerability analyses, Federal and State agencies are also developing adaptation measures on the ground and in the water. For example, the USDA Forest Service and National Park Service in Olympic National Forest and Olympic National Park are planning to target revegetation, replace road culverts, and re-target fire management based on climate change vulnerability analyses (Littell and others, 2012). The National Park Service and the University of Miami have collaborated to raise and out-plant heat-resistant local corals to restore bleached reefs in and around Biscayne Bay National Park (Lirman and others, 2010). The

**Box 3.4. An Example of Vulnerability Assessment Approaches
Applied to Northeastern Stream and River Ecosystems**

In the context of climate change impacts, vulnerability is defined as the function of the *sensitivity* of a system to climate changes, its *exposure* to those changes, and its *capacity to adapt* to those changes (IPCC, 2007). State biologists in the Northeast, who monitor the condition of stream and river ecosystems, want to understand how climate change will affect these resources, where the most vulnerable watersheds are, and what the consequences of climate change will be. A vulnerability assessment in this context examined the exposure and sensitivity components to prioritize monitoring locations in watersheds (Bierwagen, written communication 2012).

Exposure variables measure how much of a change in climate a species or system is likely to experience; while sensitivity variables measure whether and how much a species or system is likely to be affected by a change in climate. Adaptive capacity was not addressed for three reasons: first, it generally refers to human activities and institutions; second, in the ecological context it refers to biological changes in response to pressures; and third, the streams analyzed are in the best ecological condition and generally not restoration candidates. However, some insights from the vulnerability assessment can point to potential adaptation actions that can reduce vulnerability or enhance resilience for some streams. The capacity to carry out these activities then depends on the human institutions responsible for restoration or protection of ecosystems.

Changes in temperature and precipitation cause changes in hydrology and aquatic habitat. The vulnerability assessment attempts to discern which sites and watersheds will be exposed to the greatest and least amount of climate change and which will be the most and least sensitive to these changes (**Figure 3.5**). The results will inform site selection and design recommendations for a New York/New England pilot climate change monitoring network.

The monitoring network will cover seven northeastern States containing several ecoregions and stream types. In order to combine data from across the region, samples need to be sufficiently similar to minimize sources of variability. Classification procedures identify sources of variability in biological samples that are not the factors contributing to the effects that the monitoring network is designed to detect (Gerritsen and others, 2000; Hawkins and others, 2000). If the background variability is recognized, attributed to a source, and controlled by assigning sites to recognized types, then the variability observed within site types can be related to some other variable of interest, such as climate change factors. The first step in reducing variability is to isolate those sites with the least amount of disturbance caused by human activity (Hughes and others, 1986). In these reference sites, natural differences among site types are the basis of site classification.

Non-metric multidimensional scaling (NMDS) ordination and cluster analysis are techniques that group biological data (stream macroinvertebrates) into several types. Further multivariate analyses, using principal components analysis (PCA), discriminant function analysis (DFA), and correlations with NMDS axes, discern the environmental factors that may structure the biological groupings. Catchment size and stream slope are the principle factors related to the biological groupings when comparing results of the different analyses. This classification defines three stream types in the Northeast: High-gradient, small catchment (slope greater than 0.02 m/m and catchment less than 100 km^2); Moderate-gradient, small catchment (slope less than 0.02 m/m and catchment less than 100 km^2); and Low-gradient, large catchment (slope less than 0.005 m/m or catchment greater than 100 km^2).

Impacts of Climate Change on Biodiversity, Ecosystems, and Ecosystem Services |
Technical Input to the 2013 National Climate Assessment

Chapter 3
Ecosystems

Box 3.4, continued.

Climate change is projected to make the Northeast warmer and wetter overall, with more precipitation in the winter and hotter and drier summers (Frumhoff and others, 2007; Hayhoe and others, 2007). These climatic changes translate to more streamflow in the winter and spring, and lower summer low flows. Consequently, the relevant exposures due to climate change include (1) a shift in the timing of winter/spring runoff; (2) changes in peak flow events; and (3) changes in low-flow events and warmer water temperatures. Model projections that address these exposures include data on snow-water equivalents (SWE) and drought severity. No data were available on changes in peak flow events, and therefore the landscape has an equal chance of exposure.

There is a suite of environmental variables that determine whether a catchment is more or less sensitive to a particular exposure variable. Sensitivity variables that may mediate peak flow events include the amount of impervious surface, undeveloped floodplain, wetland area, and open water in the catchment, and mean catchment slope. Catchment aspect, whether the orientation is more northerly or southerly, influences the timing of winter/spring runoff and therefore SWE. Aspect, baseflow, and the amount of shading in a catchment are important components in determining drought severity.

The combination of data for each exposure pathway and the sensitivity variables determine the vulnerability score for each catchment. Least vulnerable catchments are defined as having the lowest exposure and sensitivity scores; moderately vulnerable catchments have one or more medium score; and most vulnerable catchments have one or more high exposure or sensitivity score (**Figure 3.6**).

Monitoring sites can be selected from each vulnerability category. By monitoring the highest vulnerability sites the data can be used to test hypotheses on the vulnerability of specific stream types and of climate change impacts on stream ecosystems.

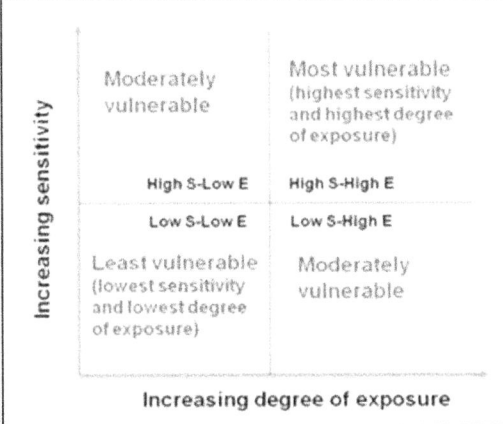

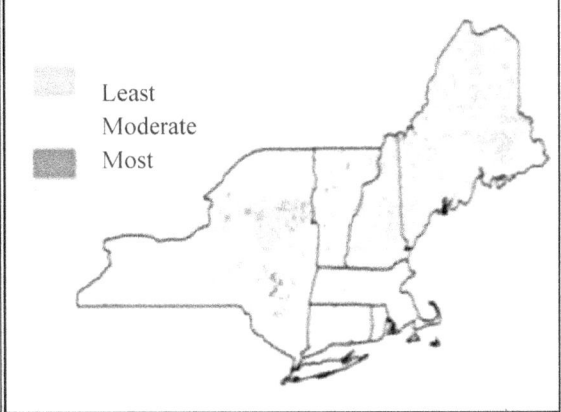

Figure 3.5: *Relative vulnerability quadrants resulting from degree of exposure (E) and sensitivity (S).*

Figure 3.6: *Relative vulnerability of catchments to low flow events and warming temperatures.*

United States Fish and Wildlife Service is using local sediment to raise and restore wetlands inundated by rising sea level at Blackwater National Wildlife Refuge and building up oyster reefs and planting flood-tolerant trees in coastal areas vulnerable to sea-level rise at Alligator River National Wildlife Refuge.

In addition to adaptation planning, State and Federal natural resource agencies are reducing greenhouse gas emissions from agency operations. In the Federal government, Executive Order no. 13514 (2009) requires all Federal agencies to create sustainability plans. The plans for the Fish and Wildlife Service and the National Park Service, for example, include greenhouse-gas emissions reduction targets, improved energy efficiency measures, reductions in vehicle fuel use, water conservation, and waste reduction and recycling. Resource management actions such as wetlands restoration and natural regeneration of forests naturally reduces climate change by removing carbon from the atmosphere. In California, the Legislature passed and signed Assembly Bill 32, the Global Warming Solutions Act of 2006, which set the 2020 greenhouse gas emissions reduction goal into law. It directed the California Air Resources Board to begin developing discrete early actions to reduce greenhouse gases while also preparing a scoping plan to identify how best to reach the 2020 limit.

State and Federal agencies are also working to inform the public about the projected impacts of climate change through outreach and to train staff through workshops and online courses. To better communicate climate change actions, the California Department of Fish and Game has created case studies to tell stories about how new and ongoing projects and programs are helping to plan for or minimize impacts associated with climate change (http://www.dfg.ca.gov/Climate_and_Energy/Climate_Change/Case_Studies/). These case studies highlight activities related to managing for ecosystem function, working collaboratively with partners across large landscapes, managing for priority species populations, and integrating climate change into Department functions.

3.3. SYNTHESIS: LIKELIHOOD AND CONSEQUENCE OF THE KEY IMPACTS OF CLIMATE CHANGE ON ECOSYSTEMS

The broad patterns of climate change impacts on ecosystems of the United States that were identified in the 2009 National Climate Assessment have been largely supported by more recent research, which provides more quantitative documentation and greater regional specificity of patterns of impact. New impacts have also been documented, suggesting pervasive impact of climate change on ecosystems of the United States. We summarize these impacts according to the key messages that were presented above, along with supporting evidence (see also **Figure 3.7**).

Biome shifts. Many of the shifts in species range that were identified in 2009 have now been shown to modify a wide range of ecosystem processes. We use the term biome shift to describe changes in the distribution of dominant plant species that fundamentally alter ecosystem processes. Climate change velocities in parts of Alaska, California, the Midwest, and the Southwest, measured in equivalent latitude per year, are 30-fold higher than the 20th century average (Burrows and others, 2011; Chen and others, 2011; **Figure 3.7**). This is associated with increased tree growth in forests close to latitudinal and altitudinal treelines and the movement of trees into adjacent tundra (Suarez and others, 1999; Lloyd and Fastie, 2003; Millar and others, 2004; Dial and others, 2007; Beckage and others, 2008). Modeling studies suggest that these biome shifts have increased NPP at zones of forest expansion, as a result of warming, and have reduced NPP at the boreal-temperate transition, due to drought stress. Forests may expand into grasslands or retreat, depending on changes in available moisture and frequency of wildfire. Models project that climate-driven biome shifts may occur on 5–20 percent of the area of the United States during the current century (Alo and Wang, 2008; Bergengren and others, 2011; Gonzalez and others, 2010; Sitch and others, 2008). *At those locations where biome shifts are occurring (forest-tundra and forest-grassland transitions), there is a high vulnerability to continued change, with important consequences for productivity, wildlife habitat, disturbance frequency, and impacts on society. There is low certainty, but potentially profound consequences, of biome shifts in places where biome shifts have not yet been observed. Ecological monitoring of transition zones between ecosystems may provide early warning of potential biome shifts.*

Ecosystem state change. Many of the biome shifts described above are stabilized by ecosystem feedbacks that maintain these ecosystems in their new state, making it difficult to reverse these changes. Movement of trees into tundra or grassland, for example, tends to shade out short-statured plants and to alter rates of carbon and nutrient cycling in ways that support the persistence of forest. If these changes are widespread, the increased energy absorption by these taller plants contributes to regional warming (Chapin and others, 2005; McGuire and others, 2006; Bonan, 2008), further supporting their persistence. Similarly, chronic drought in deserts reduces grass cover and contributes to the spread of shrubs like creosotebush, a ubiquitous desert shrub. Shrubs exert a positive feedback on climate by altering surface energy balance and promoting higher nighttime temperatures (D'Odorico and others 2010), just as described for shrub and tree encroachment into tundra. *Ecosystem state changes are difficult to predict, although the feedbacks that maintain altered states are increasingly understood. When ecosystem state changes occur, they are difficult to reverse, with large consequences for ecosystem processes and the services provided to society.*

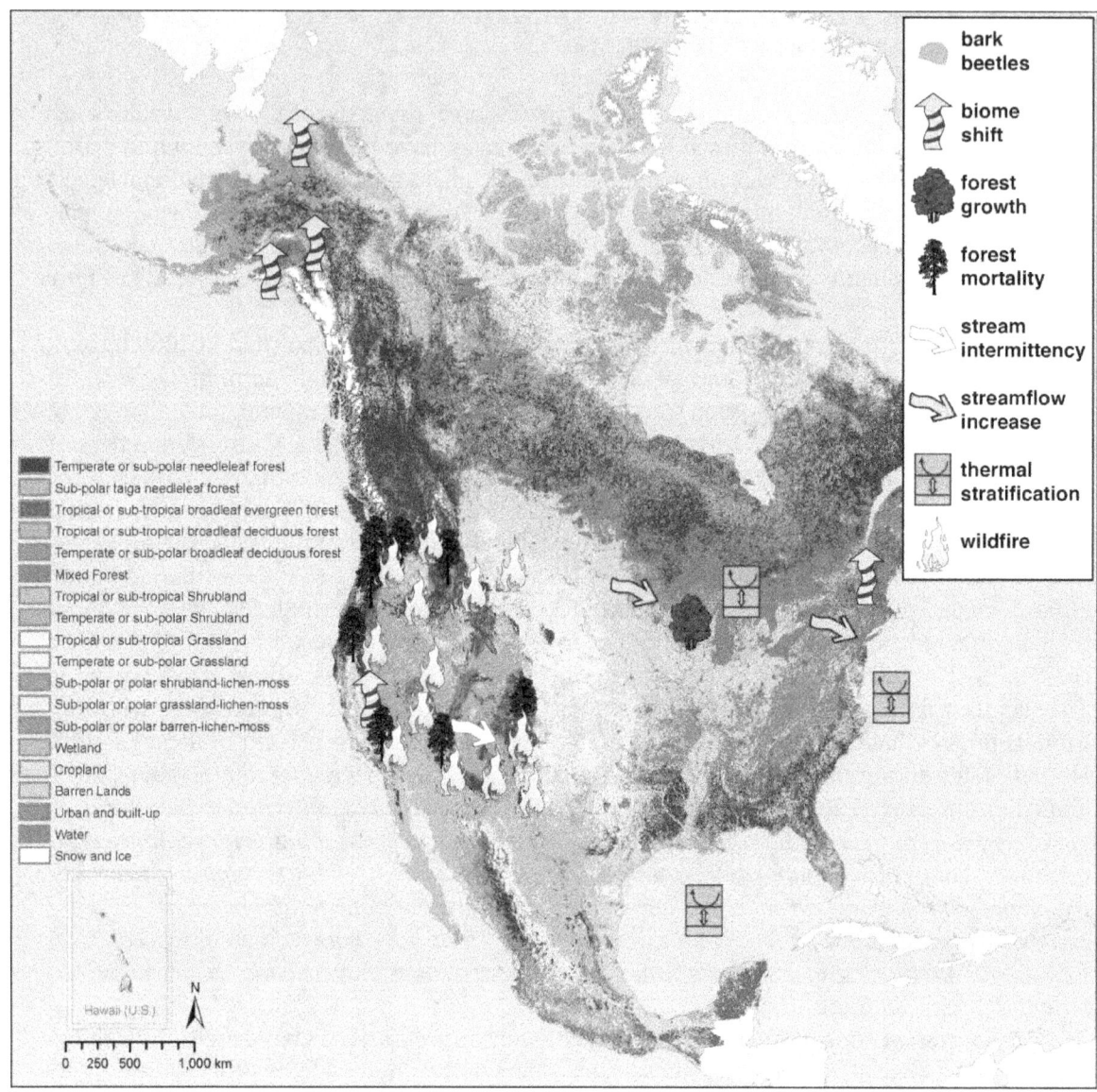

Figure 3.7. Locations in the United States of major historical changes at the ecosystem level attributed to climate change, including bark beetle infestations (Raffa and others, 2008; Logan and others, 2010), biomes shifts (Suarez and others, 1999; Lloyd and Fastie, 2003; Millar and others, 2004; Dial and others, 2007; Beckage and others, 2008), increased forest growth (Cole and others, 2010), forest mortality (van Mantgem and others, 2009), stream intermittency (Barnett and others, 2008), increased streamflow and accelerated nutrient flushing (Jones, 2011; Lins and others, 1999; Raymond and others, 2008), thermal stratification (Adrian and others, 2009), wildfire (Little and others, 2009). Land Cover: North American Land Cover 2005. Natural Resources Canada, United States Geological Survey, Insituto Nacional de Estadística y Geografía, Comisión Nacional para el Conocimiento y Uso de la Biodiversidad, and Comisión Nacional Forestal. Icons: Integration and Application Network, Center for Environmental Science, University of Maryland.

Forest growth, mortality, pests and wildfire. Climate change has increased the extent of insect outbreaks through a combination of increased plant drought stress, increased insect overwinter survival, and a shortening of the time required for insects to complete development and reproduction (Raffa and others, 2008). This has led to the most extensive insect outbreaks in western forests in the last 125 years. Warmer and drier conditions have also led to more extensive and severe wildfires. Climate has been the dominant factor controlling burned area during the 20th century, even during periods of human fire suppression (Littell and others, 2009; Westerling and others, 2011). Together, these disturbances have caused widespread reductions in forest productivity, increased tree mortality, and increased opportunities for colonization by plants that initiate biome shifts and changes in ecosystem state (**Figure 3.7**). If trends continue, baseline tree mortality rates in western forests are projected to double every 17–29 years (van Mantgem and others, 2009). In more humid areas, where these disturbances are less frequent, warming has caused an increase in forest productivity (McMahon and others, 2010; Cole and others, 2010). Satellite monitoring of NDVI suggests that these trends of increasing productivity in eastern forests and declining productivity in western and northern forests are widespread in the United States. *There is high certainty that climate change is a leading cause of the increased extent of wildfire and forest insect outbreaks leading to tree mortality in the western United States, although fire suppression, land use change and species invasions are important contributors in some places. This change in disturbance regime has large consequences for the functioning of ecosystems and the services they provide to society.*

Impact of winter warming. Climate warming in the United States has been most pronounced in winter, causing a cascade of unanticipated consequences. The most direct effects have been a shortening of the snow-covered season and reduction in snow pack, which exposes soils to more frequent freezing events and alters the seasonality of water runoff to streams and reservoirs. Soil freezing in winter has caused increases in root mortality (Tierney and others, 2001; Cleavitt and others, 2008), decreases in decomposition (Christenson and others, 2010), marked increase in leaching losses of nitrogen, phosphorus and base cations (Fitzhugh and others, 2001, 2003), and increases in nitrous oxide flux (Groffman and others, 2006), indicating fundamental changes in carbon and nutrient cycling in ecosystems. At high latitudes, declining areal extent and length of the snow- and ice-covered season increases energy absorption by ecosystems and strengthens the trend in winter warming at high latitudes (Euskirchen and others, 2007). *In places with seasonal snowpack, there is high certainty that warming has caused profound changes in snowpack, seasonality of discharge, and frequency of soil freezing, with profound consequences for both terrestrial and aquatic ecosystems.*

Intensification of the hydrologic cycle. Stream discharge has increased, particularly in New England, mid-Atlantic, Midwest and south-central regions of the United States (**Figure 3.7**; Lins and Slack, 1999). In contrast, stream discharge has decreased in many streams in the Pacific Northwest and Southeast and is projected to decrease in the arid Southwest (**Figure 3.7**; Miller and others, 2011). The most important driver of these trends is changing precipitation (Groisman and others, 2004). Streams that show increased discharge are transporting more nitrogen, phosphorus, and base cations, which both reduce soil fertility of upland terrestrial ecosystems and supports eutrophication and harmful algal blooms in streams, lakes, and the coastal zone (see also **Box 3.5**). The increased frequency of flooding that is associated with high discharge increases erosion and the delivery of sediments, contaminants, and disease organisms. *There is moderate certainty that discharge has increased in wet climates and declined in dry climates.*

Box 3.5. Nitrogen Regulation for Rivers and the Coastal Zone

Watershed nitrogen retention is an important ecosystem function underlying water-quality regulating services that protect coastal ecosystems from eutrophication. Because growth of algae in most salt water ecosystems is "limited" by nitrogen, nitrogen inputs associated with human activities—such as fertilization of agricultural fields and sewage treatment discharges—often result in overgrowth of algae, which leads to hypoxia and general ecosystem decline in many parts of the world (Conley and others, 2009). The ecosystem function, nitrogen retention, is quantified by measuring inputs (fertilizer, atmospheric deposition, import of food and feed, etc) and outputs (stream/river flow, export of food or feed) from defined watershed areas. Many studies have shown that nitrogen retention in most watersheds is high enough to absorb the majority of anthropogenic nitrogen inputs. Without this retention, human impact on coastal ecosystems would be much greater than it is currently. Even with this valuable process in place, 10 percent of US lakes and 4 percent of US rivers are impaired by nitrogen (US EPA, most recent data).

Many studies have evaluated the effects of climate on watershed nitrogen retention and provide a basis for assessing how this function may be altered by climate change. A recent comprehensive assessment (Howarth and others, 2012) shows that nitrogen flux in rivers is most strongly driven by nitrogen inputs, but that climate is also important, with a higher percentage of nitrogen inputs to the landscape being exported downstream in rivers when discharge and precipitation are higher and temperatures are lower (**Figure 3.8**). The impact of climate will vary regionally, depending on the set of processes affected most. Increases in precipitation that lead to increases in stream or river discharge will decrease nitrogen retention (increase export; **Figure 3.8a**). Changes in the frequency or timing of several seasonal processes including increased soil freezing (Brooks and others, 2011) and earlier spring and later fall will also decrease nitrogen retention. In other regions, increases in temperature will increase nitrogen retention (**Figure 3.8c**), as will reduced discharge (**Figure 3.8a**). For example, in the Southwest, projections for a drier future (Cayan *and others* 2010) suggest increased nitrogen retention as a result of decreased streamflow.

In places where the net effects of climate change reduce nutrient retention, increased nitrogen exports can have costs for society (**Figure 3.9**; Compton and others, 2011). At least five major social benefits from natural systems decline at high nitrogen loads (**Figure 3.9**). For example, in Mobile Bay (Alabama), each additional kilogram of nitrogen inflow from rivers will mean a $56 (2008 dollars) loss to the shrimp and crab fisheries through eutrophication and habitat loss (Compton and others, 2011). Climate change projections for the Mississippi Basin indicate a 20 percent increase in river discharge that will lead to higher nitrogen loads and a 50 percent increase in primary production in the Gulf, a 30-60 percent decrease in deep-water dissolved oxygen concentration, and an expansion of the dead zone (Justic and others, 1996), indicating that such economic losses are likely in this region under future climate. However, for the highly productive agricultural systems of the breadbasket, which are the source of nitrogen for the Gulf of Mexico deadzone, increases in precipitation will interact with agricultural management to determine the change in watershed retention (McIsaac and others, 2002; Raymond and others, 2012). This interaction emphasizes the opportunity for adaptation through changes in agricultural practices. Nitrogen loading strongly affects nitrogen retention (Howarth and others, 2012) so decreases in nitrogen fertilizer application rates and the use of tile drainage, or increases in the use of cover crops in this region could help decrease the costs of climate change through the loss of watershed nitrogen retention.

Impacts of Climate Change on Biodiversity, Ecosystems, and Ecosystem Services |
Technical Input to the 2013 National Climate Assessment

Chapter 3
Ecosystems

Box 3.5, continued.

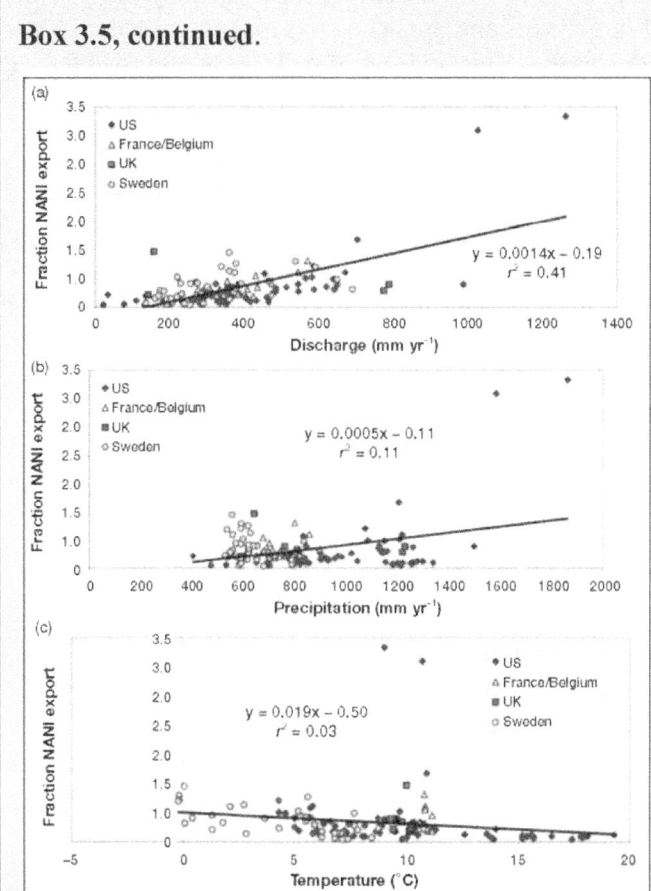

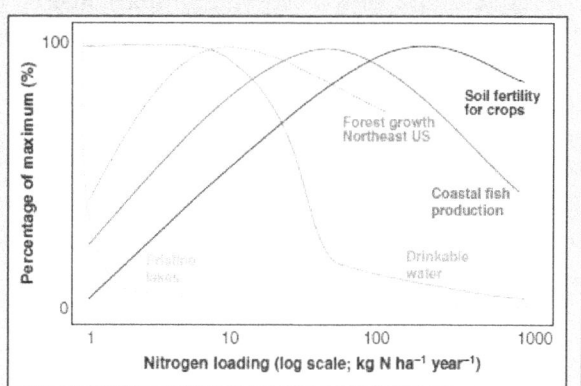

Figure 3.9. *Ecological production functions linking nitrogen and ecosystem services (pristine lakes critical threshold for changes in algae – Baron, 2006; Northeast forest growth – Thomas and others, 2010; Magill and others, 2004; Fish production – Breitburg and others, 2009). Pristine lakes and forest lines represent air deposition N loads. Reproduced from Compton and others, 2011.*

Figure 3.8. *Fraction of net anthropogenic nitrogen inputs (NANI) exported in riverine nitrogen flux as a function of (a) discharge, (b) precipitation and (c) temperature for watersheds greater than 250 km². The strongest relationship is with discharge in individual watersheds. Reproduced from Howarth and others, 2012.*

Changes in discharge, where they occur, can have profound effects on terrestrial and aquatic ecosystems through transfer of nutrients, organic matter, and sediments. There is evidence for increased discharge-dependent transfer of nutrients, contaminants and disease organisms from terrestrial to aquatic ecosystems.

    Climate effects on lakes and the coastal zone. Since 1970, the summer extent of Arctic sea ice has continued to decline, with a record low in 2007. The Arctic Ocean is projected to be ice-free in late summer by the mid-20th century, radically changing patterns of marine productivity, which is associated with ice edges (Wang and Overland 2009). In the Southern Hemisphere the population size of krill, a key crustacean food of whales and other marine vertebrates, is positively correlated with the extent of sea ice (Atkinson and others, 2004). Large

Impacts of Climate Change on Biodiversity, Ecosystems, and Ecosystem Services |
Technical Input to the 2013 National Climate Assessment

Chapter 3
Ecosystems

lakes are warming at twice the rate of the regional air-temperature warming (Moore and others, 2009; Schneider and others, 2009; Schneider and Hook, 2010), increasing the frequency of harmful algal blooms. Thermal stratification of large lakes and the ocean has also increased, reducing the upward mixing of nutrient-rich deep waters. This appears to be largely responsible for recently observed decreases in primary productivity of up to 1 percent per year in 8 out of 10 of the major ocean basins of the world (Boyce and others, 2010). *There is very high certainty that climate warming has reduced the extent of Arctic summer sea ice and the duration of winter ice-cover on inland waters, as well as the snow-covered season on land. Collectively there is high certainty that these snow and ice changes contribute to high-latitude climate warming. There is also high certainty that climate warming has increased thermal stratification in lakes and oceans, with moderate certainty that this has reduced chlorophyll concentrations in many open oceans.*

Feedbacks from ecosystem function to climate. Many of the responses of ecosystems to climate change feedback to amplify or reduce the rate of climate change. For example, the world's permafrost soils contain about 1600 Pg C, which is roughly 50 percent of the total organic C reservoir and equivalent to twice the amount of CO_2 in the atmosphere (Schuur and others, 2011). As permafrost thaws, C becomes exposed to microbial attack and decomposition, with the potential to be released as either CO_2 or CH_4. The rate and proportion of these gases released is sensitive to hydrologic changes, which are currently uncertain in the permafrost zone. Most land areas in the continental United States are a net sink for CO_2, although several processes, including declines in N deposition from air pollution and increases in wildfire and forest dieback have weakened the terrestrial C sink in many places, and dryland areas may be C sources. The equivalent of about 20-25 percent of the land-based sink for CH_4 and CO_2 are released back to the atmosphere by inland waters. *There is very high certainty that warming-induced loss of sea ice and seasonal snow cover has amplified the rate of climate warming at high latitudes. Changes in net CO_2 and CH_4 emissions are regionally variable, resulting in low certainty on the degree to which warming-induced changes in these emissions for United States ecosystems as a whole have altered the rate of climate change.*

Integration of climate change information into management and adaptation planning. State and Federal agencies and non-governmental organizations are incorporating climate change information into policy and management and working to inform the public about the impacts of climate change through outreach. *There has been a rapid increase in the use of climate change information in natural resource planning, and this information is beginning to be implemented in resource-management decisions and in outreach to the public.*

As our understanding of climate change evolves, so does our understanding of the resources values at risk and the monetary consequences of inaction. Extreme climate and weather events are one way to put a true cost on the impacts associated with a changing climate, as they comprise more than 90 percent of natural disasters in the United States (Changnon and Easterling, 2000). NOAA announced 2011 as a record year in the United States for the number of climate- and weather-related disasters exceeding $1 billion (http://www.noaa.gov/extreme2011/). In fact, there were 14 such disasters costing an estimated total of more than $53 billion, 800 lives lost, to say nothing of the social toll those deaths, of additional injuries, and of the thousands homes devastated. The year 2011 in disasters broke the previous record of nine climate- and weather-related disasters, set in 2008. Although the issues of how insurance companies are responding to climate and whether climate change is really the cause of higher rates remain controversial (http://green.blogs.nytimes.com/2011/10/06/insurance-against-the-future/), some States are now requiring insurance companies to disclose how they assess risk from climate change-related damage (for example, NY Times article: http://www.nytimes.com/2012/02/02/business/energy-environment/three-states-tell-insurers-to-disclose-responses-to-climate-change.html).

3.4. SOCIETAL RESPONSE: MANAGING CHANGE—IT CAN BE DONE. ADAPTATION AND MITIGATION RESPONSES FOR ECOSYSTEMS

In addition to regulatory mechanisms (i.e., statute and regulations) to slow the rate of emissions, climate change adaptation is a new priority that can address some climate change impacts resulting from past, current, and future GHG emissions (Interagency Climate Change Adaptation Task Force, 2010). IPCC (2007) defines adaptation as "Adjustment in natural or human systems in response to actual or expected climatic stimuli or their effects, which moderates harm or exploits beneficial opportunities." Implicit in the definition of adaptation is the assumption that climate change will likely foreclose the possibility of return to a pre-existing ecosystem state, i.e., to exceed the resilience of the system.

For Federal and State natural resource agencies, responding to climate change is ultimately about science-based ecological and biological conservation, which provides the foundation for addressing the threats climate change poses to the country's natural resources. Maintaining and restoring ecosystem functioning is a cornerstone of natural resource adaptation planning because it is essential for creating healthy ecosystems and ensuring the conservation of important ecosystem services. In order to maintain ecosystem functioning, natural resource managers are pursuing a variety of approaches to increase resistance to climate change, promote resilience, enable ecosystem responses, and realign restoration and management activities to reflect changing conditions (Millar and others 2007). Actions intended to **resist** climate change forestall undesired effects of change and/or manage ecosystems so they are better able to resist changes resulting from climate change. **Resilience** focuses on managing for viable ecosystems to increase the likelihood that they will accommodate gradual changes related to climate and tend to return to pre-disturbance conditions. **Response** is an intentional management action intended to accommodate change rather than resist it by actively or passively facilitating ecosystems to respond as environmental changes occur. **Realigning** management activities focuses on the idea that rather than restoring habitats to historic conditions, or managing for historic range of variability the managing entity would realign restoration and management approaches to current and anticipated future conditions (Millar and others, 2007; Magness and others, 2011).

Not all of the actions related to responding to climate change will need to be new or novel. There is a high likelihood that natural resource managers and decision makers will need to think differently about how we approach conservation-related problems in the future, and think at the different scales and time frames that will result from rapid climate change. Ultimately, the scientific community and the managers working on the ground need to have the knowledge, expertise, and commitment to respond in the face of uncertainty. More difficult is effectively pursuing an adaptive and responsive management approach that will allow managers the freedom and support to be flexible as our understanding of climate change and impacts evolves. We now have an opportunity to be proactive and fully embrace the importance of a dynamic plan for our actions. In the future we may need to be open to incorporating experimental approaches so that we can respond and adjust as climate change science evolves. Incorporating monitoring protocols into projects will be critical in order to document trends in impacted species of plants and animals. In addition, we must recognize that trade-offs and choices will frequently be necessary and it will be important to be receptive to prioritizing among needs. Given the enormity of this issue and limited funds and workforce, collaboration with partners is critical (as it is for other pressing conservation issues). Federal and State agencies are already actively engaging with a variety of traditional and non-traditional partners to pursue adaptation planning and implementation.

3.5. CRITICAL GAPS IN KNOWLEDGE, RESEARCH, AND DATA NEEDS

Although abundant examples of ecosystem impacts of climate change are documented in this report, substantial areas of uncertainty remain. This section outlines a selection of those challenges that the authors believe are most critical to advancing understanding. The greatest challenge is attribution of climate change impacts against the backdrop of numerous other stressors (see also *Chapter 5*). This challenge may not be solvable simply with more research; techniques of detection and attribution that are specific to ecosystem processes will need to be developed.

A second challenge concerns the multiple, indirect routes by which specific climate drivers affect particular ecosystem processes. For example, elevated temperature may stimulate soil respiration but only when soil water is not limiting; that is, soil drying may counteract the temperature effect. Such diverse responses are the rule for soil carbon–temperature– decomposition relationships (Davidson and Janssens 2006). Along with factors such as the decomposability (lability) of organic carbon, belowground microbial processes, and trace-gas fluxes, they also underlie inadequate performance of carbon-balance models based on primary productivity and ecosystem respiration under conditions of rapid change (Chapin and others, 2009).

Although we know that extreme climate events have increased and are projected to continue to do so (IPCC, 2007), their impacts on ecosystem functioning have only begun to be studied. Extreme events can be conceptualized to include 'extremeness' in both the driver and the response: changes go beyond the individual or population level, to include changes in species dominance, widespread species loss and (or) invasion by novel species, and subsequent large and potentially persistent effects on ecosystem structure and functioning (Smith, 2011). Are there attributes of ecosystems that make them more or less resilient to extreme events? Are collective, ecosystem-level properties more or less resilient than their component populations and communities? How can management and planning best incorporate knowledge of the likelihood of extreme events?

We need a better understanding of the conditions under which different ecosystems act as carbon sinks or sources; specifically, this concerns how climate change differentially affects gross primary production (GPP) and ecosystem respiration (R_E) owing to soil moisture, heat stress, and other controls. There is still disagreement regarding carbon sinks and the magnitude of net fluxes (Wang and others, 2011). While many ecosystems in North America are carbon sinks, evidence from measurements of net ecosystem exchange along a New Mexico transect suggests that drier, warmer conditions may shift sinks to sources (Anderson-Teixeira and others, 2011).

The causes and warning signs of regime shifts – changes in ecosystem state that occur when thresholds are crossed (**Box 3.1**) – are beginning to be understood but much more research is needed on this complex phenomenon. A related challenge is understanding cross-scale and cross-systems interactions. Ecosystem connectivity—the degree to which ecosystems are linked via the vectors of water, wind, organisms, and people (Peters and others, 2008)—may be an important source of either ecological surprise or unexpected resilience. We have an incomplete knowledge of the impact of connections between ecosystems and their responses to climate and other global environmental changes.

Lakes are effective indicators of climate change that provide physical, chemical, and biological signals with short response times but in some cases persistent ecosystem effects (Williamson and others, 2009a,b). More quantitative indicators are needed to assess impending regime changes such as the development of hypoxia and nutrient loading - early warning signals of collapse and transformation of healthy inland and coastal waters to anoxic dead zones and harmful algal blooms. Non-linear processes and thresholds that are exceeded at various scales of space and time may drive some of the most important changes that will affect humans. What are the "best" indicators (signal to noise ratio) and what are they telling us about the effects of changes in temperature vs. precipitation in both terrestrial ecosystems (carbon and nutrient loss from soils, primary productivity of fields and forests) and aquatic ecosystems (lakes, reservoirs, streams, rivers, wetlands)? Some of the strongest yet least well understood responses of lakes are the pronounced changes in dissolved organic carbon (DOC) concentrations that have been observed across North America and Europe. In some of these systems reductions in sulfate deposition appears to be the dominant driver of higher DOC (Monteith and others, 2007; Delpla and others, 2009), though climate is a key regulator that alters DOC concentrations (Weyhenmeyer and Karlsson, 2009). These increases in DOC not only influence the structure and function of aquatic ecosystems through changes in water transparency and chemistry (Williamson and others, 1999), but also threaten drinking water supplies (Chow and others, 2003; Kaplan and others, 2006; Haaland and others, 2010).

In order to understand the impacts of changes in DOC, we need to quantify the importance of water transparency in regulating the impact of solar radiation on epidemics of parasites and pathogens in wildlife and human populations. We know that many of these parasites and pathogens are highly sensitive to inactivation by solar UV, but neither the prevalence of this phenomenon nor its importance relative to other environmental controls of inactivation is known. Increased DOC concentrations may also create a need for fundamental changes in water-treatment systems to maintain low levels of disease in humans (for example, higher DOC necessitates longer exposure to UV, more chlorine or ozone in disinfection systems) as well as maintain low levels of carcinogenic chlorination byproducts that are produced when chlorine interacts with DOC from natural sources (Williamson and others, 1999; Chow and others, 2003; Kaplan and others, 2006; Haaland and others, 2010).

Feedbacks from ecosystem processes to climate change, primarily via the enhanced production of greenhouse gases, need further study. What ecosystem factors control the processes that generate greenhouse gases, and how are these factors themselves affected by climate change? Key variables appear to be soil moisture and its correlates (water-holding capacity, water-filled pore space) and redox, which are influenced by soil water in terrestrial ecosystems but subject to feedbacks from productivity in aquatic ecosystems. These relationships bear further scrutiny with climate change in mind.

3.5.1. Observational networks for documenting ecosystem change

Understanding and adapting to the impacts of climate change on ecosystems, ecosystem services, and biodiversity present important challenges for natural resource managers and policy makers. Currently available and future observations of ecosystems are important to addressing these intertwined challenges. Data from observational networks provide important tools for improving our understanding of the links between climate change and the functioning, structure, and health of ecosystems; ecosystem services; and biodiversity. In particular, observational networks can help to:

- Improve the understanding of ecological functioning. Examples of processes that are incompletely understood include the carbon and nitrogen cycles, predator-prey dynamics, and behavior of invasive species.
- Validate and further develop ecological models, especially with respect to models' ability to project or predict future changes.
- Understand interactions across ecological systems, geographic scales, and temporal scales as climate changes. Examples include investigating the role of disturbances in longer-term ecological change, and the dynamics at the interface between different types of ecosystems.
- Identify non-linear ecological responses to climate forcing, regime shifts, or "tipping point" behavior. Examples include large-scale die-off events.

With respect to adapting to the impacts of climate change, the use of observational networks for management and policy takes on a specialized role. In most cases, the knowledge base for identifying, selecting, and evaluating strategies to reduce the vulnerability of ecosystems to impacts is relatively immature (America's Climate Choices: Panel on Adapting to the Impacts of Climate Change; National Research Council, 2010). Consequently, initial implementation of adaptation efforts will likely be somewhat experimental, with measurements from observational networks providing the basis for further evaluation by managers and decision makers. Observational networks are thus an important ingredient in iterative adaptive management.

In addition, adaptation decision making can benefit from specialized observational networks. The impacts of climate change involve a complex interplay between regional-scale (or larger scale) changes in temperature and precipitation (or other aspects of climate) and more local stressors related to factors such as land use, pollution, and forms of resource consumption (for example, fishing, logging). Although adaptation measures will often be implemented on a relatively local scale, the evaluation and monitoring of these measures may require an understanding of multiple stresses across a range of spatial scales. To address this need, observational networks can provide information that can be appropriately "nested," such that large-scale climate information can be delivered to a wide range of local managers and policy makers who can readily combine such climate information with local-scale measurements that capture the effects of other important stressors.

3.6. LITERATURE CITED

Adrian R, O'Reilly CM, Zagarese H, Baines SB, Hessen DO, Keller W, Livingstone DM, Sommaruga R, Straile D, Van Donk E, Weyhenmeyer GA, and Winder M. 2009. Lakes as sentinels of climate change. *Limnology and Oceanography* **54**: 2283-2297.

Albani M, Moorcroft PR, Ellison AM, Orwig DA, and Foster DR. 2010. Predicting the impact of hemlock woolly adelgid on carbon dynamics of eastern United States forests. *Canadian Journal of Forest Research* **40**: 119-133.

Allan RP. 2011. Climate change human influence on rainfall. *Nature* **470**: 344-345.

Allen CD, and Breshears DD. 1998. Drought-induced shift of a forest-woodland ecotone: Rapid landscape response to climate variation. *Proceedings of the National Academy of Sciences* **95**: 14839-14842.

Allison MA, Dellapenna TM, Gordon ES, Mitra S, and Petsch ST. 2010. Impact of Hurricane Katrina (2005) on shelf organic carbon burial and deltaic evolution. *Geophysical Research Letters* **37**: L21605, 5 p.

Alo CA, and Wang G. 2008. Potential future changes of the terrestrial ecosystem based on climate projections by eight general circulation models. *Journal of Geophysical Research-Biogeosciences* **113**: G01004, 16 p.

America's Climate Choices: Panel on Adapting to the Impacts of Climate Change; National Research Council. 2010. America's Climate Choices, Adapting to the Impacts of Climate Change. National Academies Press. 292 p.

Anderson-Teixeira KJ, Delong JP, Fox AM, Brese DA, and Litvak ME. 2011. Differential responses of production and respiration to temperature and moisture drive the carbon balance across a climatic gradient in New Mexico. *Global Change Biology* **17**: 410-424.

Atkinson A, Siegel V, Pakhomov E, and Rothery P. 2004. Long-term decline in krill stock and increase in salps within the Southern Ocean. *Nature* **432**: 100-103.

Austnes K, and Vestgarden LS. 2008. Prolonged frost increases release of C and N from a montane heathland soil in southern Norway. *Soil Biology & Biochemistry* **40**: 2540-2546.

Bala G, Caldeira K, Wickett M, Phillips TJ, Lobell DB, Delire C, and Mirin A. 2007. Combined climate and carbon-cycle effects of large-scale deforestation. *Proceedings of the National Academy of Sciences* **104**: 6550-6555.

Barnett TP, Pierce DW, Hidalgo HG, Bonfils C, Santer BD, Das T, Bala G, Wood AW, Nozawa T, Mirin AA, Cayan DR, and Dettinger MD. 2008. Human-induced changes in the hydrology of the western United States. *Science* **319**: 1080-1083.

Barnett TP, and Pierce DW. 2009. Sustainable water deliveries from the Colorado River in a changing climate. *Proceedings of the National Academy of Sciences* **106**: 7334-7338.

Baron JS, Gunderson L, Allen CD, Fleishman E, McKenzie D, Meyerson LA, Oropeza J, and Stephenson N. 2009. Options for national parks and reserves for adapting to climate change. *Environmental Management* **44**: 1033-1042.

Baron JS. 2006. Hindcasting nitrogen deposition to determine an ecological critical load. *Ecological Applications* **16**: 433–439.

Barrett K, McGuire AD, Hoy EE, and Kasischke ES. 2011. Potential shifts in dominant forest cover in interior Alaska driven by variations in fire severity. *Ecological Applications* **21**: 2380-2396.

Bastviken D, Tranvik LJ, Downing JA, Crill PM, and Enrich-Prast A. 2011. Freshwater methane emissions offset the continental carbon sink. *Science* **331**: 50-50.

Beck PSA, and Goetz SJ. 2011. Satellite observations of high northern latitude vegetation productivity changes between 1982 and 2008: Ecological variability and regional differences. *Environmental Research Letters* **6**: 045501, 15 p.

Beck PSA, Juday GP, Alix C, Barber VA, Winslow SE, Sousa EE, Heiser P, Herriges JD, and Goetz SJ. 2011. Changes in forest productivity across Alaska consistent with biome shift. *Ecology Letters* **14**: 373-379.

Beckage B, Osborne B, Gavin DG, Pucko C, Siccama T, and Perkins T. 2008. A rapid upward shift of a forest ecotone during 40 years of warming in the Green Mountains of Vermont. *Proceedings of the National Academy of Sciences* **105**: 4197-4202.

Bentz BJ, Regniere J, Fettig CJ, Hansen EM, Hayes JL, Hicke JA, Kelsey RG, Negron JF, and Seybold SJ. 2010. Climate change and bark beetles of the western United States and Canada: direct and indirect effects. *BioScience* **60**: 602-613.

Berg EE, Henry JD, Fastie CL, De Volder AD, and Matsuoka SM. 2006. Spruce beetle outbreaks on the Kenai Peninsula, Alaska, and Kluane National Park and Reserve, Yukon Territory: Relationship to summer temperatures and regional differences in disturbance regimes. *Forest Ecology and Management* **227**: 219-232.

Bergengren JC, Waliser DE, and Yung YL. 2011. Ecological sensitivity: a biospheric view of climate change. *Climatic Change* **107**: 433-457.

Bond WJ, and Keeley JE. 2005. Fire as a global 'herbivore': the ecology and evolution of flammable ecosystems. *Trends in Ecology and Evolution* **20**: 387-394.

Bonan GB, 2008. Forests and climate change: Forcings, feedbacks, and the climate benefits of forests. *Science* **320**:1444-1449.

Boutin R, and Robitaille G. 1995. Increased soil nitrate loss under mature sugar maple trees affected by experimentally-induced deep frost. *Canadian Journal of Forest Research* **25**: 588-602.

Boyce DG, Lewis MR, and Worm B. 2010. Global phytoplankton decline over the past century. *Nature* **466**: 591-596.

Bradley RS, Keimig FT, and Diaz HF. 2004. Projected temperature changes along the American Cordillera and the planned GCOS network. *Geophysical Research Letters* **31**(16): L16210.

Breitburg DL, Hondorp DW, Davias LA, and Diaz RJ. 2009. Hypoxia, nitrogen, and fisheries: integrating effects across local and global landscapes. *Annual Reviews in Marine Science* **1**: 329–349.

Breshears DD, Cobb NS, Rich PM, Price KP, Allen CD, Balice RG, Romme WH, Kastens JH, Floyd ML, Belnap J, Anderson JJ, Myers OB, and Meyer CW. 2005. Regional vegetation die-off in response to global-change-type drought. *Proceedings of the National Academy of Sciences* **102**: 15144-15148.

Bringezu S, O'Brien M, and Schuetz H. 2012. Beyond biofuels: Assessing global land use for domestic consumption of biomass A conceptual and empirical contribution to sustainable management of global resources. *Land Use Policy* **29**: 224-232.

Brock WA, and Carpenter SR. 2006. Variance as a leading indicator of regime shift in ecosystem services. *Ecology and Society* **11**(2): 9, 15 p.

Brooks PD, Schmidt SK, and Williams MW. 1997. Winter production of CO_2 and N_2O from Alpine tundra: Environmental controls and relationship to inter-system C and N fluxes. *Oecologia* **110**: 403-413.

Brooks PD, Grogan P, Templer PH, Groffman PM, Oquist MG, and Schimel J. 2011. Carbon and nitrogen cycling in snow-covered environments. *Geography Compass* **5**: 682-699.

Bunn AG, Goetz SJ, Kimball JS, and Zhang K. 2007. Northern high-latitude ecosystems respond to climate change. *EOS Transactions of the American Geophysical Union* **88**: 333-340.

Burns DA, Klaus J, and McHale MR. 2007. Recent climate trends and implications for water resources in the Catskill Mountain region, New York, USA. *Journal of Hydrology* **336**: 155-170.

Burrows MT, Schoeman DS, Buckley LB, Moore P, Poloczanska ES, Brander KM, Brown C, Bruno JF, Duarte CM, Halpern BS, Holding J, Kappel CV, Kiessling W, O'Connor MI, Pandolfi JM, Parmesan C, Schwing FB, Sydeman WJ, and Richardson AJ. 2011. The pace of shifting climate in marine and terrestrial ecosystems. *Science* **334**: 652-655.

Campbell JL, Driscoll CT, Eagar C, Likens GE, Siccama TG, Johnson CE, Fahey TJ, Hamburg SP, Holmes RT, Bailey AS, and Buso DC. 2007. Long-term trends from ecosystem research at the Hubbard Brook Experimental Forest. U.S. Department of Agriculture, Forest Service, Northern Research Station, Newtown Square, PA.

Campbell JL, Rustad LE, Boyer EW, Christopher SF, Driscoll CT, Fernandez IJ, Groffman PM, Houle D, Kiekbusch J, Magill AH, Mitchell MJ, and Ollinger SV. 2009. Consequences of climate change for biogeochemical cycling in forests of northeastern North America. *Canadian Journal of Forest Research* **39**: 264-284.

Campbell JL, Ollinger SV, Flerchinger GN, Wicklein H, Hayhoe K, and Bailey AS. 2010. Past and projected future changes in snowpack and soil frost at the Hubbard Brook Experimental Forest, New Hampshire, USA. *Hydrological Processes* **24**: 2465-2480.

Carlisle DM, Wolock DM, and Meador MR. 2011. Alteration of streamflow magnitudes and potential ecological consequences: a multiregional assessment. *Frontiers in Ecology and the Environment* **9**: 264-270.

Carpenter SR. 2003. Regime shifts in lake ecosystems: Pattern and variation. Oldendorf/Luhe, Germany: Ecology Institute.

Carpenter SR, Cole JJ, Pace ML, Batt R, Brock WA, Cline T, Coloso J, Hodgson JR, Kitchell JF, Seekell DA, Smith L, and Weidel B. 2011. Early warnings of regime shifts: A whole-ecosystem experiment. *Science* **332**: 1079-1082.

Carroll A, Taylor L, Regniere J, and Safranyik L. 2004. Effects of climate change on range expansion by the mountain pine beetle in British Columbia. *In* Shore TL, Brooks JE, and Stone JE (Eds), Mountain pine beetle symposium: challenges and solutions. Natural Resources Canada, Canadian Forest Service, Pacific Forestry Centre, Victoria, British Columbia, Information Report BC-X-399, Kelowna, British Columbia, Canada. 223-232 p.

Castaneda-Moya E, Twilley RR, Rivera-Monroy VH, Zhang K, Davis SE, III, and Ross M. 2010. Sediment and nutrient deposition associated with Hurricane Wilma in mangroves of the Florida coastal Everglades. *Estuaries and Coasts* **33**: 45-58.

Cayan DR, Das T, Pierce DW, Barnett TP, Tyree M, and Gershunov A. 2010. Future dryness in the southwest U.S. and the hydrology of the early 21st century drought. *Proceedings of the National Academy of Sciences* **107**: 21271-21276.

Changnon SA, and Easterling DR. 2000. Disaster management -U.S. policies pertaining to weather and climate extremes. *Science* **289**(5487): 2053-2055.

Chapin FS, Shaver GR, Giblin AE, Nadelhoffer KJ, and Laundre JA. 1995. Responses of Arctic tundra to experimental and observed changes in climate. *Ecology* **76**: 694-711.

Chapin FS, Sturm M, Serreze MC, McFadden JP, Key JR, Lloyd AH, McGuire AD, Rupp TS, Lynch AH, Schimel JP, Beringer J, Chapman WL, Epstein HE, Euskirchen ES, Hinzman LD, Jia G, Ping CL, Tape KD, Thompson CDC, Walker DA, and Welker JM. 2005. Role of land-surface changes in Arctic summer warming. *Science* **310**: 657-660.

Chapin FS, Kofinas GP, and Folke C, editors. 2009. Principles of ecosystem stewardship: resilience-based natural resource management in a changing world. New York, New York, USA: Springer.

Chapin FS, Carpenter SR, Kofinas GP, Folke C, Abel N, Clark WC, Olsson P, Smith DMS, Walker B, Young OR, Berkes F, Biggs R, Grove JM, Naylor RL, Pinkerton E, Steffen W, and Swanson FJ. 2010a. Ecosystem stewardship: sustainability strategies for a rapidly changing planet. *Trends in Ecology and Evolution* **25**: 241-249.

Chapin FS, McGuire AD, Ruess RW, Hollingsworth TN, Euskirchen ES, Johnstone JF, Jones JB, Jorgenson MT, Kasischke ES, Kielland K, Kofinas GP, Lloyd AH, Mack MC, Taylor DL, Turetsky MR, and Yarie J. 2010b. Resilience and change in Alaska's boreal forest. *Canadian Journal of Forest Research* **40**.

Chapin FS, Matson PA, and Vitousek PM, editors. 2011. Principles of terrestrial ecosystem ecology, 2nd edition: Springer Science Business Media, LLC.

Chen IC, Hill JK, Ohlemuller R, Roy DB, and Thomas CD. 2011. Rapid range shifts of species associated with high levels of climate warming. *Science* **333**: 1024-1026.

Chow AT, Tanji KK, and Gao SD. 2003. Production of dissolved organic carbon (DOC) and trihalomethane (THM) precursor from peat soils. *Water Research* **37**: 4475-4485.

Christenson LM, Mitchell MJ, Groffman PM, and Lovett GM. 2010. Winter climate change implications for decomposition in Northeastern forests: Comparisons of sugar maple litter to herbivore fecal inputs. *Global Change Biology* **16**: 2589-2601.

Cleavitt NL, Fahey TJ, Groffman PM, Hardy JP, Henry KS, and Driscoll CT. 2008. Effects of soil freezing on fine roots in a northern hardwood forest. *Canadian Journal of Forest Research* **38**: 82-91.

CNR (California Natural Resource Agency). 2009. 2009 California Climate Adaptation Strategy Discussion Draft: A Report to the Governor of the State of California in Response to Executive Order S-13-2008. Available at: http://www.energy/ca/gov/2009publications/

Cole CT, Anderson JE, Lindroth RL, and Waller DM. 2010. Rising concentrations of atmospheric CO(2) have increased growth in natural stands of quaking aspen (*Populus tremuloides*). *Global Change Biology* **16**: 2186-2197.

Cole KL, Ironside K, Eischeid J, Garfin G, Duffy PB, and Toney C. 2011. Past and ongoing shifts in Joshua tree distribution support future modeled range contraction. *Ecological Applications* **21**: 137-149.

Collins SL, Carpenter SR, Swinton SM, Orenstein DE, Childers DL, Gragson TL, Grimm NB, Grove JM, Harlan SL, Kaye JP, Knapp AK, Kofinas GP, Magnuson JJ, McDowell WH, Melack JM, Ogden LA, Robertson GP, Smith MD, and Whitmer AC. 2011. An integrated conceptual framework for long-term social–ecological research. *Frontiers in Ecology and the Environment* **9**: 351-357.

Compton JE, Harrison JA, Dennis RL, Greaver TL, Hill BH, Jordan SJ, Walker H, and Campbell HV. 2011. Ecosystem services altered by human changes in the nitrogen cycle: a new perspective for US decision making. *Ecology Letters* **14**(8): 804-815.

Conley DJ, Paerl HW, Howarth RW, Boesch DF, Seitzinger SP, Havens KE, Lancelot C, and Likens GE. 2009. Controlling eutrophication: nitrogen and phosphorus. *Science* **323**: 1014-1015.

Connelly SJ, Wolyniak EA, Williamson CE, and Jellison KL. 2007. Artificial UV-B and solar radiation reduce in vitro infectivitiy of the human pathogen *Cryptosporidium parvum*. *Environmental Science and Technology* **41**: 7101-7106.

Curriero FC, Patz JA, Rose JB, and Lele S. 2001. The association between extreme precipitation adn waterborne disease outbreaks in the United States, 1948-1994. *American Journal of Public Health* **91**: 1194-1199.

Danz NP, Reich PB, Frelich LE, and Niemi GJ. 2011. Vegetation controls vary across space and spatial scale in a historic grassland-forest biome boundary. *Ecography* **34**: 402-414.

Davison JE, Graumlich LJ, Rowland EL, Pederson GT, and Breshears DD. 2012. Leveraging modern climatology to increase adaptive capacity across protected area networks. *Global Environmental Change* **22**: 268-274.

Davidson EA, and Janssens IA. 2006. Temperature sensitivity of soil carbon decomposition and feedbacks to climate change. *Nature* **440**: 165-173.

Davidson EA, Janssens IA, and Luo YQ. 2006. On the variability of respiration in terrestrial ecosystems: moving beyond Q(10). *Global Change Biology* **12**: 154-164.

Decker KLM, Wang D, Waite C, and Scherbatskoy T. 2003. Snow removal and ambient air temperature effects on forest soil temperatures in northern Vermont. *Soil Science Society of America Journal* **67**: 1234-1242.

Delpla I, Jung AV, Baures E, Clement M, and Thomas O. 2009. Impacts of climate change on surface water quality in relation to drinking water production. *Environment International* **35**: 1225-1233.

Desjardins RL, Pattey E, Smith WN, Worth D, Grant B, Srinivasan R, MacPherson JI, and Mauder M. 2010. Multiscale estimates of N2O emissions from agricultural lands. *Agronomy and Forest Meteorology* **150**: 817–824.

Dial RJ, Berg EE, Timm K, McMahon A, and Geck J. 2007. Changes in the alpine forest-tundra ecotone commensurate with recent warming in southcentral Alaska: Evidence from orthophotos and field plots. *Journal of Geophysical Research-Biogeosciences* **112**: G04015, 15 p.

Diaz S, Lavorel S, de Bello F, Quetier F, Grigulis K, and Robson M. 2007. Incorporating plant functional diversity effects in ecosystem service assessments. *Proceedings of the National Academy of Sciences* **104**: 20684-20689.

Dietzel R, Wolfe D, and Thies JE. 2011. The influence of winter soil cover on spring nitrous oxide emissions from an agricultural soil. *Soil Biology and Biochemistry* **43**: 1989–1991.

Dlugokencky EJ, Nisbet EG, Fisher R, and Lowry D. 2011. Global atmospheric methane in 2010: Budget, changes and dangers. *Philosophical Transactions of the Royal Society A* **369**: 2058–2072.

Doak DF, and Morris WF. 2010. Demographic compensation and tipping points in climate-induced range shifts. *Nature* **467**: 959-962.

Dodds W, Bouska WW, Eitzmann JL, Pilger TJ, Pitts KL, Riley AJ, Schloesser JT, and Thornbrugh DJ. 2009. Eutrophication of U.S. freshwaters: Analysis of potential economic damages. *Environmental Science & Technology* **43**: 12-19.

D'Odorico P, Laio F, and Ridolfi L. 2010. Does globalization of water reduce societal resilience to drought? *Geophysical Research Letters* **37**.

Impacts of Climate Change on Biodiversity, Ecosystems, and Ecosystem Services | Technical Input to the 2013 National Climate Assessment

Chapter 3
Ecosystems

Dukes JS, Pontius J, Orwig D, Garnas JR, Rodgers VL, Brazee N, Cooke B, Theoharides KA, Stange EE, Harrington R, Ehrenfeld J, Gurevitch J, Lerdau M, Stinson K, Wick R, and Ayres M. 2009. Responses of insect pests, pathogens, and invasive plant species to climate change in the forests of northeastern North America: What can we predict? *Canadian Journal of Forest Research* **39**: 231-248.

Ellis EC, and Ramankutty N. 2008. Putting people in the map: anthropogenic biomes of the world. *Frontiers in Ecology and the Environment* **6**: 439-447.

Ellison AM, Bank MS, Clinton BD, Colburn EA, Elliott K, Ford CR, Foster DR, Kloeppel BD, Knoepp JD, Lovett GM, Mohan J, Orwig DA, Rodenhouse NL, Sobczak WV, Stinson KA, Stone JK, Swan CM, J. Thompson B, Holle v, and Webster JR. 2005. Loss of foundation species: consequences for the structure and dynamics of forested ecosystems. *Frontiers in Ecology and the Environment* **3**: 479-486.

Epstein HE, Gill RA, Paruelo JM, Lauenroth WK, Jia GJ, and Burke IC. 2002. The relative abundance of three plant functional types in temperate grasslands and shrublands of North and South America: effects of projected climate change. *Journal of Biogeography* **29**: 875-888.

Euskirchen ES, McGuire AD, and Chapin FS. 2007. Energy feedbacks of northern high-latitude ecosystems to the climate system due to reduced snow cover during 20th century warming. *Global Change Biology* **13**: 2425-2438.

Euskirchen ES, McGuire AD, Chapin FS, III, Yi S, and Thompson CC. 2009. Changes in vegetation in northern Alaska under scenarios of climate change, 2003-2100: implications for climate feedbacks. *Ecological Applications* **19**: 1022-1043.

Euskirchen ES, McGuire AD, Chapin FS, and Rupp TS. 2010. The changing effects of Alaska's boreal forests on the climate system. *Canadian Journal of Forest Research* **40**: 1336-1346.

Evans CD, Chapman PJ, Clark JM, Monteith DT, and Cresser MS. 2006. Alternative explanations for rising dissolved organic carbon export from organic soils. *Global Change Biology* **12**: 2044-2053.

Eviner VT, and Chapin FS. 2003. Functional matrix: A conceptual framework for predicting multiple plant effects on ecosystem processes. *Annual Review of Ecology Evolution and Systematics* **34**: 455-485.

Findlay SEG. 2005. Increased carbon transport in the Hudson River: unexpected consequence of nitrogen deposition? *Frontiers in Ecology and the Environment* **3**: 133-137.

Fitzhugh RD, Driscoll CT, Groffman PM, Tierney GL, Fahey TJ, and Hardy JP. 2001. Effects of soil freezing disturbance on soil solution nitrogen, phosphorus, and carbon chemistry in a northern hardwood ecosystem. *Biogeochemistry* **56**: 215-238.

Fitzhugh RD, Driscoll CT, Groffman PM, Tierney GL, Fahey TJ, and Hardy JP. 2003. Soil freezing and the acid-base chemistry of soil solutions in a northern hardwood forest. *Soil Science Society of America Journal* **67**: 1897-1908.

Flechard CR, Ambus P, Skiba U, Rees RM, Hensen A, van Amstel A, Pol-van Dasselaar AV, Soussana JF, Jones M, Clifton-Brown J, Raschi A, Horvath L, Neftel A, Jocher M, Ammann C, Leifeld J, Fuhrer J, Calanca P, Thalman E, Pilegaard K, Di Marco C, Campbell C, Nemitz E, Hargreaves KJ, Levy PE, Ball BC, Jones SK, van de Bulk WCM, Groot T, Blom M, Domingues R, Kasper G, Allard V, Ceschia E, Cellier P, Laville P, Henault C, Bizouard F, Abdalla M, Williams M, Baronti S, Berretti F, and Grosz B.

2007. Effects of climate and management intensity on nitrous oxide emissions in grassland systems across Europe. *Agriculture Ecosystems & Environment* **121**: 135-152.

Foley JA, DeFries R, Asner GP, Barford C, Bonan G, Carpenter SR, Chapin FS, Coe MT, Daily GC, Gibbs HK, Helkowski JH, Holloway T, Howard EA, Kucharik CJ, Monfreda C, Patz JA, Prentice IC, Ramankutty N, and Snyder PK. 2005. Global consequences of land use. *Science* **309**: 570-574.

Frelich LE, and Reich PB. 2010. Will environmental changes reinforce the impact of global warming on the prairie-forest border of central North America? *Frontiers in Ecology and the Environment* **8**: 371-378.

Frumhoff PC, McCarthy JJ, Melillo JM, Moser SC, and Wuebbles DJ. 2007. Confronting climate change in the U.S. Northeast: Science, impacts, and solutions. Union of Concerned Scientists (UCS), Cambridge, MA.

Garbulsky M, Peñuelas J, Papale D, Ardö J, Goulden M, Kiely G, Richardson A, Rotenberg E, Veenendaal E, and Filella I. 2010. Patterns and controls of the variability of radiation use efficiency and primary productivity across terrestrial ecosystems. *Global Ecology and Biogeography* **19**: 253–267.

Gerritsen J, Barbour MT, and King K. 2000. Apples, oranges, and ecoregions: on determining pattern in aquatic assemblages. *Journal of the North American Benthological Society* **19**: 487-496.

Gibson K, Skov K, Kegley S, Jorgensen C, Smith S, and Witcosky J. 2008. Mountain pine beetle impacts in high-elevation five-needle pines: current trends and challenges. USDA Forest Service, Forest Health Protection, Missoula, Montana, USA. Available at: www.fs.usda.gov

Godsey SE, Kirchner JW, and Clow DW. 2009. Concentration-discharge relationships reflect chemostatic characteristics of U.S. catchments. *Hydrological Processes* **23**: 1844-1864.

Goetz SJ, Bunn AG, Fiske GJ, and Houghton RA. 2005. Satellite-observed photosynthetic trends across boreal North America associated with climate and fire disturbance. *Proceedings of the National Academy of Sciences* **102**: 13521-13525.

Gonzalez P. 2011. Science for natural resource management under climate change. *Issues in Science and Technology* **27**: 65-74.

Gonzalez P, Neilson RP, Lenihan JM, and Drapek RJ. 2010. Global patterns in the vulnerability of ecosystems to vegetation shifts due to climate change. *Global Ecology and Biogeography* **19**: 755-768.

Griffith B, Scott JM, Adamcik R, Ashe D, Czech B, Fischman R, Gonzalez P, Lawler J, McGuire AD, and Pidgorna A. 2009. Climate change adaptation for theU.S. National Wildlife Refuge system. *Environmental Management* **44**: 1043-1052.

Griffin JM, Turner MG, and Simard M. 2011. Nitrogen cycling following mountain pine beetle disturbance in lodgepole pine forest of Greater Yellowstone. *Forest Ecology and Management* **261**: 1077–1089.

Grime JP, Fridley JD, Askew AP, Thompson K, Hodgson JG, and Bennett CR. 2008. Long-term resistance to simulated climate change in an infertile grassland. *Proceedings of the National Academy of Sciences* **105**: 10028-10032.

Grimm NB, Faeth SH, Golubiewski NE, Redman CL, Wu J, Bai X, and Briggs JM. 2008. Global change and the ecology of cities. *Science* **319**: 756-760.

Groffman PM, Hardy JP, Driscoll CT, and Fahey TJ. 2006. Snow depth, soil freezing, and fluxes of carbon dioxide, nitrous oxide and methane in a northern hardwood forest. *Global Change Biology* **12**: 1748-1760.

Groffman PM, Hardy JP, Fashu-Kanu S, Driscoll CT, Cleavitt NL, Fahey TJ, and Fisk MC. 2011. Snow depth, soil freezing and nitrogen cycling in a northern hardwood forest landscape. *Biogeochemistry* **102**: 223-238.

Groisman PY, Knight RW, Karl TR, Easterling DR, Sun BM, and Lawrimore JH. 2004. Contemporary changes of the hydrological cycle over the contiguous United States: Trends derived from in situ observations. *Journal of Hydrometeorology* **5**: 64-85.

Groisman PY, Knight RW, Easterling DR, Karl TR, Hegerl GC, and Razuvaev VN. 2005. Trends in intense precipitation in the climate record. *Journal of Climate* **18**: 1326-1350.

Gunther KA, Haroldson MA, Frey K, Cain SL, Copeland J, and Schwartz CC. 2004. Grizzly bear–human conflicts in the Greater Yellowstone ecosystem, 1992–2000. *Ursus* **15**: 10–22.

Greater Yellowstone Coordinating Committee Whitebark Pine Subcommittee (GYCC). 2011. Whitebark pine strategy for the greater Yellowstone area. 41 p. Available at: www.fedgycc.org/documents/

Haaland S, Hongve D, Laudon H, Riise G, and Vogt RD. 2010. Quantifying the drivers of the increasing colored organic matter in boreal surface waters. *Environmental Science & Technology* **44**: 2975-2980.

Haei M, Oquist MG, Buffam I, Agren A, Blomkvist P, Bishop K, Lofvenius MO, and Laudon H. 2010. Cold winter soils enhance dissolved organic carbon concentrations in soil and stream water. *Geophysical Research Letters* **37**.

Hamilton SK. 2010. Biogeochemical implications of climate change for tropical rivers and floodplains. *Hydrobiologia* **657**: 19-35.

Hamilton AJ, Basset Y, Benke KK, Grimbacher PS, Miller SE, Novotny V, Samuelson GA, Stork NE, Weiblen GD, and Yen JDL. 2010. Quantifying uncertainty in estimation of tropical arthropod species richness. *American Naturalist* **176**: 90-95.

Hare SR, and Mantua NJ. 2000. Empirical evidence for North Pacific regime shifts in 1977 and 1989. *Progress in Oceanography* **47**: 103-145.

Harms TK, and Grimm NB. 2008. Hot spots and hot moments of carbon and nitrogen dynamics in a semiarid riparian zone. *Journal of Geophysical Research-Biogeosciences* **113**.

Harte J, and Shaw R. 1995. Shifting dominance within a montane vegetation community: Results of a climate-warming experiment. *Science* **267**: 876-880.

Hawkins CP, Norris RH, Gerritsen J, Hughes RM, Jackson SK, Johnson RK, and Stevenson RJ. 2000. Evaluation of the use of landscape classifications for the prediction of freshwater biota: synthesis and recommendations. *Journal of the North American Benthological Society* **19**: 541-556.

Hayhoe K, Wake CP, Huntington TG, Luo L, Schwartz MD, Sheffield J, Wood E, Anderson B, Bradbury J, DeGaetano A, Troy TJ, and Wolfe D. 2007. Past and future changes in climate and hydrological indicators in theU.S. Northeast. *Climate Dynamics* **28**: 381-407.

Heathwaite AL. 2010. Multiple stressors on water availability at global to catchment scales: understanding human impact on nutrient cycles to protect water quality and water availability in the long term. *Freshwater Biology* **55**: 241–257.

Heisler-White JL, Knapp AK, and Kelly EF. 2008. Increasing precipitation event size increases aboveground net primary productivity in a semi-arid grassland. *Oecologia* **158**: 129-140.

Henry HAL. 2008. Climate change and soil freezing dynamics: historical trends and projected changes. *Climatic Change* **87**: 421-434.

Hentschel K, Borken W, and Matzner E. 2008. Leaching losses of nitrogen and dissolved organic matter following repeated freeze/thaw events in a forest soil. *Journal of Plant Nutrition and Soil Science* **171**: 699-706.

Hentschel K, Borken W, Zuber T, Bogner C, Huwe B, and Matzner E. 2009. Effects of soil frost on nitrogen net mineralization, soil solution chemistry and seepage losses in a temperate forest soil. *Global Change Biology* **15**: 825-836.

Hicke JA, Allen CD, Desai AR, Dietze MC, Hall RJ, Hogg EH, Kashian DM, Moore D, Raffa KF, Sturrock RN, and Vogelmann J. 2012. Effects of biotic disturbances on forest carbon cycling in the United States and Canada. *Global Change Biology* **18**: 7-34.

Hobbs RJ, Arico S, Aronson J, Baron JS, Bridgewater P, Cramer VA, Epstein PR, Ewel JJ, Klink CA, Lugo AE, Norton D, Ojima D, Richardson DM, Sanderson EW, Valladares F, Vila M, Zamora R, and Zobel M. 2006. Novel ecosystems: theoretical and management aspects of the new ecological world order. *Global Ecology and Biogeography* **15**: 1-7.

Hodgkins GA, and Dudley RW. 2006. Changes in the timing of winter-spring streamflows in eastern North America, 1913-2002. *Geophysical Research Letters* **33**.

Holmgren M, and Scheffer M. 2001. El Niño as a window of opportunity for the restoration of degraded arid ecosystems. *Ecosystems* **4**:151–159.

Holmgren M, López BC, Gutiérrez JR, and Squeo FA. 2006. Herbivory and plant growth rate determine the success of El Niño Southern Oscillation-driven tree establishment in semiarid South America. *Global Change Biology* **12**: 2263–2271.

Howarth RW, Swaney DP, Boyer EW, Marino R, Jaworski N, and Goodale C. 2006. The influence of climate on average nitrogen export from large watersheds in the Northeastern United States. *Biogeochemistry* **79**: 163-186.

Howarth R, Swaney D, Billen G, Garnier J, Hong B, Humborg C, Johnes P, Morth CM, and Marino R. 2012. Nitrogen fluxes from the landscape are controlled by net anthropogenic nitrogen inputs and by climate. *Frontiers in Ecology and the Environment* **10**: 37-43.

Huff DD, Grad G, and Williamson CE. 2004. Environmental constraints on spawning depth of yellow perch: The roles of low temperatures and high solar ultraviolet radiation. *Transactions of the American Fisheries Society* **133**: 718-726.

Hughes RM, Larsen DP, and Omernik JM. 1986. Regional reference sites: a method for assessing stream potentials. *Environmental Management* **10**: 629–635.

Huntington TG, Richardson AD, McGuire KJ, and Hayhoe K. 2009. Climate and hydrological changes in the northeastern United States: recent trends and implications for forested and aquatic ecosystems. *Canadian Journal of Forest Research* **39**: 199-212.

ICCATF (Interagency Climate Change Adaptation Task Force). 2010. Progress Report of the Interagency Climate Change Adaptation Task Force: Recommended Actions in Support of a National Climate Change Adaptation Strategy. The White House Council on Environmental Quality, Washington, D.C.

Inman DL, and Jenkins SA. 1999. Climate Change and the Episodicity of Sediment Flux of Small California Rivers. *Journal of Geology* **107**(3): 251-270.

IPCC (Intergovernmental Panel on Climate Change). 2007. Climate change 2007: Impacts, Adaptation, and Vulnerability. Cambridge University Press, Cambridge and New York, NY.

Impacts of Climate Change on Biodiversity, Ecosystems, and Ecosystem Services |
Technical Input to the 2013 National Climate Assessment

Chapter 3
Ecosystems

Iverson LR, and Prasad AM. 2001. Potential changes in tree species richness and forest community types following climate change. *Ecosystems* **4**: 186-199.

Jenny H. 1941. Factors of soil formation. New York, New York: McGraw-Hill.

Jentsch A, Kreyling J, and Beierkuhnlein C. 2007. A new generation of climate-change experiments: events, not trends. *Frontiers in Ecology and the Environment* **5**: 365–374.

Johnson JMF, Archer D, and Barbour N. 2010. Greenhouse Gas Emission from Contrasting Management Scenarios in the Northern Corn Belt. *Soil Science Society of America Journal* **74**: 396-406.

Johnstone JF, Hollingsworth TN, Chapin FS, and Mack MC. 2010. Changes in fire regime break the legacy lock on successional trajectories in Alaskan boreal forest. *Global Change Biology* **16**: 1281-1295.

Jones JB, and Rinehart A. 2010. The long-term response of stream flow to climatic warming in headwater streams of interior Alaska. *Canadian Journal of Forest Research* **40**: 1210-1218.

Jones RI, and Grey J. 2011. Biogenic methane in freshwater food webs. *Freshwater Biology* **56**: 213-229.

Jones JA, Creed IF, Hatcher KL, Warren RJ, Adams MB, Benson MH, Boose E, Brown WA, Campbell JL, Covich A, Clow DW, Dahm CN, Elder K, Ford CR, Grimm NB, Henshaw DL, Larson KL, Miles ES, Miles KM, Sebestyen SD, Spargo AT, Stone AB, Vose JM, and Williams MW. 2012. Ecosystem processes and human influences regulate streamflow response to climate change at Long-Term Ecological Research Sites. *BioScience* **62**: 390-404.

Justic D, Rabalais NN, and Turner RE. 2005. Coupling between climate variability and coastal eutrophication: Evidence and outlook for the northern Gulf of Mexico. *Journal of Sea Research* **54**: 25-35.

Kaplan LA, Newbold JD, Van Horn DJ, Dow CL, Aufdenkampe AK, and Jackson JK. 2006. Organic matter transport in New York City drinking-water-supply watersheds. *Journal of the North American Benthological Society* **25**: 912-927.

Kareiva P, Watts S, McDonald R, and Boucher T. 2007. Domesticated nature: Shaping landscapes and ecosystems for human welfare. *Science* **316**: 1866-1869.

Karl TR, Melillo JM, and Peterson TC. 2009. Global Climate Change Impacts in the United States. Cambridge University Press.

Kaushal SS, Likens GE, Jaworski NA, Pace ML, Sides AM, Seekell D, Belt KT, Secor DH, and Wingate RL. 2010. Rising stream and river temperatures in the United States. *Frontiers in Ecology and the Environment* **8**: 461-466.

Kelly AE, and Goulden ML. 2008. Rapid shifts in plant distribution with recent climate change. *Proceeding of the National Academy of Science* **105**: 11823-11826.

King BJ, Hoefel D, Daminato DP, Fanok S, and Monis PT. 2008. Solar UV reduces *Cryptosporidium parvum* oocyst infectivity in environmental waters. *Journal of Applied Microbiology* **104**: 1311-1323.

Kinnard C, Zdanowicz CM, Fisher DA, Isaksson E, de Vernal A, and Thompson LG. 2011. Reconstructed changes in Arctic sea ice over the past 1,450 years. *Nature* **479**: 509-U231.

Klein JA, Harte J, and Zhao XQ. 2004. Experimental warming causes large and rapid species loss, dampened by simulated grazing, on the Tibetan Plateau. *Ecology Letters* **7**: 1170-1179.

Klein JA, Harte J, and Zhao XQ. 2007. Experimental warming, not grazing, decreases rangeland quality on the Tibetan Plateau. *Ecological Application* **17**: 541–557.

Kling GW, Kipphut GW, and Miller MC. 1991. Arctic lakes and streams as gas conduits to the atmosphere - Implications for tundra carbon budgets. *Science* **251**: 298-301.

Knapp AK, Beier C, Briske DD, Classen AT, Luo Y, Reichstein M, Smith MD, Smith SD, Bell JE, Fay PA, Heisler JL, Leavitt SW, Sherry R, Smith B, and Weng E. 2008. Consequences of more extreme precipitation regimes for terrestrial ecosystems. *Bioscience* **58**: 811-821.

Koven CD, Ringeval B, Friedlingstein P, Ciais P, Cadule P, Khvorostyanov D, and al e. 2011. Permafrost carbon-climate feedbacks accelerate global warming. *Proceedings of the National Academy of Sciences* **108**: 14769–14774.

Krawchuk MA, Moritz MA, Parisien MA, Van Dorn J, and Hayhoe K. 2009. Global pyrogeography: the current and future distribution of wildfire. *PLoS ONE* **4**.

Kreyling J. 2010. Winter climate change: a critical factor for temperate vegetation performance. *Ecology* **91**: 1939-1948.

Kröpelin SD, Verschuren AM, Lézine H, Eggermont C, Cocquyt P, Francus JP, Cazet M, Fagot B, Rumes JM, Russell F, Darius DJ, Conley M, Schuster H, Suchodoletz , V, and Engstrom DR. 2008. Climate-driven ecosystem succession in the Sahara: The past 6000 years. *Science* **320**: 765- 768.

Kunkel KE, Easterling DR, Kristovich DAR, Gleason B, Stoecker L, and Smith R. 2010. Recent increases in U.S. heavy precipitation associated with tropical cyclones. *Geophysical Research Letters* **37**: L24706, 4 p.

Kurz WA, Dymond CC, Stinson G, Rampley GJ, Neilson ET, Carroll AL, Ebata T, and Safranyik L. 2008. Mountain pine beetle and forest carbon feedback to climate change. *Nature* **452**: 987-990.

Lacis AA, Schmidt GA, Rind D, and Ruedy RA. 2010. Atmospheric CO2: Principal control knob governing Earth's temperature. *Science* **330**: 356-359.

Lamsal S, Cobb RC, Cushman JH, Meng Q, Rizzo DM, and Meentemeyer RK. 2011. Spatial estimation of the density and carbon content of host populations for *Phytophthora ramorum* in California and Oregon. *Forest Ecology and Management* **262**: 989-998.

Lee H, Schuur EAG, Inglett KS, Lavoie M, and Chanton JP. 2012. The rate of permafrost carbon release under aerobic and anaerobic conditions and its potential effects on climate. *Global Change Biology* **18**: 515-527.

Leopold LB, Wolman MG, and Miller JP. 1964. Fluvial processes in Geomorphology: W. H. Freeman Co.

Lins HF, and Slack JR. 1999. Streamflow trends in the United States. *Geophysical Research Letters* **26**: 227-230.

Lirman D, Thyberg T, Herlan J, Hill C, Young-Lahiff C, Schopmeyer S, Huntington B, Santos R, and Drury C. 2010. Propagation of the threatened staghorn coral *Acropora cervicornis:* Methods to minimize the impacts of fragment collection and maximize production. *Coral Reefs* **29**: 729–735.

Littell JS, McKenzie D, Peterson DL, and Westerling AL. 2009. Climate and wildfire area burned in western U. S. ecoprovinces, 1916-2003. *Ecological Applications* **19**: 1003-1021.

Liu JG, Dietz T, Carpenter SR, Alberti M, Folke C, Moran E, Pell AN, Deadman P, Kratz T, Lubchenco J, Ostrom E, Ouyang Z, Provencher W, Redman CL, Schneider SH, and

Taylor WW. 2007. Complexity of coupled human and natural systems. *Science* **317**: 1513-1516.

Livingstone DM, Adrian R, Blenckner T, George G, and Weyhenmeyer GA. 2010. Lake ice phenology. *In* G. George, (Ed), The impact of climate change on European lakes. Springer, New York. 51-62 p.

Lloyd AH, and Fastie CL. 2003. Recent changes in treeline forest distribution and structure in interior Alaska. *Ecoscience* **10**: 176-185.

Loarie SR, Duffy PB, Hamilton H, Asner GP, Field CB, and Ackerly DD. 2009. The velocity of climate change. *Nature* **462**: 1052-U1111.

Logan JA, and Powell JA. 2001. Ghost forests, global warming, and the mountain pine beetle (Coleoptera: Scolytidae). *American Entomologist* **47**: 160-173.

Logan JA, and Powell JA. 2009. Ecological consequences of forest-insect disturbance altered by climate change. *In* Wagner FH (Ed), Climate warming in western North America. University of Utah Press, Salt Lake City, Utah. 98–109 p.

Logan JA, MacFarlane WW, and Willcox L. 2010. Whitebark pine vulnerability to climate change induced mountain pine beetle disturbance in the Greater Yellowstone Ecosystem. *Ecological Applications* **20**: 895-902.

MA (Millennium Ecosystem Assessment). 2005. Ecosystems and human well-being: synthesis. Washington, D.C.: Island Press.

Ma Z, Peng C, Zhu Q, Chen H, Yu G, Li W, Zhou X, Wang W, and Zhang W. 2012. Regional drought-induced reduction in the biomass carbon sink of Canada's boreal forests. *Proceedings of the National Academy of Sciences* **109**(7): 2423-2427.

MacDonald GM, Cwynar LC, and Whitlock C. 1989. Late Quaternary dynamics of pines: northern North America. *In* Richardson DM (Ed), Ecology and biogeography of *Pinus*. Cambridge University Press, New York, NY. 122 - 136 p.

MacFarlane WW, Logan JA, and Kern WR. 2009. Using the landscape assessment system (LAS) to assess mountain pine beetle-caused mortality of whitebark pine, Greater Yellowstone Ecosystem, 2009. Jackson, Wyoming.

Magill AH, Aber JD, Currie WS, Nadelhoffer KJ, Martina ME, McDowelld WH, Melillo JM, and Steudler P. 2004. Ecosystem response to 15 years of chronic nitrogen additions at the Harvard Forest LTER, Massachusetts, USA. *Forest Ecology Management* **196**: 7–28.

Magness DR, Morton JM, Huettmann F, Chapin FS, and McGuire AD. 2011. A climate-change adaptation framework to reduce continental-scale vulnerability across the United States National Wildlife Refuge System. *Ecosphere* **2**: art112.

Magnuson JJ, Robertson DM, Benson BJ, Wynne RH, Livingstone DM, Arai T, Assel RA, Barry RG, Card V, Kuusisto E, Granin NG, Prowse TD, Stewart KM, and Vuglinski VS. 2000. Historical trends in lake and river ice cover in the Northern Hemisphere. *Science* **289**: 1743-1746.

Manca M, and DeMott WR. 2009. Response of the invertebrate predator *Bythotrephes* to a climate-linked increase in the duration of a refuge from fish predation. *Limnology and Oceanography* **54**: 2506-2512.

Matonse AH, Pierson DC, Frei A, Zion MS, Schneiderman EM, Anandhi A, Mukundan R, and Pradhanang SM. 2011. Effects of changes in snow pattern and the timing of runoff on NYC water supply system. *Hydrological Processes* **25**: 3278-3288.

Matson PA. 2009. The sustainability transition. *Issues in Science and Technology* **25**: 39-42.

Mawdsley J. 2011. Design of conservation strategies for climate adaptation. *Wiley Interdisciplinary Reviews-Climate Change* **2**: 498-515.

McClain ME, Boyer EW, Dent CL, Gergel SE, Grimm NB, Groffman PM, Hart SC, Harvey JW, Johnston CA, Mayorga E, McDowell WH, and Pinay G. 2003. Biogeochemical hot spots and hot moments at the interface of terrestrial and aquatic ecosystems. *Ecosystems* **6**: 301-312.

McGuire AD, Ruess RW, Lloyd A, Yarie J, Clein JS, and Juday GP. 2010. Vulnerability of white spruce tree growth in interior Alaska in response to climate variability: Dendrochronological, demographic, and experimental perspectives. *Canadian Journal of Forest Research* **40**: 1197-1209.

McGuire AD, Anderson LG, Christensen TR, Dallimore S, Guo LD, Hayes DJ, Heimann M, Lorenson TD, Macdonald RW, and Roulet N. 2009. Sensitivity of the carbon cycle in the Arctic to climate change. *Ecological Monographs* **79**: 523-555.

McGuire AD, Chapin FS, Walsh JE, and Wirth C. 2006. Integrated regional changes in arctic climate feedbacks: Implications for the global climate system. *Annual Review of Environment and Resources* **31**:61-91.

McIsaac GF, David MB, Gertner GZ, and Goolsby DA. 2002. Relating net nitrogen input in the Mississippi River basin to nitrate flux in the lower Mississippi River: A comparison of approaches. *Journal of Environmental Quality* **31**: 1610-1622.

McMahon SM, Parker GG, and Miller DR. 2010. Evidence for a recent increase in forest growth. *Proceedings of the National Academy of Sciences* **107**: 3611-3615.

Metzger MJ, Schroter D, Leemans R, and Cramer W. 2008. A spatially explicit and quantitative vulnerability assessment of ecosystem service change in Europe. *Regional Environmental Change* **8**: 91-107.

Millar CI, Westfall RD, Delany DL, King JC, and Graumlich LJ. 2004. Response of subalpine conifers in the Sierra Nevada, California, USA, to 20th-century warming and decadal climate variability. *Arctic Antarctic and Alpine Research* **36**: 181-200.

Millar CI, Stephenson NL, and Stephens SL. 2007. Climate change and forests of the future: Managing in the face of uncertainty. *Ecological Applications* **17**: 2145-2151.

Miller WP, Piechota TC, Gangopadhyay S, and Pruitt T. 2011. Development of streamflow projections under changing climate conditions over Colorado River basin headwaters. *Hydrology and Earth System Sciences* **15**: 2145-2164.

Milner AM, Robertson AL, Brown LE, Sonderland SH, McDermott M, and Veal AJ. 2011. Evolution of a stream ecosystem in recently deglaciated terrain. *Ecology* **92**: 1924-1935.

Min SK, Zhang XB, Zwiers FW, and Hegerl GC. 2011. Human contribution to more-intense precipitation extremes. *Nature* **470**: 378-381.

Monteith DT, Stoddard JL, Evans CD, de Wit HA, Forsius M, Hogasen T, Wilander A, Skjelkvale BL, Jeffries DS, Vuorenmaa J, Keller B, Kopacek J, and Vesely J. 2007. Dissolved organic carbon trends resulting from changes in atmospheric deposition chemistry. *Nature* **450**: 537-U539.

Moore MV, Hampton SE, Izmest'eva LR, Silow EA, Peshkova EV, and Pavlov BK. 2009. Climate change and the World's "Sacred Sea"-Lake Baikal, Siberia. *BioScience* **59**: 405-417.

Nathan R, Horvitz N, He YP, Kuparinen A, Schurr FM, and Katul GG. 2011. Spread of North American wind-dispersed trees in future environments. *Ecology Letters* **14**: 211-219.

NCA (National Climate Assessment). 2009. Global Climate Change Impacts in the United States. Karl TR, Melillo JM, and Peterson TC (eds). Cambridge University Press. Available at: http://www.globalchange.gov/publications/reports/scientific-assessments/us-impacts/full-report

Neff JC, Ballantyne AP, Farmer GL, Mahowald NM, Conroy JL, Landry CC, Overpeck JT, Painter TH, Lawrence CR, and Reynolds RL. 2008. Increasing eolian dust deposition in the western United States linked to human activity. *Nature Geoscience* **1**: 189-195.

Nemani R, Keeling CD, Hashimoto H, Jolly WM, Piper SC, Tucker CJ, Myneni RB, and Running SW. 2003. Climate-Driven Increases in Global Terrestrial Net Primary Production from 1982 to 1999. *Science* **300**: 1560-1563.

Nemani R, Hashimoto H, Votava P, Melton F, Wang WL, Michaelis A, Mutch L, Milesi C, Hiatt S, and White M. 2009. Monitoring and forecasting ecosystem dynamics using the Terrestrial Observation and Prediction System (TOPS). *Remote Sensing of Environment* **113**: 1497-1509.

New M, Lister D, Hulme M, and Makin I. 2002. A high-resolution data set of surface climate over global land areas. *Climate Research* **21**: 1–25.

Niu SL, and Wan SQ. 2008. Warming changes plant competitive hierarchy in a temperate steppe in northern China. *Journal of Plant Ecology-Uk* **1**: 103-110.

NRC (National Research Council). 2010. Adapting to the Impacts of Climate Change. Washington, D.C.: The National Academies Press.

O'Reilly CM, Alin SR, Plisnier PD, Cohen AS, and McKee BA. 2003. Climate change decreases aquatic ecosystem productivity of Lake Tanganyika, Africa. *Nature* **424**: 766-768.

Olden JD, and Naiman RJ. 2010. Incorporating thermal regimes into environmental flows assessments: modifying dam operations to restore freshwater ecosystem integrity. *Freshwater Biology* **55**: 86-107.

Overholt EP, Hall SR, Williamson CE, Meikle CK, Duffy MA, and Caceres CE. 2012. Solar radiation decreases parasitism in *Daphnia*. *Ecology Letters* **15**: 47-54.

Pace ML, and Cole JJ. 2002. Synchronous variation of dissolved organic carbon and color in lakes. *Limnology and Oceanography* **47**: 333-342.

Paerl HW, and Huisman J. 2008. Climate - Blooms like it hot. *Science* **320**: 57-58.

Painter TH, Deems JS, Belnap J, Hamlet AF, Landry CC, and Udall B. 2010. Response of Colorado River runoff to dust radiative forcing in snow. *Proceedings of the National Academy of Sciences* **107**: 17125-17130.

Pall P, Aina T, Stone DA, Stott PA, Nozawa T, Hilberts AGJ, Lohmann D, and Allen MR. 2011. Anthropogenic greenhouse gas contribution to flood risk in England and Wales in autumn 2000. *Nature* **470**: 382-385.

Parent MB, and Verbyla D. 2010. The browning of Alaska's boreal forest. *Remote Sensing* **2**: 2729–2747.

Pederson GT, Gray ST, Woodhouse CA, Betancourt JL, Fagre DB, Littell JS, Watson E, Luckman BH, and Graumlich LJ. 2011. The unusual nature of recent snowpack declines in the North American Cordillera. *Science* **333**: 332-335.

Pendleton EA, Thieler ER, and Williams SJ. 2010. Importance of coastal change variables in determining vulnerability to sea- and lake-level change. *Journal of Coastal Research* **26**: 176-183.

Perkins DM, Reiss J, Yvon-Durocher G, and Woodward G. 2010. Global change and food webs in running waters. *Hydrobiologia* **657**: 181-198.

Perry LG, Andersen DC, Reynolds LV, Nelson SM, and Shafroth PB. 2012. Vulnerability of riparian ecosystems to elevated CO_2 and climate change in arid and semiarid western North America. *Global Change Biology* **18**: 821-842.

Peters DPC, Pielke RA, Bestelmeyer BT, Allen CD, Munson-McGee S, and Havstad KM. 2004. Cross-scale interactions, nonlinearities, and forecasting catastrophic events. *Proceedings of the National Academy of Sciences* **101**: 15130-15135.

Peters DPC, Groffman PM, Nadelhoffer KJ, Grimm NB, Coffins SL, Michener WK, and Huston MA. 2008. Living in an increasingly connected world: a framework for continental-scale environmental science. *Frontiers in Ecology and the Environment* **6**: 229-237.

Peters DPC, Laney CM, Lugo AE, Collins SL, Driscoll CT, Groffman PM, Grove JM, Knapp AK, Kratz TK, Ohman MD, Waide RB, and Yao JAK. 2011a. Long-term trends in climate and climate-related drivers. Washington, D.C.: USDA Agricultural Research Service Publication.

Peters DPC, Lugo AE, F. S. Chapin, III, Pickett STA, Duniway M, Rocha AV, Swanson FJ, Laney C, and Jones J. 2011b. Cross-system comparisons elucidate disturbance complexities and generalities. *Ecosphere* **2**: art81.

Peters DPC, Yao J, Sala OE, and Anderson JP. 2012. Directional climate change and potential reversal of desertification in arid and semiarid ecosystems. *Global Change Biology* **18**: 151-163.

Piao S, Ciais P, Friedlingstein P, de Noblet-Ducoudre N, Cadule P, Viovy N, and Wang T. 2009. Spatiotemporal patterns of terrestrial carbon cycle during the 20th century. *Global Biogeochem Cycles* **23**: GB4026, 16 p.

Poff NL, Pyne MI, Bledsoe BP, Cuhaciyan CC, and Carlisle DM. 2010. Developing linkages between species traits and multiscaled environmental variation to explore vulnerability of stream benthic communities to climate change. *Journal of the North American Benthological Society* **29**: 1441-1458.

Poff NL, and Zimmerman JKH. 2010. Ecological responses to altered flow regimes: a literature review to inform environmental flows science and management. *Freshwater Biology* **55**: 194-120.

Power ME, Tilman D, Estes JA, Menge BA, Bond WJ, Mills LS, Daily G, Castilla JC, Lubchenco J, and Paine RT. 1996. Challenges in the quest for keystones. *Bioscience* **46**: 609-620.

Pregitzer KS, and Euskirchen ES. 2004. Carbon cycling and storage in world forests: biome patterns related to forest age. *Global Change Biology* **10**: 2052–2077.

Preston BL, Yuen EJ, and Westaway RM. 2011. Putting vulnerability to climate change on the map: a review of approaches, benefits, and risks. *Sustainability Science* **6**: 177-202.

Raffa KF, Aukema BH, Bentz BJ, Carroll AL, Hicke JA, Turner MG, and Romme WH. 2008. Cross-scale drivers of natural disturbances prone to anthropogenic amplification: The dynamics of bark beetle eruptions. *Bioscience* **58**: 501-517.

Ranalli AJ, and Macalady DL. 2010. The importance of the riparian zone and in-stream processes in nitrate attenuation in undisturbed and agricultural watersheds - A review of the scientific literature. *Journal of Hydrology* **389**: 406-415.

Raymond PA, Oh NH, Turner RE, and Broussard W. 2008. Anthropogenically enhanced fluxes of water and carbon from the Mississippi River. *Nature* **451**: 449-452.

Raymond PA, and Saiers JE. 2010. Event controlled DOC export from forested watersheds. *Biogeochemistry* **100**: 197-209.

Raymond PA, Zappa CJ, Butman D, Bott TL, Potter J, Mulholland P, Laursen AE, McDowell WH, and Newbold D. 2012. Scaling the gas transfer velocity and hydraulic geometry in streams and small rivers. *Limnology and Oceanography Fluids and Environments* **2**: 41-53.

Richardson AJ. 2008. In hot water: zooplankton and climate change. *ICES Journal of Marine Science* **65**: 279-295.

Rood SB, Pan J, Gill KM, Franks CG, Samuelson GM, and Shepherd A. 2008. Declining summer flows of Rocky Mountain rivers: Changing seasonal hydrology and probable impacts on floodplain forests. *Journal of Hydrology* **349**: 397-410.

Rose JB, Gerba CP, and Jakubowski W. 1991. Survey of potable water-supplies for *Cryptosporidium* and *Giardia*. *Environmental Science and Technology* **25**: 1393-1400.

Ryan MG, Archer SR, Birdsey R, Dahm C, Heath L, Hicke J, Hollinger D, Huxman T, Okin G, Oren R, Randerson J, and Schlesinger W. 2008. Land resources. *In* Backlund PA, Janetos A, Schimel D, Hatfield J, Boote K, Fay P, Hahn L, Izaurralde C, Kimball BA, Mader T, Morgan J, Ort D, Polley W, Thomson A, Wolfe D, Ryan MG, Archer SR, Birdsey R, Dahm C, Heath L, Hicke J, Hollinger D, Huxman TE, Okin G, Oren R, Randerson J, Schlesinger WH, Lettenmaier D, Major D, Poff L, Running SW, Hansen L, Inouye D, Kelly BP, Meyerson L, Peterson B, and Shaw R (Eds), The effects of climate change on agriculture, land resources, water resources, and biodiversity in the United States. U.S. Department of Agriculture, Washington, D.C. 75-120 p.

Sabo JL, Finlay JC, Kennedy T, and Post DM. 2010. The role of discharge variation in scaling of drainage area and food chain length in rivers. *Science* **330**: 965-967.

Sandel B, Arge L, Dalsgaard B, Davies RG, Gaston KJ, Sutherland WJ, and Svenning JC. 2011. The influence of Late Quaternary climate-change velocity on species endemism. *Science* **334**: 660-664.

Scheffer M, Carpenter SR, Foley JA, Folke C, and Walker B. 2001. Catastrophic shifts in ecosystems. *Nature* **413**: 591-596.

Scheffer M, Bascompte J, Brock WA, Brovkin V, Carpenter SR, Dakos V, Held H, van Nes EH, Rietkerk M, and Sugihara G. 2009. Early-warning signals for critical transitions. *Nature* **461**: 53-59.

Schneider P, Hook SJ, Radocinski RG, Corlett GK, Hulley GC, Schladow SG, and Steissberg TE. 2009. Satellite observations indicate rapid warming trend for lakes in California and Nevada. *Geophysical Research Letters* **36**. pages

Schneider P, and Hook SJ. 2010. Space observations of inland water bodies show rapid surface warming since 1985. *Geophysical Research Letters* **37**: L22405, 5 p.

Scholze M, Knorr W, Arnell NW, and Prentice IC. 2006. A climate-change risk analysis for world ecosystems. *Proceedings of the National Academy of Sciences* **103**: 13116-13120.

Schuur EAG, Abbott B, and Network PC. 2011. High risk of permafrost thaw. *Nature* **480**: 32-33.

Senthilkumar S, Basso B, Kravchenko AN, and Robertson GP. 2009. Contemporary evidence of soil carbon loss in the U.S. corn belt. *Soil Science Society of America Journal* **73**: 2078-2086.

Serrat-Capdevila A, Valdés JB, Pérez JG, Baird K, Mata LJ, and Maddock T. 2007. Modeling climate change impacts – and uncertainty – on the hydrology of a riparian system: The San Pedro Basin (Arizona/Sonora). *Journal of Hydrology* **347**: 48-66.

Sherry RA, Arnone JA, Johnson DW, Schimel DS, Verburg PS, and Luo Y. 2011. Carry over from previous year environmental conditions alters dominance hierarchy in a prairie plant community. *Journal of Plant Ecology*: 1-13.

Sitch S, Huntingford C, Gedney N, Levy PE, Lomas M, Piao SL, Betts R, Ciais P, Cox P, Friedlingstein P, Jones CD, Prentice IC, and Woodward FI. 2008. Evaluation of the terrestrial carbon cycle, future plant geography and climate-carbon cycle feedbacks using five Dynamic Global Vegetation Models (DGVMs). *Global Change Biology* **14**: 2015-2039.

Smith MD. 2011. An ecological perspective on extreme climatic events: a synthetic definition and framework to guide future research. *Journal of Ecology* **99**: 656–663.

Sobota DJ, Harrison JA, and Dahlgren RA. 2009. Influences of climate, hydrology, and land use on input and export of nitrogen in California watersheds. *Biogeochemistry* **94**: 43-62.

Sponseller RA, Grimm NB, Boulton AJ, and Sabo JL. 2010. Responses of macroinvertebrate communities to long-term variability in a Sonoran Desert stream. *Global Change Biology* **16**: 2891–2900.

Stafford JM, Wendler G, and Curtis J. 2000. Temperature and precipitation of Alaska: 50 year trend analysis. *Theoretical and Applied Climatology* **67**: 33-44.

Steffen WL. 2004. Global change and the earth system: a planet under pressure. New York, New York, USA: Springer-Verlag.

Stewart IT. 2009. Changes in snowpack and snowmelt runoff for key mountain regions. *Hydrological Processes* **23**: 78-94.

Stewart IT, Cayan DR, and Dettinger MD. 2005. Changes toward earlier streamflow timing across western North America. *Journal of Climate* **18**: 1136-1155.

Strayer DL. 2010. Alien species in fresh waters: ecological effects, interactions with other stressors, and prospects for the future. *Freshwater Biology* **55**: 152-174.

Stromberg JC, Lite SJ, and Dixon MD. 2010. Effects of stream flow patterns on riparian vegetation of a semiarid river: Implications for a changing climate. *River Research and Applications* **26**: 712-729.

Sturm M, Racine C, and Tape K. 2001. Climate change - Increasing shrub abundance in the Arctic. *Nature* **411**: 546-547.

Suarez F, Binkley D, Kaye MW, and Stottlemyer R. 1999. Expansion of forest stands into tundra in the Noatak National Preserve, northwest Alaska. *Ecoscience* **6**: 465-470.

Thomas RQ, Canham CD, Weathers KC, and Goodale CL. 2010. Increased tree carbon storage in response to nitrogen deposition in the US. *Nature Geoscience* **3**: 13–17.

Thomey ML, Collins SL, Vargas R, Johnson JE, Brown RF, Natvig DO, and Friggens MT. 2011. Effect of precipitation variability on net primary production and soil respiration in a Chihuahuan Desert grassland. *Global Change Biology* **17**(4): 1505-1515.

Tierney GL, Fahey TJ, Groffman PM, Hardy JP, Fitzhugh RD, and Driscoll CT. 2001. Soil freezing alters fine root dynamics in a northern hardwood forest. *Biogeochemistry* **56**: 175-190.

Tilman D, Knops J, Wedin D, Reich P, Ritchie M, and Siemann E. 1997. The influence of functional diversity and composition on ecosystem processes. *Science* **277**: 1300-1302.

Tomback DF, and Achuff P. 2010. Blister rust and western forest biodiversity: ecology, values and outlook for white pines. *Forest Pathology* **40**: 186-225.

Tranvik LJ, Downing JA, Cotner JB, Loiselle SA, Striegl RG, Ballatore TJ, Dillon P, Finlay K, Fortino K, Knoll LB, Kortelainen PL, Kutser T, Larsen S, Laurion I, Leech DM,

McCallister SL, McKnight DM, Melack JM, Overholt E, Porter JA, Prairie Y, Renwick WH, Roland F, Sherman BS, Schindler DW, Sobek S, Tremblay A, Vanni MJ, Verschoor AM, von Wachenfeldt E, and Weyhenmeyer GA. 2009. Lakes and reservoirs as regulators of carbon cycling and climate. *Limnology and Oceanography* **54**: 2298-2314.

Tucker AJ, Williamson CE, Rose KC, Oris JT, Connelly SJ, Olson MH, and Mitchell DL. 2010. Ultraviolet radiation affects invasibility of lake ecosystems by warmwater fish. *Ecology* **91**: 882-890.

Turner MG. 2010. Disturbance and landscape dynamics in a changing world. *Ecology* **91**: 2833-2849.

Turner BL, 2nd, Matson PA, McCarthy JJ, Corell RW, Christensen L, Eckley N, Hovelsrud-Broda GK, Kasperson JX, Kasperson RE, Luers A, Martello ML, Mathiesen S, Naylor R, Polsky C, Pulsipher A, Schiller A, Selin H, and Tyler N. 2003a. Illustrating the coupled human–environment system for vulnerability analysis: Three case studies. *Proceedings of the National Academy of Sciences* **100**: 8080–8085.

Turner BL, Kasperson RE, Matson PA, McCarthy JJ, Corell RW, Christensen L, Eckley N, Kasperson JX, Luers A, Martello ML, Polsky C, Pulsipher A, and Schiller A. 2003b. A framework for vulnerability analysis in sustainability science. *Proceedings of the National Academy of Sciences* **100**: 8074-8079.

van Mantgem PJ, Stephenson NL, Byrne JC, Daniels LD, Franklin JF, Fule PZ, Harmon ME, Larson AJ, Smith JM, Taylor AH, and Veblen TT. 2009. Widespread increase of tree mortality rates in the western United States. *Science* **323**: 521-524.

Venalainen A, Tuomenvirta H, Heikinheimo M, Kellomaki S, Peltola H, Strandman H, and Vaisanen H. 2001. Impact of climate change on soil frost under snow cover in a forested landscape. *Climate Research* **17**: 63-72.

Verburg P, and Hecky RE. 2009. The physics of the warming of Lake Tanganyika by climate change. *Limnology and Oceanography* **54**: 2418-2430.

Verbyla D. 2008. The greening and browning of Alaska based on 1982-2003 satellite data. *Global Ecology and Biogeography* **17**: 547-555.

Verbyla D. 2011. Browning boreal forests of western North America. *Environmental Research Letters* **6**: 041003.

Vitousek PM. 2004. Nutrient cycling and limitation: Hawai'i as a model system: Princeton University Press.

Walker B, Holling CS, Carpenter SR, and Kinzig A. 2004. Resilience, adaptability and transformability in social–ecological systems. *Ecology and Society* **9**: 5.

Walker MD, Wahren CH, Hollister RD, Henry GHR, Ahlquist LE, Alatalo JM, Bret-Harte MS, Calef MP, Callaghan TV, Carroll AB, Epstein HE, Jonsdottir IS, Klein JA, Magnusson B, Molau U, Oberbauer SF, Rewa SP, Robinson CH, Shaver GR, Suding KN, Thompson CC, Tolvanen A, Totland O, Turner PL, Tweedie CE, Webber PJ, and Wookey PA. 2006. Plant community responses to experimental warming across the tundra biome. *Proceedings of the National Academy of Sciences* **103**: 1342-1346.

Wang MY, and Overland JE. 2009. A sea ice free summer Arctic within 30 years? *Geophysical Research Letters* **36**: L07502.

Wang DB, and Cai XM. 2010. Comparative study of climate and human impacts on seasonal baseflow in urban and agricultural watersheds. *Geophysical Research Letters* **37**: L06406.

Wang XH, Piao SL, Ciais P, Li JS, Friedlingstein P, Koven C, and Chen AP. 2011. Spring temperature change and its implication in the change of vegetation growth in North America from 1982 to 2006. *Proceedings of the National Academy of Sciences* **108**:1240-1245.

Westerling AL, Hidalgo HG, Cayan DR, and Swetnam TW. 2006. Warming and earlier spring increase western U.S. forest wildfire activity. *Science* **313**: 940-943.

Westerling AL, Turner MG, Smithwick EAH, Romme WH, and Ryan MG. 2011. Continued warming could transform Greater Yellowstone fire regimes by mid-21st century. *Proceedings of the National Academy of Sciences* **108**: 13165-13170.

Weyhenmeyer GA, and Karlsson J. 2009. Nonlinear response of dissolved organic carbon concentrations in boreal lakes to increasing temperatures. *Limnology and Oceanography* **54**: 2513-2519.

Williams MW, Brooks PD, and Seastedt T. 1998. Nitrogen and carbon soil dynamics in response to climate change in a high-elevation ecosystem in the Rocky Mountains, USA. *Arctic and Alpine Research* **30**: 26-30.

Williamson CE, Morris DP, Pace ML, and Olson AG. 1999. Dissolved organic carbon and nutrients as regulators of lake ecosystems: Resurrection of a more integrated paradigm. *Limnology and Oceanography* **44**: 795-803.

Williamson CE, Saros JE, and Schindler DW. 2009a. Climate change: Sentinels of change. *Science* **323**: 887-888.

Williamson CE, Saros JE, Vincent WF, and Smol JP. 2009b. Lakes and reservoirs as sentinels, integrators, and regulators of climate change. *Limnology and Oceanography* **54**: 2273-2282.

Wilmking M, Juday GP, Barber VA, and Zald HSJ. 2004. Recent climate warming forces contrasting growth responses of white spruce at treeline in Alaska through temperature thresholds. *Global Change Biology* **10**: 1724-1736.

Winder M, Schindler DE, Essington TE, and Litt AH. 2009. Disrupted seasonal clockwork in the population dynamics of a freshwater copepod by climate warming. *Limnology and Oceanography* **54**: 2493-2505.

Woodward FI, Lomas MR, and Kelly CK. 2004. Global climate and the distribution of plant biomes. *Philosophical Transactions of the Royal Society of London Series B-Biological Sciences* **359**: 1465-1476.

Yang HJ, Wu MY, Liu WX, Zhang Z, Zhang NL, and Wan SQ. 2011. Community structure and composition in response to climate change in a temperate steppe. *Global Change Biology* **17**: 452-465.

Yohe G. 2008. Characterizing the value of reducing greenhouse gas emissions: Creating benefit profiles tracking diminished risk. United States Environmental Protection Agency, Washington, D.C.

Zavaleta ES, Shaw MR, Chiariello NR, Thomas BD, Cleland EE, Field CB, and Mooney HA. 2003. Grassland responses to three years of elevated temperature, CO_2, precipitation, and N deposition. *Ecological Monographs* **73**: 585-604.

Zavaleta ES. 2006. Shrub establishment under experimental global changes in a California grassland. *Plant Ecology* **184**: 53-63.

Zhang K, Kimball JS, Hogg EH, Zhao MS, Oechel WC, Cassano JJ, and Running SW. 2008. Satellite-based model detection of recent climate-driven changes in northern high-latitude vegetation productivity. *Journal of Geophysical Research-Biogeosciences* **113**: G03033.

Zhang J, Hudson J, Neal R, Sereda J, Clair T, Turner M, Jeffries D, Dillon P, Molot L, Somers K, and Hesslein R. 2010. Long-term patterns of dissolved organic carbon in lakes across eastern Canada: Evidence of a pronounced climate effect. *Limnology and Oceanography* **55**: 30-42.

Zhao MS, and Running SW. 2010. Drought-Induced Reduction in Global Terrestrial Net Primary Production from 2000 Through 2009. *Science* **329**: 940-943.

Zion MS, Pradhanang SM, Pierson DC, Anandhi A, Lounsbury DG, Matonse AH, and Schneiderman EM. 2011. Investigation and Modeling of winter streamflow timing and magnitude under changing climate conditions for the Catskill Mountain region, New York, USA. *Hydrological Processes* **25**: 3289-3301.

Chapter 4. Impacts of Climate Change on Ecosystem Services

Convening Lead Authors: Peter Kareiva and Mary Ruckelshaus
Lead Authors: Katie Arkema, Gary Geller, Evan Girvetz, Dave Goodrich, Erik Nelson, Virginia
Matzek, Malin Pinsky, Walt Reid, Martin Saunders,
Darius Semmens, Heather Tallis

Key Findings

- By 2050, climate change will triple the fraction of counties in the U.S. that are at high or extremely high risk of outstripping their water supplies (from 10 percent to 32 percent). The most at risk areas in the U.S. are the West, Southwest and Great Plains regions.
- Regulation of drinking water quality will be strained as high rainfall and river discharge conditions may lead to higher levels of nitrogen in rivers and greater risk of waterborne disease outbreaks.
- Climate change will have uneven effects on timber production across the U.S. Recent increases in tree mortality due to disease and pests, and the intensity of fires and area burned will continue to destroy productive forests. On the other hand, in some regions climate change is expected to boost overall forest productivity due to longer growing seasons.
- There is a better than 50 percent chance that climate change will overwhelm the ability of natural systems to mitigate the harm to people resulting from extreme weather events (such as heat waves, heavy rains, and drought).
- Vulnerability of people and property in coastal areas is highly likely to increase dramatically – due to the effects of sea-level rise, storm surge, and the loss of habitats that provide protection from flooding and erosion. The areas at greatest risk to coastal hazards in the U.S. are the Atlantic and Gulf coasts.
- The human communities most vulnerable to climate-related increases in coastal hazards are the elderly and the poor who are less able to respond quickly before and during hazards and to respond over the long term through relocation.
- Changes in abundance and ranges of commercially important marine fish are highly likely to result in loss of some local fisheries, and increases in value for others if fishing communities and management practices can adapt.
- In recreation and tourism, the greatest negative climate impacts will continue to be felt in winter sports and beach recreation (due to coastal erosion). Other forms of recreation are highly likely to increase due to better weather, leading to a redistribution of the industry and its economic impacts, with visitors and tourism dollars shifting away from some communities in favor of others.
- Supporting, regulating, and provisioning ecosystem services all contribute to food security in the United States, and the fate of the nation's food production are very likely to depend on the interplay of these services and how the agriculture and fishery sectors respond to climate stresses.

4.1. INTRODUCTION: WHAT ARE ECOSYSTEM SERVICES AND WHY DO THEY MATTER?

Climate change will likely put at risk many of nature's benefits, or ecosystem services, that humans derive from our lands and waters. Climate-mediated loss or disruption of ecosystem functions are very likely to have repercussions for society's dependence on ecosystems for wild-caught and farmed food, recreation, nutrient cycling, waste processing, protection from natural hazards, climate regulation, and other services. One of the many advantages of nature-based services is that not only can they provide jobs and economic opportunities, but they are not subject to "economic bubbles" – in other words they can be reliably counted on as long as ecosystems are well-managed. In addition, ecosystem structures and functions typically provide multiple services; for example, the same habitats that can buffer devastating impacts of floods or storms also provide other benefits, including critical habitat for commercial and recreationally valued species, filtration of sediment and pollutants, and carbon storage and sequestration.

The social values of ecosystem services are broad and include those reflected in markets, avoided damage costs, maintenance of human health and livelihoods, and cultural and aesthetic values. Understanding how human activities and a changing climate are likely to interact to affect the delivery of these ecosystem services is of the utmost importance as we make decisions now that affect the health of terrestrial, coastal, and marine systems and their ability to sustain future generations.

There are a number of ways of accounting for the value of ecosystem services (NRC, 2005), and the literature cited in this Chapter reflects this diversity of methods. The most reliable methodology for estimating how *changes* in human or natural drivers lead to *changes* in ecosystem-derived value is production function analysis (NRC, 2005; Daily and others, 2009; Kareiva and others, 2011). Information about demand for ecosystem services (for example, the distribution of people who use the services supplied) and their social value can be combined with biophysical supply estimates to generate predictive maps of service use and value (Daily and others, 2009; Nelson and others, 2009; Tallis and others, 2011). Economic valuation methods take changes in the supply of ecosystem services as input and translate these into changes in human welfare in monetary terms (Daily and others, 2000; Arrow and others, 2004). There is a common misconception that valuing ecosystem services requires converting everything to a dollar value, when in fact this is not the case (Reyers and others, 2012). The value of ecosystem services can be effectively captured in terms of reduced risk, jobs, and human well-being, without having to convert everything to a dollar bottom line.

The state of our understanding of climate impacts on ecosystem services across the U.S. is relatively undeveloped, primarily because there is no national system for tracking the status or trends in ecosystem services for the USA (PCAST, 2011). However, there are numerous studies from which one can identify selected, albeit not comprehensive, impacts of climate change on ecosystem services.

4.2. WHAT ARE OBSERVED IMPACTS OF RECENT CLIMATE CHANGE ON ECOSYSTEM SERVICES AND THEIR VALUE?

Because there is no national assessment of ecosystem services for the USA, it is impossible to report on the overall status of all of the nation's natural assets. However, specific studies and analyses allow us to survey a range of documented impacts of recent climate changes on ecosystem services and their values. These are summarized in **Table 4.1** (table is located at the end of *Chapter 4*). There is strong evidence of negative effects on human wellbeing having already occurred due to climate change through such impacts as: increased forest wildfires, reduced carbon storage in coastal marine systems, reduced storm protection, shifting marine fish ranges and localized reduction in fish harvest, decreased trout and salmonid recreational fisheries, shortened season for winter recreation, loss of subsistence hunting for Inupiat communities, and closed campgrounds as a result of drought and wildfire risk. These highly focused studies likely reflect only a small fraction of the impacts of climate change that have already occurred, when one considers the total value of ecosystem services in the United States. By looking at specific ecosystem services it is possible to make a start on assessing the economic and employment losses due to recent climate trends.

4.2.1. Marine fishery yields

The economic value of fishery-related services from the ocean is substantial. In 2009, marine living resource industries had $116 billion in sales and contributed $48 billion in value added to the U.S. economy (NMFS, 2010). In 2010, 8.2 billion pounds of fish and other marine species were landed at U.S. ports, worth $4.5 billion in ex-vessel values (Van Voorhees and Lowther, 2011)

Although fisheries are a small fraction of the total U.S. Gross National Product, marine fishing is central to the economies and identities of hundreds of local and regional economies. For example, coral reef fisheries provided $54.7 million to American Samoa and Northern Marianas from 1982-2002 (Zeller and others, 2007); and tuna canneries provide 90 percent of total exports for American Samoa (BEA, 2010). U.S. consumers in all States like to eat seafood: we ate 15.8 pounds of fish per person in 2010, and that quantity has been slowly growing for decades (Van Voorhees and Lowther, 2011). Almost all communities within the Pacific Islands derive over 25 percent of their animal protein from fish, with some deriving up to 69 percent (NCA, 2009).

Fisheries provide a culturally important source of employment in coastal communities that often have few other economic opportunities. In 2009, 1 million people were employed in full- and part-time jobs by commercial fishing, seafood processors and dealers, seafood wholesalers and distributors, importers, and seafood retailers (NMFS, 2010). Where vibrant fishing industries exist, supporting industries are also sustained, including boat building and maintenance, shipping, processing, and service industries.

Climate change already is affecting where and how much fish biomass is available for harvest, and thus the value of fisheries for local fishers. The distributions of many fished species are shifting poleward as sea surface temperatures warm (Nye and others, 2009; Murawski, 1993; Mueter and Litzow, 2008); resulting in concomitant poleward shifts in jobs, catch and value (**Box 4.1**) (McCay and others, 2011; Pinsky and Fogarty, written communication 2012). In Alaska, salmon production increased when ocean temperatures warmed as part of the Pacific

Box 4.1. Climate Impacts on New England Groundfish Fisheries
Author: Malin Pinkski

Fishing in New England has been associated with bottom-dwelling species of fish, collectively called groundfishes, for more than 400 years and is a central part of the region's cultural identity and social fabric. Atlantic halibut (*Hippoglossus hippoglossus*), cod (*Gadus morhua*) and haddock (*Melanogrammus aeglefinus*) were among the earliest species caught, but this fishery has now expanded to include over fifteen species, including winter flounder (*Pseudopleuronectes americanus*), white hake (*Urophycis tenuis*), pollock (*Pollachius virens*), American plaice (*Hippoglossoides platessoides*), and yellowtail flounder (*Limanda ferruginea*). The fishery is pursued by both small boats (less than 50 ft) that are typically at sea for less than a day to large boats (greater than 50 ft) that fish for a day to a week at a time. These vessels use home ports spread across more than 100 coastal communities from Maine to New Jersey, and they land fish worth about $60 million at the dock each year (New England Fishery Management Council, 2011). Captains and crew are often second- or third-generation fishermen who have learned the trade from their families and who hope to pass the tradition on to their children (New England Fishery Management Council, 2011).

The documented impacts of warming temperatures on this fishery over the last few decades suggests indications of further changes ahead. From 1982-2006, sea surface temperature in the coastal waters of the northeastern U.S. warmed by 0.23°C, close to twice the global rate of warming over this period (0.13°C) (Belkin, 2009). The velocity of climate change from 1960-2009 was 20-100 km/decade in the Northeastern U.S., with spring temperatures advancing by 2-10 days/decade (Burrows and others, 2011). Long-term monitoring of bottom-dwelling fish communities in New England revealed that the abundance of warm-water species increased, while cool-water species decreased (Collie and others, 2008; Lucey and Nye, 2010). A recent study suggests that many species in this community have shifted their geographic distributions northwards by up to 200 miles since 1968, though substantial variability among species also exists (Nye and others, 2009). The northward shifts of these species are reflected in the fishery as well: landings and landed value of these species have shifted towards northern States such as Massachusetts and Maine, while southern States have declined (Pinsky and Fogarty, written communication 2012). A number of the commercially important groundfish species in the region such as cod, haddock, winter flounder and yellowtail flounder are at the southern extent of their range in the Northeast and are particularly vulnerable to temperature increase.

Climate projections for this region suggest similar trends in the future. A coarse global projection of future fisheries potential under IPCC scenario A1B (720 ppm CO_2 in 2100) suggests a 15-50 percent loss of fisheries in this region (Cheung and others, 2010). Specific projections for pollock and haddock also suggest substantial declines in this region by 2090 based on changes in temperature and salinity (Lenoir and others, 2010). Under the A1fi emissions scenario (970 ppm CO_2 in 2100), increasing temperatures suggest a substantial loss of cod in the Mid-Atlantic Bight and a decline on Georges Bank (Fogarty and others, 2007). These losses appear substantially less likely to occur under low emissions scenarios (B1, 550 ppm CO_2 by 2100). In contrast, subtropical species such as croaker (*Micropogonia undulatus*) appear likely to increase in the northeast (Hare and others, 2010). To both avoid overfishing of these declining populations and to take advantage of expanding populations, fisheries management will need to adjust exploitation levels, including benchmark measures such as maximum sustainable yield, to account for the impacts of climate change on changing species distributions (Hare and others, 2010).

Box 4.1, continued.

The economic and social impacts of these biophysical changes depend in large part on the response of the human communities in the region (McCay and others, 2011). Fishing communities have a range of strategies for coping with the inherent uncertainty and variability of fishing, including diversification among species and livelihoods, but climate change imposes both increased variability and sustained change that may push these fishermen beyond their ability to cope (Coulthard, 2009). Technology plays a role in this transition. Larger fishing boats can follow the fish to a certain extent as they shift northward, while smaller inshore boats will be more likely to leave fishing or switch to new species (Coulthard, 2009). The past decade in New England has seen dramatic changes to the groundfish industry that has already pushed boats towards larger sizes (New England Fishery Management Council, 2011). However, long-term viability of fisheries in the region is likely to ultimately depend on a transition to new species that have shifted from regions further south (Sumaila and others, 2011).

In light of these transitions, actions that enhance the flexibility of the industry in the region will be important (Coulthard, 2009). Co-management, or the sharing of regulatory decision-making between the government and fishing stakeholders, has been suggested as one mechanism for enhancing the ability of fishing communities to cope with change (McCay and others, 2011). Secure and exclusive fishing rights also promote future-oriented action that can help with difficult transitions (McCay and others, 2011). New England fisheries management includes some of these mechanisms, including fishing industry representation on the management council and a newly implemented sector management program that provides fishermen with more flexibility and responsibility for managing their resources. These measures, however, were primarily focused on ending overfishing in the region. Climate change presents a new challenge that will likely require additional effort to align individual and industry incentives with a sustainable transition to new fishing opportunities before traditional fisheries decline further under the combined impacts of climate and intensive fishing.

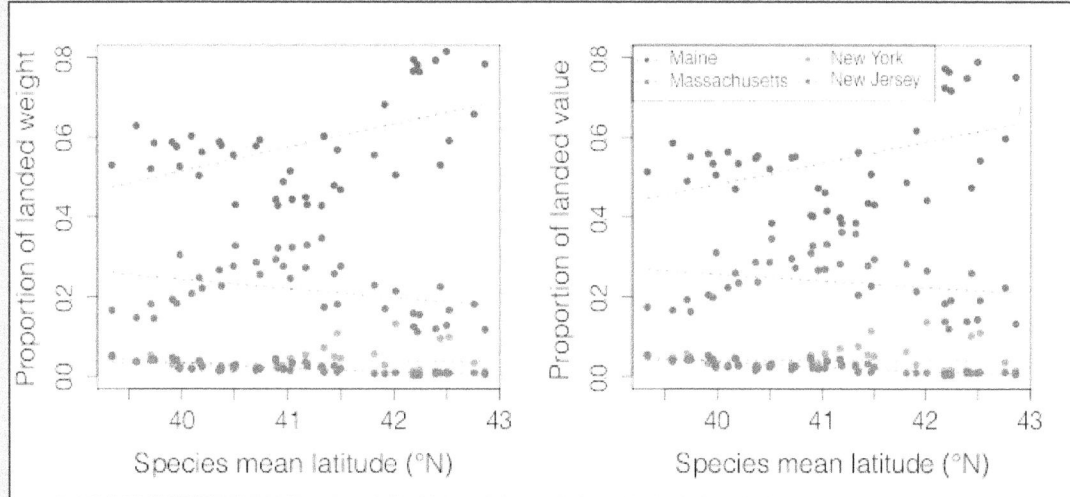

Figure 4.1. Winners and losers as a result of lobster range shifts: Northern ports (for example, Maine) land relatively more lobster by weight and by value as lobster stocks shift north (towards the right side of graph), while southern ports do worse (for example, Massachusetts). Data are from Van Vorhees and Lowther, 2011, and Nye and others, 2009.

Decadal Oscillation, while salmon production decreased in the Pacific Northwest; additional heterogeneity in stock abundance in response to climate also occurs at smaller geographic scales (Hare and others, 1999, Schindler and others, 2008). In Monterey Bay, CA, albacore tuna abundance and catch per unit effort increased during past warm periods, while Chinook salmon declined (Dalton, 2001). The overall economic impact on fishermen of recent warming temperatures was positive for tuna and negative for salmon (Dalton, 2001).

Geographic shifts in fish species in response to climate change could be due to a number of interacting factors, including physiological tolerance thresholds, phenology mismatches of competitor, predator and prey species (for example, Beaugrand and others, 2003), and through effects of climate on habitats that in turn affect fish population dynamics (Jennings and Brander, 2008). Together, these shifts are creating transitions from cold-water fish communities to a different set of warm-water species available for harvest in specific regions (Collie and others, 2008; Lucey and Nye, 2010). In some cases, new industries have developed in response to novel warm-water fish species (Pinnegar and others, 2010; McCay and others, 2011). Furthermore, warm surface water temperatures are driving some fish species deeper (Nye and others, 2009; Dulvy and others, 2008; Perry and others, 2005), which will affect harvest strategies and potentially, costs of exploitation, as fish move to deeper waters (Caputi and others, 2010).

Research is ongoing to explicitly link climate and the condition of natural habitats to fisheries production; yet numerous examples demonstrate that the relationship is often close. On the east coast of the U.S., approximately two out of every three species of economically important fish species rely on estuaries for shelter and resources when young (nursery habitat) (Able and Fahay, 1998). Gulf of Mexico shrimp support the largest crustacean fishery in the U.S., and up to 66 percent of their production may rely on salt marshes (Zimmerman and others, 2000). Similarly, about a quarter of the Gulf's blue crab fishery may be dependent on salt marshes (Zimmerman and others, 2000). The supporting value of marshes for the blue crab fishery in the Gulf is $0.19 to $1.89/acre (Freeman, 1991). Climate impacts on marsh and other habitats affecting fishery production are well documented.

4.2.2. Nature-dependent tourism

Climate change is known to impact opportunities for outdoor recreation by increasing beach erosion, reducing winter snows, increasing wildfire risk, threatening coral reefs, and decreasing valuable cold-water fisheries, among other impacts (**Table 4.1**). To date, the evidence for current climate change impacts on recreation are mostly anecdotal or indirect; for instance, in summer 2008, as a result of tree die-offs related to drought and beetle infestations in the West, Colorado and Wyoming closed 38 campgrounds (Robbins 2008). However, the size of the tourism and outdoor recreation industry gives a good indication of the assets may be at risk in the future.

Ocean-related tourism contributed $82 billion to the U.S. Gross Domestic Product in 2009 (NOEP, 2005); skiing and snowmobiling together contribute another $88 billion (International Snowmobile Manufacturers Association); while recreational fishing, hunting, and wildlife watching add up to $113 billion combined (US Department of the Interior (DOI), Fish and Wildlife Service (FWS), Department of Commerce (DOC), 2006). Some of these activities have profound local impacts. For instance, Hawaiian reefs allowed about 100 dive operators to make $50-60 million/year in total (van Beukering and Cesar, 2004), while Florida's east coast marshes are worth $6471/acre for their support of recreational fishing alone (Bell, 1997).

California has the nation's largest ocean economy, valued at approximately $43 billion annually, with about 80 percent of this coming from tourism and recreation (NOEP, 2005).

Demand for recreation is sensitive to improvements and declines in the health of the ecosystem. For instance, implementation of a beach replenishment policy in North Carolina to increase beach width by 100ft was expected to increase the average number of trips by visitors from 11 to 14, with beach goers willing to pay $166/trip or $1574 per visiting household per year (Landry and Liu, 2009). Another study of North Carolina beaches found that widening beach width increases the consumer surplus of visitors by $7/trip (Whitehead and others, 2009). Conversely, economists have estimated that a single catastrophic fire in New Mexico would reduce forest visits by 7 percent, resulting in a loss of 1,900 jobs and $81,000,000 (Starbuck and others, 2006).

4.2.3. Hazard Reduction: Coastal protection services

Nationwide, more than one-third of the U.S. population currently lives in the coastal zone; and 14 of the 20 largest U.S. urban centers are located along the coast. As population and development along our coasts continue to increase (Crossett and others, 2004), so will their vulnerability to coastal hazards such as storms and sea-level rise. A 17ft storm surge from Hurricane Andrew cost $26.5 billion worth of damage to Miami residents in 1992. In 2005, Hurricane Katrina caused $85.6B worth of damage, with New Orleans taking the brunt of the economic and social damage (First American, 2010). Following Hurricane Katrina and international disasters such as the Indian Ocean Tsunami of 2004, attention has been focused on the ability of coastal ecosystems, such as wetlands and mangroves, to provide protection from ocean-related hazards (Danielsen and others, 2005; Kathiresan and Rajendran, 2005; Das and Vincent, 2009; Koch and others, 2009; Wamsley and others, 2010). A variety of these coastal habitats border the edges of the U.S. shoreline, reducing the vulnerability of people and property to coastal hazards. But marine and coastal ecosystems that provide protection are at risk from coastal development, pollution, destructive fishing practices, aquaculture, marine transportation and other ocean uses. Loss of these ecosystems and the protection they provide could prove devastating for U.S. coastal communities. For example, reduced coastal protection due to salt marsh loss and degradation is thought to have contributed to the extent of the disaster caused by Hurricanes Katrina and Rita in the Gulf of Mexico, which caused over 1,500 deaths (Day and others, 2007). Here we focus on risks of coastal communities to climate impacts and the documented role of protective habitats in ameliorating impacts of sea level rise and storms to people.

Some regions of the U.S. are experiencing more dramatic climate-related coastal hazards. The two primary biophysical processes affecting risk to coastlines and people from climate change are (1) erosion from sea-level rise and storm-induced waves and (2) flooding from sea-level rise and storm surges (**Table 4.1**). Long-term data (greater than 30 yrs) from tide stations indicate that the greatest increases in sea level are occurring along the Atlantic coast from New York south to Virginia (3-6 mm/yr) and in the Gulf of Mexico from Louisiana to Texas (3-12 mm/yr). The majority of the U.S. coast is experiencing a rise of 1-3mm/yr (NOAA, 2011). Furthermore, wave heights from hurricanes (greater than 3m, during the summer months) have increased by 0.7-1.8 m during the last 30 years, increasing erosion processes. The observed increases in wave heights have been greater in higher latitudes (Allan and Komar, 2006; Komar and Allan, 2008); but whether such increases are due to climate change or background environmental variability remains unclear (Komar and others, 2009).

Some of the observed geographic variation in coastal climate impacts in the U.S. is caused by heterogeneity in the distribution of habitats such as wetlands, marshes, mangroves, seagrasses, coral reefs, and dunes that can offer protection from flooding by attenuating storm surges and protection from erosion by dampening wave heights (Barbier and others, 2008). For example, estimates suggest that 0.4 million ha of salt marsh has been lost in North America over the last 200 yrs (Sifleet and others, 2011). It is not known how much of this loss has been due to climate change. Some studies have found that salt marshes in the U.S. are keeping pace with the current long term rate of relative sea-level rise (for example, in North Carolina (Morris and others, 2002); yet other studies show the opposite (Craft and others, 2009; Gedan and others, 2011). In the Chesapeake Bay, satellite imagery suggests that more than half of the tidal marsh area has been degraded by erosion since 1000 AD; and erosion rates have increased from 0.5mm per year to more than 3.2 mm per year during the 20th century (Stevenson and others, 2002). This erosion has caused marsh loss—for example, from 1849 to 1992, the land area of one of the large saltmarsh islands in the Bay decreased by 579 acres or 26 percent of the area (Downs and others, 1994). The documented loss of protective habitats to climate change, human activities, and natural disasters is putting more people and property at risk from coastal hazards. For example, salt marshes along the central Louisiana coast are estimated to reduce storm surges by 3 inches (0.25 feet) per mile of marsh (USACE, 2006). Many years of coastal erosion coupled with Hurricane Katrina's damages to the estuaries surrounding New Orleans have reduced the natural storm defenses around the city by more than 500 square miles (USACE, 2006).

Vulnerability to erosion hazards depends both on physical and social characteristics of coastlines. A social vulnerability index accounting for such attributes as poverty status, race, gender, development density and infrastructure reliance calculated for the U.S. found that social and physical vulnerabilities to erosion hazards from storms are not uniformly distributed (Boruff and others, 2005). For example, the vulnerability of the Gulf coast to erosion is more a product of social than physical characteristics because of the relatively high prevalence of low-income communities along the coast. The reverse is true for the Pacific and Atlantic counties, where physical characteristics are more influential in determining erosion-hazard vulnerability (Boruff and others, 2005).

The value to people of the protection offered by coastal habitats is impressive. For example, marshes are worth an estimated $8235/yr/ha in reduced hurricane damages to the U.S. (Costanza and others, 2008). An analysis of the economic damages associated with 34 major hurricanes striking the United States coast since 1980 found that the additional storm protection value per unit area of coastal wetlands from a specific hurricane ranged from a minimum of U.S. $23 per hectare for Hurricane Bill to a maximum of U.S. $463,730 per hectare for Hurricane Opal, with a median value of just under U.S. $5,000 per hectare (Costanza and others, 2008).

4.2.4. Fire Regulation

The risk of severe wildfires is a function of climate, forest composition and management practices in that forest or grassland. Wildfires in the U.S. damage hundreds of homes in the U.S. each year and annual fire-fighting expenditures alone regularly exceed $1billion dollars per year (Whitlock, 2004). The incidence of large forest fires in the western U.S. increased nearly four-fold in the 1980s onward, and the total area burned by fires six-fold (Westerling and others, 2006). Most of this increase can be explained by increased spring and summer temperatures (Westerling and others, 2006). However, management of forests, grazing regimes, and thinning can dramatically impact the spread and risk of wildfires. For example the Arizona Wallow fire of

Impacts of Climate Change on Biodiversity, Ecosystems, and Ecosystem Services |
Technical Input to the 2013 National Climate Assessment

Chapter 4
Ecosystem Services

2011, which was Arizona's largest fire on record, did not burn ridges where there had been previous thinning of the forests. The thinning effort in portions of Arizona was a forest stewardship project aimed to reduce fire risk and to create jobs. It did both (BIA, 2011). Thus, well-managed forests provide the auxiliary service of fire risk reduction—a service whose importance increases as warming trends can exacerbate background propensity for severe fires. The nexus of climate and forest fires is a flashpoint for several other pathways towards degraded ecosystems services such as water supply and quality (**Box 4.2**).

Box 4.2. Climate Impacting Fire Risk, Water Supply, Recreation, and Flood Risk in Western U.S. Forests
Authors: Evan Girvetz, Dave Goodrich, Darius Semmens, Carolyn Enquist

The 2009 National Climate Change Assessment (CCSP, 2009) documented the broad-scale forest dieback as a threshold response to climate change in the Southwestern United States (Fagre and others, 2009) and noted this can be a precursor to high severity wildfires. Since that assessment, in the summer of 2011 the largest recorded wildfires in Arizona (Wallow - greater than 538,000 acres with 15,400 acres in New Mexico; greater than$100 million in suppression costs) and New Mexico (Las Conchas - ~156,600 acres) occurred. Both fires had significant impacts on a range of ecosystem processes, individual species, and a number of ecosystem services provided by these systems.

The Las Conchas fire in northern New Mexico burned over 63 residences, 1100 archeological sites, more than sixty percent of Bandelier National Monument (BNM), and over 80 percent of the forested lands of the Santa Clara Native American Pueblo (16,600 acres), and was severe enough to cause forest stand replacement scale damage over broad areas. Following the fire, heavy rain storms led to major flooding and erosion throughout the fire area. Scientific modeling found that this type of storm (25-year event) would lead to river runoff approximately 2.5 times greater and sediment yield three times greater due to this fire in the main canyon of Bandelier National Monument (Semmens and others, 2008; **Table 4.1**).

Climate change a likely contributing factor: There is good evidence for warmer temperatures, reduced snowpack, and earlier onset of springtime leading to already observed increased wildfires in the western U.S (Westerling, 2006). The National Research Council (2011) projected 2 to 6 times increase in areas in the West burned by wildfires given a 1°C increase. Recent research employing paleodata and an ensemble of climate models projects that the frequency of droughts, which cause broad-scale forest die-back may occur approximately 50 times per century by 2100, far beyond the range of variability of the driest centuries in the past millennium (Williams and others, 2012).

Other Stressors Exacerbating Fire: Forest management practices and invasive insect pests contributed to catastrophic wildfire occurring in these systems. Even-aged second growth forests much denser than natural occur in the West, remove more water out of the soil and increase the likelihood of catastrophic crown fires. In addition, naturally occurring bark beetles breed more frequently and successfully under conditions that are projected to become more frequent with climate change (Jonsson and others, 2009; Schoennagel and others, 2011). Outbreaks of bark beetles and associated tree mortality have increased in severity in recent years, suggesting a

Box 4.2, continued

possible connection between large fires and the changing fuel conditions caused by beetle outbreaks. In turn, the dead trees left behind by bark beetles can make crown fires more likely (Hoffman and others, 2010; Schoennagel and others, 2011).

Impacts to species and biodiversity: The catastrophic crown fire conditions during the Las Conchas fire undoubtedly had a devastating impact on above-ground wildlife (McCarthy, 2012). Relatively few animals living above ground likely survived. In addition, the mid-elevation areas of all the major canyon systems of Bandelier National Monument experienced extensive to near complete mortality of all tree and shrub cover while leaving dead trees standing. Mexican Spotted Owls (*Strix occidentalis lucida*) nesting and roosting habitat has been altered, potentially affecting its suitability for this species (Jenness and others, 2004). The Jamez salamander is an endangered species whose population was put in further danger due to this fire (McCarthy, 2012).

Impacts to recreation: Post-fire localized thunderstorms on a single day resulted in at least ten debris flows originating from the north slopes of a single canyon in Bandelier National Monument. Popular recreation areas in the Monument were evacuated for four weeks and flash floods damaged the newly-renovated multi-million dollar National Park Service visitor center. In addition, other recreation areas managed by the U.S. Forest Service, U.S. Army Corps of Engineers, and the Bureau of Land Management closed down recreation areas due to the fire, and associated flooding and erosion.

Impacts to Urban water supply: The increased sediment and ash eroded by the floods in the wake of the fire were transported to downstream streams and rivers, including the Rio Grande, a major source of drinking water for New Mexico and 50 percent of the drinking water supply for Albuquerque. The sediment and ash led to Albuquerque's water agency to turn off all water supplies from the Rio Grande for a week, and reducing water withdrawals in the subsequent months due to increased cost of treatment (Albuquerque Journal, September 2, 2011 http://www.abqjournal.com/main/2011/09/02/news/2-agencies-curtail-rio-grande-draws.html)

An adaptation effort is needed: Safeguarding against fire related impacts and adaptation to change will require innovative solutions, large-scale action and engagement among a variety of different stakeholders. The Southwest Climate Change Initiative (SWCCI), led by The Nature Conservancy, is an example of this type of adaptation planning effort. SWCCI is a public-private partnership developed in 2009 with the University of Arizona Climate Assessment for the Southwest, Wildlife Conservation Society, National Center for Atmospheric Research, and Western Water Assessment along with government agency partners with the goal of providing information and tools to build resilience in ecosystems and communities of the southwestern U.S. The SWCCI is currently leading efforts across the Southwest, including adjacent to the Las Conchas fire area, to identify and implement adaptation solutions that help prevent these types of catastrophic events. Some of the solutions being considered include forest restoration activities such as non-commercial mechanical thinning of small-diameter trees, controlled burns to reintroduce the low-severity ground fires that historically maintained forest health, and comprehensive ecological monitoring to determine effects of these treatments on forest and stream habitats, plants, animals, habitats and soils.

4.2.5 Carbon storage and sequestration

Carbon accumulates in soil and biomass (for example, vegetation), and represents a greater pool of carbon than is present in the atmospheric pool (Lal, 2004a). When carbon is released from the earth during cultivation, deforestation, fire, and other land use practices, it binds with other chemicals to form greenhouse gases (GHG) in the atmosphere and accelerates global climate change (Lal, 2004b). The conservation of carbon sinks or pools is therefore important to mitigate GHG levels. Property owners and land managers can influence the pace of global climate change and related impacts through climate-smart land use decisions that maintain, rather than perturb or destroy carbon sources (Post and Kwon, 2008). Carbon sequestration and other actions that reduce emissions have become valued goods and services that benefit and potentially reduce global economic damage from climate change (Conte and others, 2011). Estimates of the global economic value created by each ton of carbon that is sequestered or reduced through lowering emissions ranges from $25 to $675 (Tol, 2009). This large range in values is in part explained by uncertainties in climate change projections, mitigation actions, climate change adaptation, and the resilience of ecological systems to future changes (Aldy and others, 2010).

Because carbon sequestration and reduced emissions can create an economic value, society is willing to pay to encourage it. Carbon markets are a manifestation of this willingness to pay. Several mandated and voluntary markets that pay landowners to sequester carbon have been created in the last decade (Canadell and Raupach, 2008; Arriagada and Perrings, 2011). The carbon market price and the policy infrastructure that supports the carbon market is likely to be an important determinant for U.S. landowners to remove or prevent emissions to the atmosphere (Lubowski and others, 2006). A well-functioning market can approximately equate the carbon price with the global value created by a ton of sequestered. If climate changes reduce the capacity of ecosystems to sequester carbon, the ability to mitigate global economic damages caused by climate change is likely to decrease.

Forest carbon

Climate change-induced perturbations in forest distribution, growth rates, and risk of wildfire, invasive species, and disease are impacting the rates of carbon sequestration and expectations for length of storage. Dry, warm conditions over the last 10 years across 20 million hectares in western North America have led to extensive insect outbreaks and mortality of diverse tree species, including oaks in the Midwest and southeastern U.S. (Allen and others, 2010). Although these tree mortality rates are higher than any observed in 50 years, greater than 99 percent of forest species inventory available for harvest remains unaffected (Oswalt and others, 2009). Governments at all levels and private landowners are investing significant sums to protect forests from further damage. For example, the cost to Federal agencies for fire suppression now exceeds $1 billion annually (U.S. Government Accountability Office, 2006).

An extrapolation of current economic dynamics in the conterminous U.S. suggests that forested areas could increase by 10 to 14 million hectares from 2001 to 2051 (Radeloff and others, 2012), resulting in about 220 million hectares of forest across the conterminous U.S. by 2051. This same study suggests that a combination of payments for landowners converting to forest lands and taxes on those who cut their trees could increase the area of forest in 2051 by an additional 30 million hectares, resulting in forest carbon storage levels that are orders of magnitude larger than storage levels under the current baseline. Payments for landowners who decide not to deforest are beginning (for example, through the United Nations Collaborative initiative on Reducing Emissions from Deforestation and forest Degradation (REDD) policies), and the potential for management incentives to change forest area is great (Canadell and Raupach, 2008; Arriagada and Perrings, 2011).

Marine Carbon

Research on carbon storage and sequestration has focused predominantly on terrestrial forest and deep ocean ecosystems. Vegetated coastal ecosystems are not part of either ecosystem type, creating a gap in estimates of global carbon storage and sequestration capacity estimates (Mcleod and others, 2011). Coastal ecosystems dominated by plants such as mangroves, salt marshes and seagrasses, sequester and store carbon in the short term in biomass and over the long term in sediments (Duarte and others, 2005; Mcleod and others, 2011). The annual burial of carbon in mangroves, salt marshes, and seagrass beds across the world is estimated to be 31–34 teragrams (Tg), 5–87 Tg, and 48–112 Tg C per year, respectively (Mcleod and others, 2011). The carbon storage and sequestration potential of these marine habitats is impressive. In just the first meter of coastal and nearshore sediments, soil organic carbon averages 500 - 4966 t carbon dioxide equivalent (CO_2e)/ha for sea grasses, 917 t CO_2e/ha for salt marshes, 1060 t CO_2e/ha for estuarine mangroves, and nearly 1800 t CO_2e/ha for marine mangroves (Murray and others, 2011).

Approximately 0.4 million hectares of salt marsh has been lost in North America over the last 200 yrs (Sifleet and others, 2011). Currently, 1.9 million hectares of salt marsh in the U.S. store and sequester carbon. Most annual estimates of salt marsh carbon sequestration fall below 2.2 Mg per hectare (Sifleet and others, 2011). Most U.S. studies on carbon storage and sequestration in salt marshes are from the northeastern States.

Estimates of carbon sequestration rates in Floridian mangroves range from 0.03-3.8 Mg of C per hectare (Sifleet and others, 2011 and citations therein). Annual carbon sequestration rates have been calculated for 39 mangrove sites worldwide. Values range from 0.03 to 6.54 Mg of carbon per hectare. However, most estimates fall below 1.9 Mg per hectare per year (Sifleet and others, 2011 and citations therein). Annual carbon sequestration data are available for 377 seagrass sites worldwide. Values range from -21 to 23.2 Mg of C per hectare. A large number of estimates show annual net losses of carbon (Sifleet and others, 2011). Most estimates of annual seagrass bed sequestration show 1.9 Mg of C per hectare.

Soil carbon

Climate change induced perturbations in nutrient cycling and precipitation is very likely to impact the ability of soil to sequester and store carbon. Currently, soil carbon levels are most influenced by rates of land use change. In general, switching from cropland to grassland and forest increases carbon levels in the soil (Post and Kwon, 2000; Powlson and others, 2011). How much additional soil is conserved in such transitions is open to debate (Dlugofl and others, 2010; Syswerda and others, 2011; Rumpel and Kogel-Knabner, 2010; Powlson and others, 2011). Further, the soil carbon sequestration benefits created by various less intense land use management practices are in doubt; for example, benefits from reduced tillage are relatively small, and increased N_2O emissions observed in some cases could offset increases in stored carbon (Powlson and others, 2011).

4.3. HOW WILL CLIMATE CHANGE AFFECT ECOSYSTEM SERVICES AND HUMAN WELL BEING OVER THE NEXT 50 TO 100 YEARS?

The status of ecosystem services summarized above point to regional, species- and habitat-based differences in the current distribution of services and their impacts on human well-being. Below we summarize information on the vulnerability of ecosystem services under future climate conditions (**Table 4.1**). In some cases, ecosystem service delivery and value will increase; and in others, there is a high likelihood that the benefits from ecosystem processes to humans will be severely reduced under projected future climate. Vulnerability in ecosystem services and the impacts on human communities are likely to vary in the future due to where people are located, or because of particular susceptibility of habitats or species upon which the service values depend. Here we briefly highlight ecosystem services that are particularly vulnerable to climate change or that have not been previously summarized (**Table 4.2;** table is located at the end of *Chapter 4*)).

4.3.1 Marine fishery yields
The range and abundance of economically important marine fish already are shifting due to climate change and they are highly likely to continue to change; some local fisheries are very likely to cease to be viable, whereas other fisheries may increase in value if the fishing community can adapt to the changes. Globally, fish species are projected to shift 45-49 km/decade poleward under the A1B future climate scenario (Cheung and others, 2009), and thus the abundance and availability of fish are projected to decline (Cheung and others 2011). Fisheries potential is projected to decline under future climate in coastal lower 48 States, but increase in parts of Alaska (Cheung and others, 2010). In the northeastern U.S., Atlantic croaker are likely to increase, while pollock, haddock, and cod decrease (Hare and others 2010; Fogarty and others, 2007; Lenoir and others, 2010) (**Box 4.1**). In the NE Atlantic, fish distributions are projected to shift 5.1 m/decade deeper under future climate (A1B) (Cheung and others, 2011). Salmon ocean habitat is projected to disappear from the Gulf of Alaska (Abdul-Aziz and others, 2011). Not all marine species can move quickly in response to climate. Some fishes and invertebrates spend little time dispersing as larvae and move little as adults (Kinlan and Gaines, 2003; Shanks, 2009). Whether these and other species will keep up with climate change remains an important question. Similarly, fishery-based industries are likely to bear increased costs due to transitioning to new species, relocation of processing plants and fishing jobs poleward (NCA, 2009; Sumaila and others, 2011), but these socio-economic impacts have not been well studied.

4.3.2 Nature-based recreation and tourism
Climate change impacts on outdoor recreation are projected to be most profound in winter sports and in beach recreation (**Tables 4.1 and 4.2**). There is a high probability of abbreviated ski seasons in many parts of North America. The California ski season is expected to shorten by 49-103 days, potentially missing the Christmas-New Year's week (Hayhoe and others, 2004). Snow seasons are very likely to shorten by 5-60 percent in various parts of the Northeast (Scott and others, 2006; Dawson and Scott, 2007; Scott and others, 2008). In the Pacific Northwest, 12.5 percent of ski areas in the Cascades and 60 percent of ski areas in the Olympic range are at risk due to increasingly frequent warm winters (Nolin and Daly, 2006), and Arizona resorts may be unable to forestall losses to the ski season after 2050, due to

insufficiently cold temperatures for snowmaking (Bark and others, 2010). If drier conditions lead to a greater frequency of dust storms, windblown dust on snow will also increase rates of snowmelt, shortening the ski season and increasing evapo-transpiration, resulting in reduced water flows to the Colorado River (Painter and others, 2010). Snowmobile areas will be particularly vulnerable to economic losses because snowmaking is not practical on the terrain exploited by snowmobile enthusiasts (Scott and others, 2008). In addition to economic losses from lower visitation and increased costs of snowmaking at ski areas, homeowners in winter sports resort areas are expected to suffer declines in home value (Butsic and others, 2011).

Beach recreation losses will result from loss of beach width due to the combined effects of sea level rise and erosion. Narrower beaches make it harder to access fishing sites for anglers, and are less attractive to sunbathers. An analysis of projected losses due to beach erosion from 2006 to 2080 in North Carolina estimates losses of over $1 billion due to reduced recreation (Whitehead and others, 2009); a similar analysis for Southern California projects negative impacts of climate change on beaches, amounting to $63 million annually (Pendleton and others, 2011). However, beach user days may increase with warmer, drier weather, possibly resulting in economic gains in some areas (Loomis and Crespi, 1999).

The potential for longer stretches of more pleasant weather for enjoying the outdoors may actually increase some recreation opportunities, or simply shift others to new areas. For these activities, it is unclear what the net effect in human well-being will be; for instance, one study found that visitation to Rocky Mountain National Park would increase with higher temperatures (Richardson and Loomis, 2005), while other parks are projected to lose visitors if catastrophic fires result from drier conditions (Starbuck and others, 2006). "Winter sun" and "summer cool" destinations for retirees will redistribute around North American cities (Scott and others, 2004), whale-watching outfitters will have to shift locations to improve the reliability of their sightings (Lambert and others, 2010), and some recreational anglers will have to switch from cold-water species like salmon and trout to warm-water fish like bass and perch (Pendleton and Mendelsohn 1998). Golfing and boating are projected to increase with good weather (Loomis and Crespi, 1999; Shaw and Loomis, 2008); diving and snorkeling may experience losses due to declines in coral reef habitat.

Recreation is considered an ecosystem service not only because it has economic value, but also because it contributes to cultural well-being. Another cultural service at risk from future climate change is traditional subsistence hunting by indigenous people of the Arctic. Among coastal Inupiat people, hunting is a substantial contributor to dietary protein, a source of cash income, and a cultural touchstone (Gearheard and others, 2006). Climate change is decreasing the extent of sea ice and breaking up the sea ice earlier (Gearheard and others, 2006), changing the abundance and migratory patterns of wildlife (Kruse and others, 2004), decreasing the predictability of weather conditions (Ford and others 2006), increasing storminess and windiness (Ford and others, 2006; Hinzman and others, 2005), and generally increasing hazards to traditional hunters (Ford and others, 2006; Ford and others, 2008). Indigenous hunters in Alaska are projected to spend less time hunting (Berman and Kofinas, 2004), suffer decreased wildlife harvests (Hinzman and others, 2005; Kruse and others 2004) and the obsolescence of the traditional ecological knowledge that has guided weather prediction and risk assessment for centuries (Ford and others, 2006).

4.3.3 Hazard reduction by coastal habitats

Climate change has a very high likelihood of increasing property loss and vulnerability of people to coastal hazards (**Table 4.1**). With the projected accelerated rise in sea level and increased storm intensity in some areas, the conflicts between development along the coast and the protective value of natural processes will likely increase, causing negative economic and societal impacts (Titus and others, 2009). Modeling of future storm surges suggests that the number of people affected by flooding worldwide will increase five-fold by 2080 (Nicholls and others, 1999). Rising sea level is making populations in low-lying coastal areas increasingly vulnerable to catastrophic floods and coastal erosion from storms (McGranahan and others, 2007; Fitzgerald and others, 2008). In summary, over the next 50 to 100 years the vulnerability of people and property in coastal areas is highly likely to increase dramatically – due to the effects of sea-level rise, storm surge, and the loss of habitats that provide protection from flooding and erosion.

Some regions of the U.S. are particularly at risk from climate-related coastal hazards (**Table 4.2; Box 4.3**). The Atlantic and Gulf of Mexico coasts are most vulnerable to the loss of coastal protection services provided by wetlands and coral reefs. A prime example is the Gulf of Mexico coast, where a combination of sea level rise (SLR), exposure to large storms, coastal development, large river systems, and engineered coastlines puts thousands of people and acres of property at risk from flooding and erosion from storm surge flooding (**Box 4.3**). Along the California coast, a 1.4 m sea level rise would put an anticipated 480,000 people at risk of a 100-yr coastal flooding event, and cause nearly $100 Billion in damages (Heberger and others, 2009). In addition, large sections of the Pacific Coast are vulnerable to erosion – which would accelerate with sea level rise. Such erosion is projected to result in a loss of 41 sq. miles from the California coast by 2100, affecting more than 14,000 people who currently live in the area (Heberger and others, 2009). In the northeast, a Long Island example indicates that even modest sea level rise (0.5 m by 2080) would dramatically increase the number of people (47 percent increase in persons affected) and property loss (73 percent increase) impacted by storm surge (Shepard, 2011). Similarly, approximately 1 percent to 3 percent of the land area of New Jersey would be permanently inundated over the next century under modest sea-level rise scenarios (0.61m-1.22m) (Cooper and others, 2008). As a result, coastal storms coming ashore in New Jersey could temporarily flood low-lying areas up to 20 times more frequently as marsh and other protective habitats are inundated (Cooper and others, 2008).

In addition to direct increases in inundation and erosion through sea-level rise, loss of protective coastal habitats places certain regions at particular risk of greater damages in the future. Effects of climate change on coastal hazards will depend both on changes in wave and storm events, and on effects of sea level rise and other climate-related variables on coastal habitats (for example, coastal forests, wetlands, dunes, and corals). Climate impacts on these habitats will likely include increases in the intensity and frequency of storms, sea level rise, salt water intrusion, warming temperature, and ocean acidification, and human modification of the shoreline in response to rising seas. The ability of coastal ecosystems to provide protection from future climate-related hazards depends upon their ability to adapt to changing conditions (Alongi, 2008). Wetlands are extremely vulnerable to sea-level rise and can maintain their elevation and viability only if sediment accumulation (both mineral and organic matter) keeps pace with sea-level rise and tidal range is not too extreme (Morris and others, 2002; Temmerman and others, 2004; Stevenson and Kearney, 2009). Controversy exists about whether wetlands,

Box 4.3: Climate Impacts on Coastal Hazards in the Gulf of Mexico
Author: Katie Arkema

The Gulf coast region is especially vulnerable to a changing climate because of its relatively flat topography, rapid rates of coastal lands subsidence, land and waterway engineering, coastal development and exposure to large storms. Sea level rise is likely to increase the vulnerability of Gulf coast communities by increasing flooding during storm events. For example, Katrina and Rita were the fourth and fifth most powerful storms to strike the Mississippi Delta since 1893 with respect to maximum wind speed at landfall, but they both were more devastating for the hundreds of kilometers of the coast affected by a storm surge exceeding 3 m. Climate models project that sea level will rise by 0.3 to 1.0 m along the Gulf Coast in the next century (Twilley, 2007). Because of high rates of land subsidence in the Mississippi Delta, relative sea-level rise – the combination of absolute sea level rise and subsidence – is about 1 cm/yr in contrast to eustatic sea level rise of 1.5 mm/yr (Day and others, 2007).

In addition to the direct effects of sea level rise and storms, vulnerability of the Gulf coast to climate change also accrues through indirect processes, through the loss of protective salt marshes and coastal forests caused by a combination of rising ocean temperatures, ocean acidification, flooding and salt-water intrusion (Craft and others, 2009). Simulations from numerical models (Wamsley and others, 2010) and empirical observations (USACE, 2006) have highlighted the importance of coastal wetlands for providing the Gulf coast with protection from flooding and storms. Yet, some regions of the Gulf coast, such as the Mississippi River delta and Florida Everglades are experiencing some of the highest wetland loss rates of the country (Twilley, 2007). Nearly 5,000 km^2 of wetlands have been lost from coastal Louisiana at rates as high as 100 km^2/year (Gagliano and others, 1981; Britsch and Dunbar, 1993). Coastal development and engineering can increase the vulnerability of these wetlands to climate change and diminish their ability to provide protection for surrounding areas in the future. Large restoration efforts are underway to restore the functioning of the system (Day and others, 2007), but climate change will likely also affect watersheds that feed coastal ecosystems. Hydrology will depend on effects on precipitation, evaporation and management of water resources, which could lead to periods of drought as well as flooding. For example, a 25-month drought, interacting with other environmental stresses, is considered the main cause of a severe dieback of 100,000 acres of salt marsh in coastal Louisiana in 2000 (Twilley, 2007).

The Gulf Coast is vulnerable to climate related coastal hazards for social as well as physical reasons (Boruff and others, 2005). Relatively high vulnerability of the Gulf Coast to erosion hazards is due primarily to the percent of the population over 65 years old, followed by birth rate, sea-level rise, mean wave height, and median age of the population. More generally, the effects of hurricanes may be indicative of the potential consequences of rising sea levels and changes in wave height under future climate scenarios. Communities unprotected by levees or where levees failed were inundated during hurricanes Rita and Katrina. More than 1500 people died as a direct or indirect result of Hurricane Katrina, almost 1100 of them in Louisiana (Day and others, 2007). Sea level rise would increase costs incurred due to storm surge flooding. For example, the economic damages resulting from Hurricanes Carla (1961), Beulah (1967), and Bret (1999) in Corpus Christi, Texas would increase by $30-$1,100 million under a 2080 climate scenario (Frey and others, 2010). Furthermore, the area of land flooded and the number of people affected in the projected storms would increase with respect to those impacts in the original storm (Frey and others, 2010).

Box 4.3, continued.

Climate adaptation planning is underway at the State, county, and local government levels along the Gulf coast (NOAA, 2011). These efforts are varied, ranging from assessments of the effects of rising sea levels on infrastructure, transportation systems, and property rights and using ecosystem protection as a means of reducing hazard risks in Louisiana.

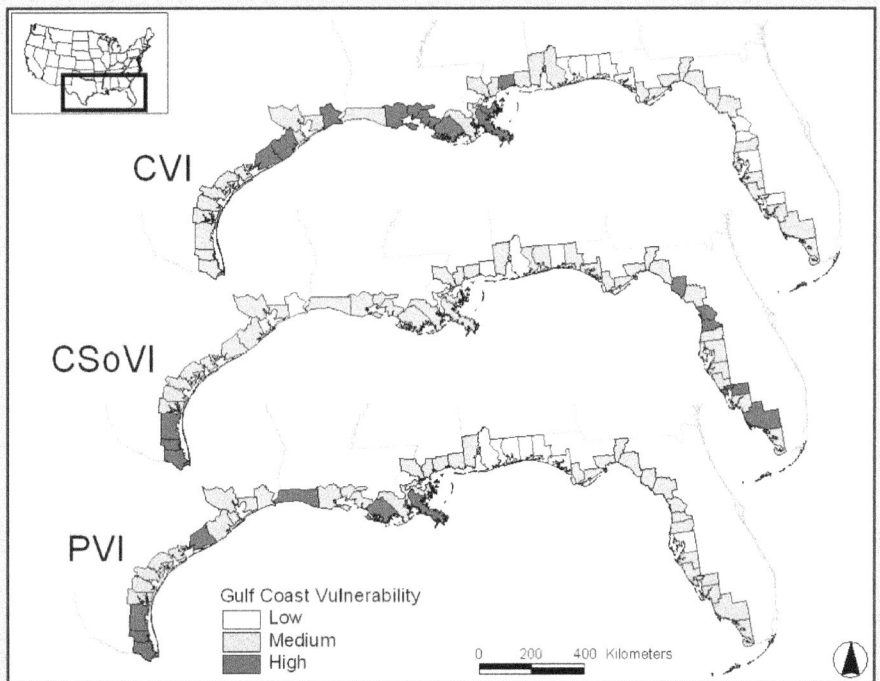

Figure 4.2. Vulnerability of Gulf coastal counties based on physical (CVI) and social (CSoVI) indicators and their integration into place vulnerability (PVI) (From: Boruff and others, 2005).

and in particular U.S. marshes, can accrete and keep up with sea level rise or be lost to open water (Craft and others, 2009; Morris and others, 2002; Gedan and others, 2011). For example, the Atlantic coast of North America may experience one of the world's largest losses in wetlands due to projected sea-level rise (Nicholls and others, 1999). On the other hand, simulations of mangrove forest dynamics along the southwest coast of Florida suggest that forests will change in structure and composition; although diminished in height, future mangrove forests will likely be able to adapt to sea level rise and migrate inshore (Doyle and others, 2003, 2010).

There is a high likelihood that coral reefs will suffer much damage from climate impacts. Roughly one third of all reef-building corals are estimated to be at elevated risk of extinction due to projected climate change (Carpenter and others, 2008). Coral cover in Hawaii, Florida and the Gulf is likely to decrease, as warming and acidifying seas are very likely to compromise coral reef carbonate accretion worldwide (Hoegh-Guldberg and others, 2007). Degradation of other protective habitats, such as barrier islands along the Texas coast, combined with sea level rise may lead to increased flooding from even intermediate hurricane events (Irish and others, 2010; Frey and others, 2010).

Vulnerability and loss of protective habitats will be greater for those populations lacking the social and economic means to cope with the short and long-term consequences of coastal hazards. One study that projected storm surge inundation showed that for Hampton, Virginia, the most vulnerable regions to storm surge are those areas where the most socially vulnerable populations live (Kleinosky and others, 2007). In Alaska, 86 percent of Alaskan Native villages are already affected by flooding and erosion, due in part to rising temperatures (US General Accounting Office (USGAO), 2003; **Figure 4.3**). Further warming is projected to lead to greater loss of sea ice, which provides some protection from winter storms. As many of these villages do not qualify for flood and erosion control projects, the only option would be relocation (USGAO, 2003).

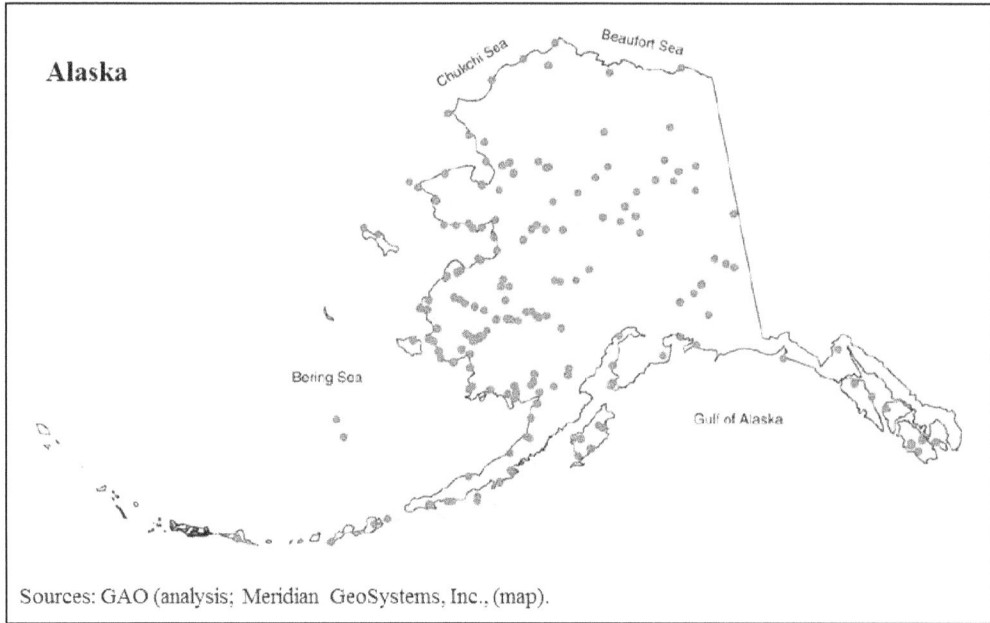

Sources: GAO (analysis; Meridian GeoSystems, Inc., (map).

Figure 4.3. Location of the 184 out of 213 Alaska Native villages already affected by flooding and erosion, due in part to rising temperatures (USGAO, 2003).

4.3.4. Water supply and water quality under future climate

It is widely appreciated that water scarcity and water quality could become a significant problem for the United States. Some of this is driven simply by human population growth and human activities. However, climate is modifying the hydrological cycle in a way that makes water supply in some places increasingly subject to flash floods, and enhances evaporation and (or) evapo-transpiration (**Table 4.1**).

Much of the Western U.S. is projected to experience decreasing water yield under a number of future climate scenarios, especially the Southwestern U.S., Great Basin, and California (Walker and others, 2011). Snow pack driven systems are especially susceptible to changes in hydrology, with these river systems experiencing earlier peak flows and a reduction in dry season base flows throughout the western U.S. (Hamlet and others, 2005). Snowpack water storage has already been reduced in much of the U.S., with a greater percentage of precipitation falling as rain, and future projections for 2040 springtime (March-April) snow water equivalent indicate a reduction in all of the conterminous U.S (**Figure 4.4**) (Mote and others, 2005; Adam and others, 2009). To compound the problem, decreases in runoff—particularly during the dry season—may be coupled with increased flooding in some parts of country (Bukovsky and Karoly, 2011).

An increase in the number of U.S. counties with water sustainability risk by 2050 is projected as a consequence of climate change (**Figure 4.4**; Roy, 2012). Using a county-level water supply sustainability index based on attributes of susceptibility to drought, increase in water withdrawal, increased need for storage, and groundwater use, this research found that by 2050 climate change is projected to double the percent of counties with moderate or higher water sustainability risk (35 percent to 70 percent). Even more striking, the number of counties with high or extreme water sustainability risk (10 percent to 32 percent) would triple, and the number of counties with extreme risk is projected to increase 14-fold. The most at risk areas in the U.S. are the West, Southwest and Great Plains regions.

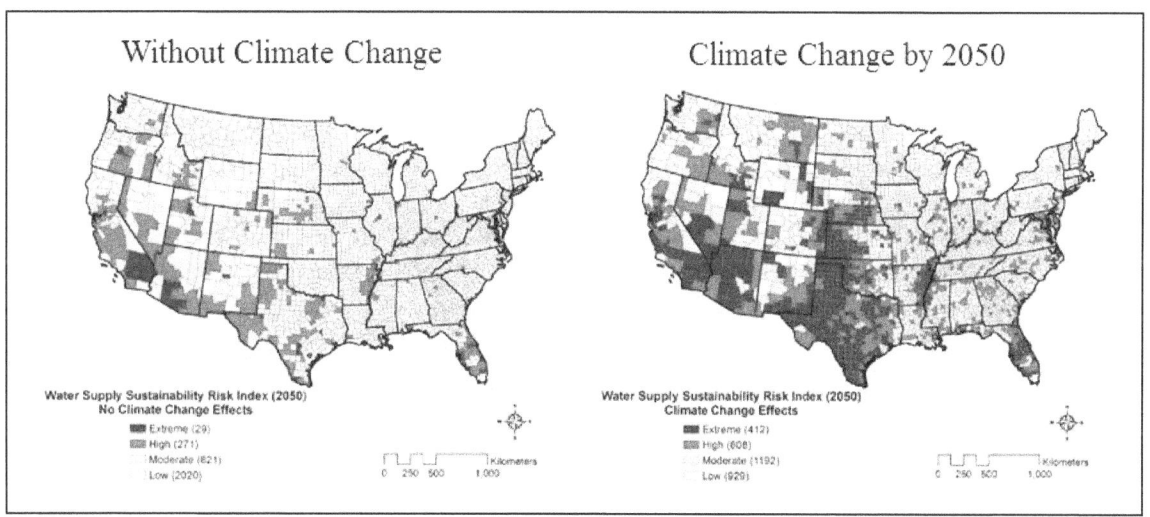

Figure 4.4. The number of U.S. counties with water sustainability risk by 2050 with and without climate change (Roy and others, 2012).

As the climate continues to warm and soil moisture deficits accumulate beyond historical levels, a consensus among climate model simulations suggests that sustaining water supplies in parts of the Southwest will be a challenge (Cayan and others, 2010). If this happens, an array of impacts could affect the American Southwest, including more dust storms that affect human health and traffic safety, and reduced soil fertility that affects agricultural yields and food security.

Some of these changes in climate and hydrology are expected to cause changes in water quality. The links between precipitation, temperature and nitrogen retention are well described (Vitousek and others, 1997). The flux of nitrogen from watersheds and exported to coastal waters is correlated with high rainfall and river discharge conditions (Howarth and others, 2012). Similarly, extreme precipitation and river discharge events are positively correlated with waterborne disease outbreaks (Curriero and others, 2001). Higher water temperatures can be associated with increases in nitrogen retention, but the relationship is weaker than the relationship of nitrogen with precipitation and discharge (Howarth and others, 2012).

Although these links with water quality have been observed under current climate conditions, few studies have projected the impacts of climate change on water quality. Several studies state that waterborne illness is likely to increase because extreme precipitation events increase the loading of contaminants to waterways (Rose and others, 2001; Curriero and others, 2001; Ebi and others, 2006). One regional study estimates the impacts of climate change on nutrient retention and the downstream impacts on the coastal ocean. Climate change projections for the Mississippi Basin (under doubled CO_2) indicate a 20 percent increase in river discharge that will lead to higher nitrogen loads and a 50 percent increase in primary production in the Gulf of Mexico, a 30-60 percent decrease in deep water dissolved oxygen concentration and an expansion of the dead zone (Justic and others, 1996).

4.4. WHAT RESPONSE STRATEGIES COULD ADDRESS THE MOST HARMFUL IMPACTS OF CLIMATE CHANGE ON ECOSYSTEM SERVICES?

Climate adaptation approaches will need to be implemented across all sectors of the U.S. economy—we highlight several by way of example here (**Table 4.2**). To combat expected negative yield impacts from climate change, the U.S. agriculture sector can improve the soils they crop on, both by reserving the best soils for agriculture and improving the marginal soils already used. Farmers could also better adapt to projected climate change by using irrigation water more strategically and becoming more flexible in management and planting decisions. Soil conservation will become particularly important as several global forces increase the pressure to cultivate more marginal lands, resulting in the accompanying risk of increased erosion and decreases in sequestered soil carbon and soil fertility (**Box 4.4**). Farmers can also enhance existing soil quality for agriculture by establishing major drainage facilities, building levees or flood-retarding structures, providing water for irrigation, removing stones, or grading gullied land (USDA, 2012). The first pressure point is likely to come from the strong growth in food demand due to a growing and increasingly richer world (Foley and others, 2011; Tilman and others, 2011).

Impacts of Climate Change on Biodiversity, Ecosystems, and Ecosystem Services |
Technical Input to the 2013 National Climate Assessment

Chapter 4
Ecosystem Services

Box 4.4. Adapting to Climate Change By Maximizing a Supporting Service: Soil Quality
Author: Erik Nelson

Projected climate change is very likely to require adaptation in crop production processes in the U.S. within the next 100 years. Farmers are likely to use technology and adaptive management (for example, different crop and variety choices, different input use, changing planting and harvesting dates) to maintain profits in the face of climate change. One significant pathway to adaptation could be shifting crops to the most productive soils, or improving the quality of existing soils.

The benefits of adaptation through improved soils can be estimated with a statistical model that describes variation in corn yield in Illinois, Indiana, Iowa, Minnesota, Michigan, and Ohio counties as a function of time, county growing season weather, and distribution of soil capabilities (USDA-NASS, 2011; CRU, 2010; Radeloff and others, 2012). The model uses annual 1950 to 2008 data as well as data on percent of county land used for corn, soybeans, wheat, and all other land use types. Counties are grouped according to their soil quality profile; counties with the most capable soil profiles are in the soil class 5 group, counties with slightly less productive soils are in the soil class 4 group, and so on. Soil class 1 includes the counties with the least capable soils (**Figure 4.5**). **Table 4.3** presents the expected average annual yield on a typical acre in each soil class using 2000 to 2008 data on average crop type distribution and growing season weather. The estimated yields from **Table 4.3** are plotted in **Figure 4.6**.

Table 4.3. Predicted annual corn yield from 2000 to 2008[3]

Soil Class	Estimated yield (bu / acre)	Avg. annual growing degree days (GDD)	Avg. annual growing season precipitation (mm)	Average annual share of class corn production across 6 States
5	156	2,301	521	34 percent
4	147	2,292	503	26 percent
3	141	2,391	512	22 percent
2	134	2,427	517	15 percent
1	121	2,178	499	3 percent

Table 4.3 Notes: Temperature only adds to GDD if it is 5 degrees Celsius or greater for corn growing seasons defined in Sacks and others (2010). Only precipitation that occurs during the growing season is counted. Counties with significant missing data on soil capabilities are dropped from the dataset used to estimate the model.

As expected, after controlling for growing season weather from 2000 to 2008 and distribution of land uses across counties, a typical acre in soil class $c + 1$ is predicted to generate higher yield than a typical acre in class c. The results in **Table 4.3** indicate how much U.S. corn production could increase under current weather conditions if corn production was shifted from lesser soils to better soils (and the associated change in management practices associated with farming on better soils). There is significant capacity to do this right now without negatively affecting the production of other crops. The number of acres available for cropland use in the

---

[3] Data and statistical model code can be found at (http://www.bowdoin.edu/faculty/e/enelson/index.shtml); data used for predictions are from 2000-2008

Box 4.4, continued.

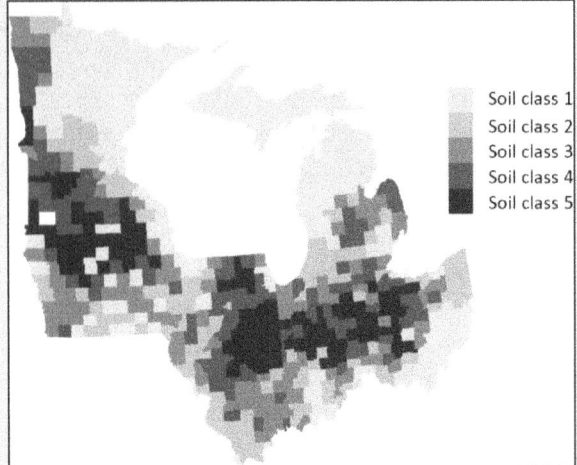

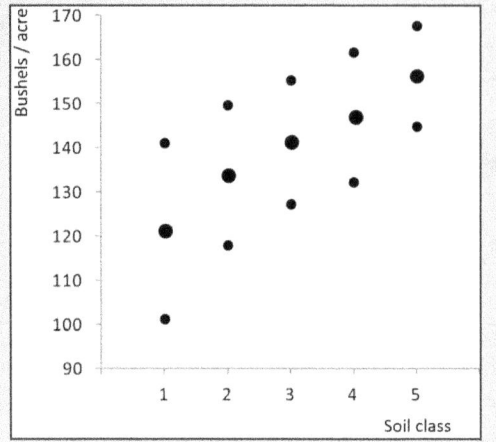

Figure 4.5. A map of soil classes.

Figure 4.6. Estimated average corn yield from 2000-2008 by soil class. The smaller dots indicate estimated yield plus and minus 1 standard deviation.

most capable soils (the type of soils found in the typical acre in soil class 5) in each soil class as of 2001 is given in **Table 4.4**.

The use of these better soils would come at an ecosystem service cost, however, as much of this soil is under forest and other natural land covers (for example, restored prairie) and conversion to cropland would result in a reduction in stored carbon, habitat for some species, water regulation capacity, and recreational lands.

Another management strategy for increasing current corn production with little to no ecosystem service loss would be to increase the soil capability on a typical acre in soil class $c - 1$ such that it mimicked the soil capability of a typical acre in soil class c (and adopted the higher class' typical management practices as well). **Table 4.5** reports expected contemporaneous yield gains given recent weather trends (2000 through 2008) for an acre in soil class $c - 1$ that mimics the soil capacity of an acre in class c.

Table 4.4. Acres available for cropping on the best soils as of 2001

Soil Class	Acres of undeveloped acres in the most capable soils as of 2001	Average number of acres used for corn harvest from 2000 to 2008
5	2,088,003	12,843,674
4	3,362,076	10,089,931
3	4,240,432	8,295,655
2	5,414,875	5,575,536
1	19,183,846	1,314,666

Table 4.4 Notes: Data in the "Acres of undeveloped acres in the most capable soils as of 2001" column comes from Radeloff and others (2012). Undeveloped acres available in the most capable soils for cropping include protected cropland and protected and unprotected pasture, forest, and range in the land capability classes 1 and 2 (USDA-NRCS, 2012).

Box 4.4, continued.

Table 4.5. Contemporaneous yield impact of marginal soil improvement

Soil Class Improvement	Increase in expected yield due to soil improvement (bu /acre)	Average number of acres harvested for corn from 2000 to 2008 in the original soil class (acres)	Gain in corn production all else equal (bu)
4 → 5	4.2	10,089,931	42,377,710
3 → 4	-1.1	8,295,655	-9,125,221
2 → 3	2.0	5,575,536	11,151,072
1 → 2	15.1	1,314,666	19,851,457
Total			64,255,018

Table 4.5 Notes: These results use the observed weather from class $c - 1$. For example, the predicted increase in expected yield due to improving the corn soil typically found in counties in class 1 to corn soil typically found in counties in class 2 uses the observed weather from soil class 1.

By multiplying the typical number of corn acres in a class "Average number of acres harvested for corn from 2000 to 2008 in the original soil class (acres)" by the expected gain in yield due to soil improvement, the productive value of a uniform one-soil-class improvement across the 6 States is determined "Gain in corn production all else equal (bushels)". Using this number as a baseline, this uniform improvement in soil capabilities across all classes would increase bushel production across the six State area by 1 percent, all else being equal.

Climate change

Measured climate change, especially change in GDD, over corn acres in the six States was relatively minor from 1950 to 2008. **Table 4.6** reports the percentage change in average annual GDD and growing season precipitation by soil class between the periods of 1950–1958 and 2000–2008.

Most climate models predict much more rapid climate change over these six States in the next 50 years. **Table 4.7** presents predicted average corn yield in the period 2050–2058 by soil class assuming that average annual GDD and growing season precipitation increase 10 percent between the periods of 2000–2008 and 2050–2058 across the entire study area.

Table 4.6. Change in average annual corn GDD and growing season precipitation between the periods of 1950–1958 and 2000 –2008

Soil Class	Change in average annual GDD	Change in average annual growing season precipitation
5	0.9 percent	13.2 percent
4	0.9 percent	13.2 percent
3	0.2 percent	14.0 percent
2	1.0 percent	11.5 percent
1	4.5 percent	10.7 percent

Even with accelerated climate change, average corn yields are predicted to be much higher in 50 years than they are today over all soil classes (see **Table 4.3** for comparison).

Box 4.4, continued.
Much of the expected gain in yield corn as reported in **Table 4.7** is due to the extrapolation of past technological rates of change into the future. In **Table 4.8**, we predict average yields between 2050–2058 with uniform 10 percent climate change, but now assume that technological improvements in corn farming occur at half the rate that they did in the past.

Table 4.7. Predicted average corn yield in the period 2050–2058 assuming that average annual GDD and growing season precipitation increase 10 percent between the periods of 2000-2008 and 2050 –2058 across the entire study area.

Predicted 2050 – 2058				
Soil Class	Average annual GDD	Average annual growing season precipitation	Predicted yield (bu /acre)	Percentage increase in yield between 2000-2008 and 2050 – 2058
5	2,531	573	235	50.6 percent
4	2,521	553	222	51.0 percent
3	2,630	564	209	48.2 percent
2	2,670	569	204	52.2 percent
1	2,395	549	184	52.1 percent

Table 4.8. Predicted average corn yield in the period 2050–2058 assuming that average annual GDD and growing season precipitation increase 10 percent between the periods of 2000-2008 and 2050–2058 across the entire study area but technological improvements in corn farming occur at half the rate that they did in the past.

Predicted 2050–2058				
Soil Class	Average annual GDD	Average annual growing season precipitation	Predicted yield (bu. /acre)	Percentage increase in yield between 2000-2008 and 2050–2058
5	2,531	573	191	22.4 percent
4	2,521	553	181	23.1 percent
3	2,630	564	171	21.3 percent
2	2,670	569	164	22.4 percent
1	2,395	549	151	24.8 percent

A more pessimistic scenario would include more rapid climate change. **Table 4.9** shows the results from such a scenario—specifically an across the board GDD and growing season precipitation increase of 20 percent from 2000–2008 to 2050–2058 and technological progress slowing to half its historic rate.

Under this last scenario of accelerated climate change and slowing technological progress, there is great opportunity for adaptation by improving the most marginal corn soils (**Table 4.10**). Specifically, an extra 23 bushels could be obtained per acre by improving the soil quality of the most marginal corn land (and adopting the management practices typical on slightly better soils).

Box 4.4, continued.

These analyses of soil supporting services in conjunction with climate change show that better selection of high quality soils, and improving lower quality soils will likely provide a strong capacity for adaptation. Examples of management changes to improve soil quality include establishing major drainage facilities, building levees or flood-retarding structures, providing water for irrigation, removing stones, or large-scale grading of gullied land (USDA 2012). Previous analyses of ecosystem services have focused on the direct impacts of climate change on provisioning and regulating services. One hypothesis suggested by analyses of soil supporting services is that better management of supporting services in general could provide substantial adaptive capacity for the negative impacts of climate change on other services.

Table 4.9: Predicted average corn yield in the period 2050–2058 assuming that average annual GDD and growing season precipitation increase 20 percent between the periods of 2000-2008 and 2050–2058 across the entire study area but technological improvements in corn farming occur at half the rate that they did in the past.

	Predicted 2050–2058			
Soil Class	Average annual GDD	Average annual growing season precipitation	Predicted yield (bu. /acre)	Percentage increase in yield between 2000-2008 and 2050–2058
5	2,761	625	172	10.3 percent
4	2,750	603	166	12.9 percent
3	2,869	615	156	10.6 percent
2	2,912	621	151	12.7 percent
1	2,613	599	143	18.2 percent

Table 4.10. Potential improvements by improving marginal corn soils.

Soil Class Improvement	Marginal gain in expected yield due to investment in soil (bu. /acre)
4 → 5	2.1
3 → 4	-4.5
2 → 3	-3.0
1 → 2	23.3

Other agriculture management approaches could help address climate impacts on nitrogen retention. The main driver of nitrogen pollution in U.S. waterways is anthropogenic input (Howarth and others, 2012). Reducing fertilizer application rates could reduce pollution directly. Many current practices, such as tile drains and leaving fields fallow without cover crops, circumvent the ability of natural capital to retain nitrogen before it reaches riverways (Raymond and others, 2012). Reducing the use of tile drains and increasing the use of cover crops could increase nitrogen retention on the landscape.

For timber production, private forest managers have the financial incentive and the flexibility to protect against extensive loss from climate-related impacts. They can use several existing management techniques: short rotations to reduce the length of time that a tree is influenced by unfavorable climate conditions; planting improved varieties developed through selection, breeding, or genetic engineering to reduce vulnerability; and thinning, weeding, managing pests, irrigating, improving drainage, and fertilizing to improve general vigor. Such actions are likely to reduce the probability of moisture stress and secondary risks from fire, insects, and disease.

Strategies to secure food and secondary feed supplies from fisheries can use existing management approaches. Stock assessments that form the basis of regulated catch limits increasingly incorporate modeled climate-driven shifts in fish spatial distributions (Barange and others, 2011; Ianelli and others, 2011); and protection and restoration of habitats for nursery and other life stages can bolster stock resilience to environmental change (Hughes, 2007; Perry and others, 2010; McGilliard and others, 2011). However, the more rapid the rate of climate change, the more it may strain the ability of ecosystems to support the supply of crops, timber, or fish (Oswalt and others, 2009; Lobell and others, 2011; Perry and others, 2010). A faster rate of warming also may limit species constrained by slow dispersal rates and/or habitat fragmentation, or those that are already stressed by other factors, such as pollution.

Developing alternative livelihood options as part of climate adaptation strategies for food and timber producing sectors can help avoid surprises under future climate (Marschke and Berkes, 2006; Coulthard and others, 2011). These strategies can help identify conditions under which fishing- or timber-based communities should be encouraged to undergo livelihood diversification, shift the location of their fishing and timber harvest, or change livelihoods.

Assessments show that where ecological resilience is high (for example, habitat heterogeneity and connectivity among habitats is maintained), marine and terrestrial systems will be better equipped to respond to climate-related changes in storms, freshwater runoff, harvest pressures, and other potential stressors (Adger and others, 2005; Gaines and others, 2010; Howes and others, 2010). There is promise in using restoration of key habitats to provide a broad suite of benefits ameliorating climate impacts with relatively little ongoing maintenance costs. For example, if an oyster reef or mangrove restoration strategy included consideration of not only sea level rise, but also fish habitat benefits for commercial and recreational uses and coastal protection services, the benefits to surrounding communities could multiply quickly (Aburto-Oropeza and others, 2008; Das and Vincent, 2009). Although restoration strategies are less certain—and often more expensive—than protection of intact ecosystems, in many parts of the world protection alone will be insufficient to ensure the provision of benefits. More work is needed to move beyond general principles and understand the cost effectiveness of alternative 'gray' versus 'green' approaches to climate adaptation and to identify conditions under which ecosystem versus technological approaches are most likely to sustain benefits.

Payments for ecosystem services are occurring through standard approaches such as wetland banking, land acquisitions for conservation (Madsen, 2011), and payments for watershed services, which totaled $1.35 billion in the U.S. in 2008, primarily through the Farm Bill (Stanton, 2010). The only ecosystem market explicitly developed to address climate concerns is for carbon. Forest carbon sequestration projects already exist and payment plans for landowners who decide not to cut their trees are beginning to come on-line (Canadell and Raupach, 2008; Arriagada and Perrings, 2011). In 2010, global prices paid for qualified sequestered forest carbon ranged from $4.30 to $47.50 per ton (Diaz and others, 2011).

Further, innovative approaches to adjusting user-fees to account for maintenance and protection costs of valuable, natural habitats are growing in popularity. For example, destructive fishing in coral reefs has high initial economic value, but the combined sustainable fishing, tourism and coastal protection benefits of more protected reefs have higher value for climate adaptation over time (WB, 2010).

Ecosystem services do not vary independently of one another, and as a result, one general strategy for responding to harmful reductions in one ecosystem service is to boost another ecosystem service, or to reduce interacting stressors. One hypothesis suggested by analyses of soil-supporting services is that better management of supporting services in general could provide substantial adaptive capacity for the negative impacts of climate change on other services. A second general principle is that policies and incentives aimed at getting people to behave differently, or change the location and type of livelihoods they engage in, may be necessary. For example, paying farmers to increase soil carbon and retain nitrogen could compensate for the negative impacts of climate change on water quality and on carbon sequestration.

4.5. CRITICAL GAPS IN KNOWLEDGE, RESEARCH, AND DATA NEEDS FOR CLIMATE IMPACTS ON ECOSYSTEM SERVICES

Among the numerous gaps in our scientific understanding of how ecosystem services will respond to climate change, a few stand out as critical to answer in the next 5-10 years if society is to be able to reduce the human and economic costs of the climate disruption we are already observing:

- What are the likely effects of climate change on rates of carbon storage and sequestration in soils and vegetation? Are there farming practices that can be implemented to substantially enhance soil carbon in a predictable manner?
- What are likely effects of climate change on water quality regulation in freshwater streams and rivers?
- How can fishery management best respond to climate impacts in a way that maintains harvest and jobs without putting the resource base at risk?
- "Green" energy use in the U.S. is increasing in part as a response to climate change. What impact will an increasing reliance on "green" energy have on ecosystem services? For example, how do windmills, solar panel arrays, and land area and water used to create biofuel feedstocks affect service delivery and value?
- What specific incentives, regulations, management strategies, or investments can be implemented to allow fishing, farming, timber, agricultural and aquaculture communities to adapt to changing and more variable climate conditions?
- What is the relative cost-effectiveness of engineered versus ecosystem-based approaches to reducing vulnerability of communities to coastal hazards?
- What is the current distribution and abundance of coastal habitats that provide protection from coastal hazards? Where could restoration of these habitats deliver the greatest value to coastal communities?
- How can vulnerable communities get specific information about projected climate change impacts at local and regional scales that would be useful in planning for hazards and promoting resilience?

4.6. LITERATURE CITED

Abdul-Aziz O, Mantua N, Myers K, and Bradford M. 2011. Potential climate change impacts on thermal habitats of Pacific salmon (*Oncorhynchus spp.*) in the North Pacific Ocean and adjacent seas. *Canadian Journal of Fisheries and Aquatic Sciences* **68**: 1660-1680.

Aburto-Oropeza O, Ezcurra E, Danemann G, Valdez V, Murray J, and Sala E. 2008. Mangroves in the Gulf of California increase fishery yields. *Proceedings of the National Academy of Sciences* **105**(30): 10456-10459.

Adam J, Hamlet A, and Lettenmaier D. 2009. Implications of global climate change for snowmelt hydrology in the twenty-first century. *Hydrological Processes* **23**: 962-972.

Adger W, Hughes T, Folke C, Carpenter S, and Rockstrom J. 2005. Social-ecological resilience to coastal disasters. *Science* **309**: 1036-1039.

Aldy JE, Krupnick AJ, Newell RG, Parry IWH, and Pizer WA. 2010. Designing Climate Mitigation Policy. *Journal of Economic Literature* **48**(4): 903-934.

Allan J, and Komar P. 2006. Climate controls on U.S. West Coast erosion processes. *Journal of Coastal Research* **22**: 511-529.

Allen C, Macalady A, Chenchouni H, Bachelet D, McDowell N, Vennetier M, Kitzberger T, Rigling A, Breshears D, Hogg E, Gonzalez P, Fensham R, Zhang Z, Castro J, Demidova N, Lim J, Allard G, Running S, Semerci A, and Cobb N. 2010. A global overview of drought and heat-induced tree mortality reveals emerging climate change risks for forests. *Forest Ecology and Management* **259**: 660-684.

Alongi DM. 2008. Mangrove forests: Resilience, protection from tsunamis, and responses to global climate change. *Estuarine, Coastal and Shelf Science* **76** (1): 1-13.

Arriagada R, and Perrings C. 2011. Paying for international environmental public goods. *Ambio* **40** (7): 798-806.

Arrow K, Dasgupta P, Goulder L, Daily G, Ehrlich P, Heal G, Levin S, Mäler KG, Schneider S, Starrett D, and Walker B. 2004. Are we consuming too much? *Journal of Economic Perspectives* **18**(2): 147-172.

Barange M, Allen I, Allison E, Badjeck M-C, Blanchard J, Drakeford B, Dulvy NK, Harle J, Holmes R, Holt J, Jennings S, Lowe J, Merino G, Mullon C, Pilling G, Tompkins E, and Werner F. 2011. Predicting the impacts and socio-economic consequences of climate change on global marine ecosystems and fisheries: the QUEST_Fish framework. *In* Ommer R, Cochrane KL, Cury P (Eds), World Fisheries: A Social-Ecological Analysis. Wiley-Blackwell.

Barbier EB, Hacker SD, Kennedy CJ, Koch EW, Stier AC, and Silliman BR. 2011. The Value of Estuarine and Coastal Ecosystem Services. *Ecological Monographs* **81**(2): 169-193.

Bark R, Colby H, Dominguez B, Dominguez G, and Dominguez F. 2010. Snow days? Snowmaking adaptation and the future of low latitude, high elevation skiing in Arizona, USA. *Climatic Change* **102**: 467-491.

Beaugrand G, and Reid PC. 2003. Long-term changes in phytoplankton, zooplankton and salmon related to climate. *Global Change Biology* **9**: 801-817.

Belkin IM. 2009. Rapid warming of Large Marine Ecosystems. *Progress in Oceanography* **81**: 207-213.

Bell FW. 1997. The economic valuation of saltwater marsh supporting marine recreational fishing in the southeastern United States. *Ecological Economics* **21**: 243–254.

Berman M, and Kofinas G. 2004. Hunting for models: grounded and rational choice approaches to analyzing climate effects on subsistence hunting in an Arctic community. *Ecological Economics* **49**: 31-46.

Bin O, Dumas C, Poulter B, and Whitehead J. 2007. Measuring the impacts of climate change on North Carolina coastal resources. Report prepared for National Commission on Energy Policy. Available at: http://libres.uncg.edu/ir/uncw/listing.aspx?id=743

BIA. 2011. Bureau of Indian Affairs. [cited 2/21/2012. Available from http://www.frontrangeroundtable.org/External_Documents.html]

Boruff BJ, Emrich C, and Cutter SL. 2005. Erosion hazard vulnerability of U.S. coastal counties. *Journal of Coastal Research* **215**: 932-942.

Britsch LD, Dunbar JB. 1993. Land Loss Rates: Louisiana Coastal Plain. *Journal of Coastal Research* **9**(2): 324-338.

Bukovsky MS, and Karoly DJ. 2011. A regional modeling study of climate change impacts on warm-season precipitation in the Central United States. *Journal of Climate* **24**: 1985-2002.

Bureau of Economic Analysis (BEA). 2010. The Bureau of Economic Analysis (BEA) Releases estimates of the major components of gross domestic product for American Samoa. U.S. Department of Commerce, Bureau of Economic Analysis. Available at: http://www.bea.gov/newsreleases/general/terr/2010/asgdp_051010.htm

Burrows MT, Schoeman DS, Buckley LB, Moore P, Poloczanska ES, Brander KM, Brown C, Bruno JF, Duarte CM, Halpern BS, Holding J, Kappel CV, Kiessling W, O'Connor MI, Pandolfi JM, Parmesan C, Schwing FB, Sydeman WJ, and Richardson AJ. 2011. The pace of shifting climate in marine and terrestrial ecosystems. *Science* **334**: 652-655.

Butsic V, Hanak E, and Valletta RG. 2011. Climate change and housing prices: hedonic estimates for ski resorts in western North America. *Land Economics* **87**: 75-91.

Canadell JG, and Raupach MR. 2008. Managing forests for climate change mitigation. *Science* **320**: 1456-1457.

Caputi N, Melville-Smith R, de Lestang S, Pearce A, and Feng M. 2010. The effect of climate change on the western rock lobster (*Panulirus cygnus*) fishery of Western Australia. *Canadian Journal of Fisheries and Aquatic Sciences* **67**: 85-96.

Carpenter KE, Abrar M, Aeby G, Aronson RB, Banks S, Bruckner A, Chiriboga A, Cortes J, Delbeek JC, DeVantier L, Edgar GJ, Edwards AJ, Fenner D, Guzman HM, Hoeksema BW, Hodgson G, Johan O, Licuanan WY, Livingstone SR, Lovell ER, Moore JA, Obura DO, Ochavillo D, Polidoro BA, Precht WF, Quibilan MC, Reboton C, Richards ZT, Rogers AD, Sanciangco J, Sheppard A, Sheppard C, Smith J, Stuart S, Turak E, Veron JEN, Wallace C, Weil E, and Wood E. 2008. One-third of reef-building corals face elevated extinction risk from climate change and local impacts. *Science* **321**: 560-563.

Cayan DR, Das T, Pierce DW, Barnett TP, Tyree M, and Gershunov A. 2010. Future dryness in the southwest U.S. and the hydrology of the early 21st century drought. *Proceedings of the National Academy of Sciences* **107**: 21271-21276.

Cesar HSJ, and van Beukering P. 2004. Economic valuation of the coral reefs of Hawai'i. *Pacific Science* **58**: 231-242.

Cheung WWL, Lam VWY, Sarmiento JL, Kearney K, Watson R, and Pauly D. 2009. Projecting global marine biodiversity impacts under climate change scenarios. *Fish and Fisheries* **10**: 235-251.

Cheung WWL, Lam VWY, Sarmiento JL, Kearney K, Watson R, Zeller D, and Pauly D. 2010. Large-scale redistribution of maximum fisheries catch potential in the global ocean under climate change. *Global Change Biology* **16**: 24-35.

Cheung WWL, Dunne J, Sarmiento J, and Pauly D. 2011. Integrating ecophysiology and plankton dynamics into projected maximum fisheries catch potential under climate change in the Northeast Atlantic. *ICES Journal of Marine Science* **68**: 1-11.

Christian-Smith J, Cooley H, and Gleick PH. 2012. Potential water savings associated with agricultural water efficiency improvements: a case study of California, USA. *Water Policy* 14(2): 194–213.

Collie JS, Wood AD, and Jeffries HP. 2008. Long-term shifts in the species composition of a coastal fish community. *Canadian Journal of Fisheries and Aquatic Sciences* **65**: 1352-1365.

Cooley SR, and Doney SC. 2009. Anticipating ocean acidification's economic consequences for commercial fisheries. *Environmental Research Letters* 4: 024007.

Compton JE, Harrison JA, Dennis RL, Greaver TL, Hill BH, Jordan SJ, Walker H, and Campbell HV. 2011. Ecosystem services altered by human changes in the nitrogen cycle: a new perspective for US decision making. *Ecology Letters* **14**(8): 804-815.

Conte M, Nelson E, Carney K, Fissore C, Olwero N, Plantinga AJ, Stanley B, and Ricketts T. 2011. Terrestrial carbon sequestration and storage. *In* Karvera P, Tallis H, Ricketts TH, Daily GC and Polasky S (Eds), Natural Capital. Theory and Practice of Mapping Ecosystem Services. Oxford University Press, Oxford. 111–128 p.

Cooper M, Beevers M, and Oppenheimer M. 2008. The potential impacts of sea level rise on the coastal region of New Jersey, USA. *Climatic Change* **90**: 475-492.

Costanza R, Pérez-Maqueo O, Martinez ML, Sutton P, Anderson SJ, and Mulder K. 2008. The value of coastal wetlands for hurricane protection. *Ambio* **37**: 241-248.

Coulthard S. 2009. Adaptation and conflict within fisheries: insights for living with climate change. *In* Adger WN, Lorenzoni I, and O'Brien KL (Eds), Adapting to Climate Change: Thresholds, Values, Governance. Cambridge University Press. Cambridge, UK. 255-268 p.

Coulthard S, Johnson D, and McGregor JA. 2011. Poverty, sustainability and human wellbeing: A social wellbeing approach to the global fisheries crisis. *Global Environmental Change* **21**: 453-463.

Craft C, Clough J, Ehman J, Joyce S, Park R, Pennings S, Guo H, and Machmuller M. 2009. Forecasting the effects of accelerated sea-level rise on tidal marsh ecosystem services. *Frontiers in Ecology and the Environment* **7**: 73-78.

Crossett KM, Culliton TJ, Wiley PC, and Goodspeed TR. 2004. Population trends along the coastal United States. 1980-2008. Coastal Trends Report Series. NOAA, National Ocean Service, Management and Budget Office, Special Projects. 47 p.

CRU (University of East Anglia Climatic Research Unit) [Jones P, and Harris I]. 2010. CRU Time Series (TS) high resolution gridded datasets. NCAS British Atmospheric Data Centre. Available at: http://badc.nerc.ac.uk/view/badc.nerc.ac.uk__ATOM__dataent_1256223773328276

Curriero FC, Patz JA, Rose JB, and Lele S. 2001. The association between extreme precipitation and waterborne disease outbreaks in the United States, 1948-1994. *American Journal of Public Health* **91**: 1194-1199.

Daily GC, Söderqvist T, Aniyar S, Arrow K, Dasgupta P, Ehrlich PR, Folke C, Jansson AM, Jansson BO, Kautsky N, Levin S, Lubchenco J, Mäler KG, Simpson D, Starrett D, Tilman D, and Walker B. 2000. The Value of Nature and the Nature of Value. *Science* 289(5478): 395-396.

Daily GC, Polasky S, Goldstein J, Kareiva PM, Mooney HA, Pejchar L, Ricketts TH, Salzman J, and Shallenberger R. 2009. Ecosystem services in decision making: time to deliver. *Frontiers in Ecology and the Environment* 7: 21–28.

Dalton MG. 2001. El Niño, expectations, and fishing effort in Monterey Bay, California. *Journal of Environmental Economics and Management* 42: 336-359.

Danielsen F, Sorensen MK, Olwig MF, Selvam V, Parish F, Burgess ND, Hiraishi T, Karunagaran VM, Rasmussen MS, Hansen LB, Quarto A, and Suryadiputra N. 2005. The Asian tsunami: A protective role for coastal vegetation. *Science* 310: 643.

Das S, and Vincent JR. 2009. Mangroves protected villages and reduced death toll during Indian super cyclone. *Proceedings of the National Academy of Sciences* 106: 7357-7360.

Dawson J, and Scott D. 2007. Climate change vulnerability of the Vermont ski tourism industry. *Annals of Leisure Research* 10: 550-572.

Day JW, Boesch DF, Clairain EJ, Kemp PG, Laska SB, Mitsch WJ, Orth K, Mashriqui H, Reed DJ, Shabman L, Simenstad CA, Streever BJ, Twilley RR, Watson CC, Wells JT, and Whigham DF. 2007. Restoration of the Mississippi delta: lessons from Hurricanes Katrina and Rita. *Science* 315: 1679-1684.

De Silva SS, and Soto D. 2009. Climate change and aquaculture: potential impacts, adaptation and mitigation. *In* Cochrane K, De Young C, SotoD, and Bahri T (Eds), Climate change implications for fisheries and aquaculture: overview of current scientific knowledge. Food and Agriculture Organization of the United Nations, Rome. 151-212 p.

Diaz D, Hamilton K, and Johnson E. 2011. State of the forest carbon markes 2011: From canopy to currency. Ecosystem Marketplace. 93 p. Available at: www.ecosystemmarketplace.com

Dlugofl V, Fiener P, and Schneider K. 2010. Layer-specific analysis and spatial prediction of soil organic carbon using terrain attributes and erosion modeling. *Soil Science Society of America Journal* 74: 922-935.

Doney SC, Ruckelshaus M, Duffy JE, Barry JP, Chan F, English CA, Galindo HM, Grebmeier JM, Hollowed AB, Knowlton N, Polovina J, Rabalais NN, Sydeman WJ, and Talley LD. 2012. Climate change impacts on marine ecosystems. *Annual Reviews in Marine Science* 4: 4.1–4.27.

Downs LL, Nicholls RJ, Leatherman SP, and Hautzenroder J. 1994. Historic Evolution of a Marsh Island: Bloodsworth Island, Maryland. *Journal of Coastal Research* 10(4): 1031-1044.

Doyle TW, Girod GF, and Books MA. 2003. Modeling Mangrove Forest Migration Along the Southwest Coast of Florida Under Climate Change. *In* Ning ZH, Turner RE, Doyle TW, and Abdollahi K (Eds), Integrated Assessment of the Climate Change Impacts on the Gulf Coast Region. Louisiana State University Graphic Services, Baton Rouge, LA, 211–221 p.

Doyle TW, Krauss KW, Conner WH, and From AS. 2010. Predicting the retreat and migration of tidal forests along the northern Gulf of Mexico under sea-level rise. *Forest Ecology and Management* 259: 770–777.

Duarte CM, Middelburg JJ, and Caraco N. 2005. Major role of marine vegetation on the oceanic carbon cycle. *Biogeosciences* **2**: 1-8.

Dulvy NK, Rogers SI, Jennings S, Stelzenmuller V, Dye SR, and Skjoldal HR. 2008. Climate change and deepening of the North Sea fish assemblage: a biotic indicator of warming seas. *Journal of Applied Ecology* **45**: 1029-1039.

Ebi KL, Mills DM, Smith JB, and Grambsch A. 2006. Climate chagne and human health impacts in the United States: an update on the results of theU.S. National Assessment. *Environmental Health Perspectives* **114**: 1318-1324.

Fagre DB, Charles CW. 2009. Thresholds of climate change in ecosystems. U.S. Geological Survey, Department of the Interior, Washington, DC.

Feely RA, Sabine CL, Hernandez-Ayon JM, Ianson D, and Hales B. 2008. Evidence for upwelling of corrosive "acidified" water onto the continental shelf. *Science* **320**: 1490-1492.

Fitzgerald DM, Fenster MS, Argow BA, and Buynevich IV. 2008. Coastal impacts due to sea-level rise. *Annual Review of Earth and Planetary Sciences* **36**: 601–647.

Foley JA, Ramankutty N, Brauman KA, Cassidy ES, Gerber JS, Johnston M, Mueller ND, O'Connell C, Ray DK, West PC, Balzer C, Bennett EM, Carpenter SR, Hill J, Monfreda C, Polasky S, Rockstrom J, Sheehan J, Siebert S, Tilman, D, and Zaks DPM. 2011. Solutions for a cultivated planet. *Nature* **478**: 337–342.

Frey AE, Olivera F, Irish JL, Dunkin LM, Kaihatu JM, Ferreira CM, and Edge BL. 2010. Potential impact of climate change on hurricane flooding inundation, population affected and property damages in Corpus Christi. *JAWRA Journal of the American Water Resources Association* **46**(5): 1049-1059.

First American. 2010. First American Storm Surge Report 2010. Residential storm surge exposure estimates for 13 U.S. Cities. First American Corporation. Available at: http://www.stormadjuster.com/Home/News/tabid/56/articleType/ArticleView/articleId/686/2010-First-American-Storm-Surge-Report-on-Residential-Exposure.aspx

Fogarty MJ, Incze L, Hayhoe K, Mountain D, and Manning J. 2007. Potential climate change impacts on Atlantic cod (*Gadus morhua*) off the northeastern USA. *Mitigation and Adaptation Strategies for Global Change* **13**: 453-466.

Ford JD, Smit B, and Wandel J. 2006. Vulnerability to climate change in the Arctic: A case study from Arctic Bay, Canada. *Global Environmental Change* **16**: 145-160.

Ford JD, Smit B, Wandel J, Allurut M, Shappa K, Ittusarjuat H, and Qrunnut K. 2008. Climate change in the Arctic: current and future vulnerability in two Inuit communities in Canada. *The Geographical Journal* **174**: 45-62.

Freeman A. 1991. Valuing environmental resources under alternative management regimes. *Ecological Economics* **3**: 247-256.

Gagliano SM, Meyer-Arendt KJ, and Wicker KM. 1981. Land Loss in the Mississippi River Deltaic Plain. *Gulf Coast Association of Geological Societies Transactions* **31**: 295-300.

Gaines SD, White C, Carr MH, and Palumbi S. 2010. Designing marine reserve networks for both conservation and fisheries management. *Proceedings of the National Academy of Sciences* **107**(43): 18286-18293.

Gearheard S, Matumeak W, Angutikjuaq I, Maslanik J, Huntington HP, Leavitt J, Kagak DM, Tigullaraq G, and Barry RG. 2006. "It's not that simple": A collaborative comparison of sea ice environments, their uses, observed changes, and adaptations in Barrow, Alaska, USA, and Clyde River, Nunavut, Canada. *Ambio* **35**: 203-211.

Gedan KB, Kirwan M, Barbier E, Wolinksi E, and Silliman BR. 2011. The present and future role of coastal wetland vegetation in protecting shorelines: answering recent challenges to the paradigm. *Climatic Change* **106**: 7-29.

Glenk K, and Colombo S. 2011. Designing policies to mitigate the agricultural contribution to climate change: an assessment of soil based carbon sequestration and its ancillary effects. *Climatic Change* **105**: 43-66.

Haim D, Alig RJ, Plantinga AJ, and Sohngen B. 2011. Climate change and future land use in the United States: an economic approach. *Climate Change Economics* **2**(1): 27–51.

Hamlet AF, Mote PW, Clark MP, and Lettenmaier DP. 2005. Effects of temperature and precipitation variability on snowpack trends in the western United States. *Journal of Climate* **18**: 4545-4561.

Hare SR, Mantua NJ, and Francis RC. 1999. Inverse production regimes: Alaska and West Coast Pacific salmon. *Fisheries* 24(1): 6-14.

Hare JA, Alexander MA, Fogarty MJ, Williams EH, and Scott JD. 2010. Forecasting the dynamics of a coastal fishery species using a coupled climate-population model. *Ecological Applications* **20**: 452-464.

Hayhoe K, Cayan D, Field CB, Frumhoff PC, Maurer EP, Miller NL, Moser SC, Schneider SH, Cahill KN, Cleland EE, Dale L, Drapek R, Hanemann RM, Kalkstein LS, Lenihan J, Lunch CK, Neilson RP, Sheridan SC, and Verville JH. 2004. Emissions pathways, climate change, and impacts on California. *Proceedings of the National Academy of Sciences* **101**:12422-12427.

Heberger M, Cooley H, Herrera P, Gleick P, and Moore E. 2009. The impacts of sea-level rise on the California coast. California Climate Change Center. 101 p. Available at: www.energy.ca.gov/pier/

Hinzman L, Bettez D, Bolton N, Chapin D, Dyurgerov W, Fastie R, Griffith F, Hollister S, Hope M, Huntington B, Jensen C, Jia L, Jorgenson B, Kane R, Klein D, Kofinas A, Lynch H, Lloyd P, McGuire A, Nelson M, Oechel G, Osterkamp J, Racine T, Romanovsky D, Stone L, Stow D, Sturm R, Tweedie G, Vourlitis A, Walker H, Walker A, Webber H, Welker A, Winker D, and Yoshikawa K. 2005. Evidence and implications of recent climate change in Northern Alaska and other Arctic regions. *Climatic Change* **72**: 251-298.

Hoegh-Guldberg O, Mumby PJ, Hooten AJ, Steneck RS, Greenfield P, Gomez E, Harvell CD, Sale PF, Edwards AJ, Caldeira K, Knowlton N, Eakin CM, Iglesias-Prieto R, Muthiga N, Bradbury RH, Dubi A, and Hatziolos ME. 2007. Coral reefs under rapid climate change and ocean acidification. *Science* **318**: 1737 -1742.

Hofmann E, Ford S, Powell E, and Klinck J. 2001. Modeling studies of the effect of climate variability on MSX disease in eastern oyster (*Crassostrea virginica*) populations. *Hydrobiologia* **460**: 195-212.

Hoffman C, Parsons R, Morgan P, Mell R. 2010. Numerical simulation of crown fire hazard following bark beetle-caused mortality in lodgepole pine forests. *In* Wade DD, Robinson ML (Eds), Proceedings of 3rd Fire Behavior and Fuels Conference; 25-29 October 2010; Spokane, WA. International Association of Wildland Fire, Birmingham, AL. 1 p.

Howarth R, Swaney D, Billen G, Garnier J, Hong B, Humborg C, Johnes P, Morth CM, and Marino R. 2012. Nitrogen fluxes from the landscape are controlled by net anthropogenic nitrogen inputs and by climate. *Frontiers in Ecology and the Environment* **10**: 37-43.

Howes NC, FitzGerald DM, Hughes ZJ, Georgiou IY, Kulp MA, Miner MD, Smith JM, and Barras JA. 2010. Hurricane-induced failure of low salinity wetlands. *Proceedings of the National Academy of Sciences* **107**: 14014-14019.

Hughes T, David R. Bellwood, Carl Folke, Robert S. Steneck and James Wilson. 2007. New paradigms for supporting the resilience of marine ecosystems. *Trends in Ecology and Evolution* **20**: 380-386.

Ianelli J, Hollowed A, Haynie A, Mueter FJ, and Bond N. 2011. Evaluating management strategies for eastern Bering Sea walleye pollock (*Theragra chalcogramma*) in a changing environment. *ICES Journal of Marine Science* **68**: 1297-1304.

Irish JL, Frey AE, Rosati JD, Olivera F, Dunkin LM, Kaihatu JM, Ferreira CM, and Edge BL. 2010. Potential implications of global warming and barrier island degradation on future hurricane inundation, property damages, and population impacted. *Ocean and Coastal Management* **53**: 645-657.

Jenness JS, Beier P, and Ganey JL. 2004. Associations between Forest Fire and Mexican Spotted Owls. *Forest Science* **50**(6): 765-772.

Jennings S, and Brander K. 2008. Predicting the effects of climate change on marine communities and the consequences for fisheries. *Journal of Marine Systems* **79**: 418-426.

Jones BM, Arp CD, Jorgenson MT, Hinkel KM, Schmutz JA, and Flint PL. 2009. Increase in the rate and uniformity of coastline erosion in Arctic Alaska. *Geophysical Research Letters* **36**: L03503-L03508.

Jönsson AM, Appelberg G, Harding S, and Bärring L. 2009. Spatio-temporal impact of climate change on the activity and voltinism of the spruce bark beetle, *Ips typographus*. *Global Change Biology* **15**(2): 486–499.

Justic D, Rabalais NN, and Turner RE. 1996. Effects of cliamte change on hypoxia in coastal waters: a doubled CO_2 scenario for the northern Gulf of Mexico. *Limnology and Oceanography* **41**: 992-1003.

Kareiva P, Tallis H, Ricketts TH, Daily GC, and Polasky S. 2011. Natural Capital: Theory and Practice of Mapping Ecosystem Services. Oxford University Press. 432 p.

Kathiresan K, and Rajendran N. 2005. Coastal mangrove forests mitigated tsunami. *Estuarine, Coastal and Shelf Science* **65**: 601-606.

Kelly RP, Foley MM, Fisher WS, Feely RA, Halpern BS, Waldbusser GG, and Caldwell MR. Mitigating Local Causes of Ocean Acidification with Existing Laws. *Science* **332**: 1036-1037.

Kinlan BP, and Gaines SD. 2003. Propagule dispersal in marine and terrestrial environments: a community perspective. *Ecology* **84**: 2007-2020.

Kleinosky L, Yarnal B, and Fisher A. 2007. Vulnerability of Hampton Roads, Virginia to Ssorm-surge Flooding and sea-level rise. *Natural Hazards* **40**: 43-70.

Koch EW, Barbier EB, Silliman BR, Reed DJ, Perillo GME, Hacker SD, Granek EF, Primavera JH, Muthiga N, Polasky S, Halpern BS, Kennedy CJ, Kappel CV, and Wolanski E. 2009. Non-linearity in ecosystem services: temporal and spatial variability in coastal protection. *Frontiers in Ecology and the Environment:* **7**(1): 29–37.

Komar PD, and Allan JC. 2008. Increasing hurricane-generated wave heights along the U.S. East Coast and their climate controls. *Journal of Coastal Research* **242**: 479-488.

Komar PD, Allan JC, and Ruggiero P. 2009. Ocean wave climates: trends and variations due to Earth's changing climate. *In* Kim YC (Ed), Handbook of Coastal and Ocean Engineering, World Scientific Publishing Company. 971–975 p.

Kroeker KJ, Kordas RL, Crim RN, and Singh GG. 2010. Meta-analysis reveals negative yet variable effects of ocean acidification on marine organisms. *Ecological Letters* 13: 1419–1434.

Kruse JA, White RG, Epstein HE, Archie B, Berman M, Braund SR, Chapin FS, Charlie J, Daniel CJ, Eamer J, Flanders N, Griffith B, Haley S, Huskey L, Joseph B, Klein DR, Kofinas GP, Martin SM, Murphy SM, Nebesky W, Nicolson C, Russell DE, Tetlichi J, Tussing A, Walker MD, and Young OR. 2004. Modeling sustainability of Arctic communities: An interdisciplinary collaboration of researchers and local knowledge holders. *Ecosystems* 7: 815-828.

Kurihara H. 2008. Effects of CO_2-driven ocean acidification on the early developmental stages of invertebrates. *Marine Ecology Progress Series* 373: 275-284.

Lal R. 2004a. Soil Carbon Sequestration Impacts on Global Climate Change and Food Security. *Science* 304(5677): 1623-1627.

Lal R. 2004b. Soil carbon sequestration to mitigate climate change. *Geoderma* 123(1-2): 1-22.

Lambert E, Hunter C, Pierce GJ, and MacLeod CD. 2010. Sustainable whale-watching tourism and climate change: towards a framework of resilience. *Journal of Sustainable Tourism* 18: 409-427.

Landry CE, and Liu H. 2009. A semi-parametric estimator for revealed and stated preference data—An application to recreational beach visitation. *Journal of Environmental Economics and Management.* 57(2): 205–218.

Latta G, Temesgen H, Adams D, and Barrett T. 2010. Analysis of potential impacts of climate change on forests of the United States Pacific Northwest. *Forest Ecology and Management* 259: 720–729.

Lenoir S, Beaugrand G, and Lecuyer A. 2010. Modelled spatial distribution of marine fish and projected modifications in the North Atlantic Ocean. *Global Change Biology* 17: 115-129.

Liu Y, Stanturf J, and Goodrick S, Trends in global wildfire potential in a changing climate, *Forest Ecology and Management* 259(4): 685-697.

Lobell DB, Schlenker W, Costa-Roberts J. 2011. Climate trends and global crop production since 1980. *Science* 333 (6042): 616-620.

Loomis J, and Crespi J. 1999. Estimated effects of climate change on selected outdoor recreation activities in the United States. *In* The Impact of Climate Change on the United States Economy. Cambridge University Press, Cambridge, UK. 289-314 p.

Lubowski RN, Plantinga AJ, and Stavins RN. 2006. Land-use change and carbon sinks: Econometric estimation of the carbon sequestration supply function. *Journal of Environmental Economics and Management* 51: 135-152.

Lucey SM, and Nye JA. 2010. Shifting species assemblages in the Northeast U.S. continental shelf large marine ecosystem. *Marine Ecology Progress Series* 415: 23-33.

Madsen B, Carroll N, Kandy D, and Bennett G. 2011. Update: State of biodiversity markets. Washington, DC: Forest Trends, 2011., Available at: http://www.ecosystemmarketplace.com/reports/2011_update_sbdm.

Marschke MJ, and Berkes F. 2006. Exploring strategies that build livelihood resilience: A case from Cambodia. *Ecology and Society* 11: 42.

McCarthy PD. 2012. Climate Change Adaptation for People and Nature: A Case Study from the U.S. Southwest. *Advances in Climate Change Research* 3(1): 22-37.

McCay BJ, Weisman W, and Creed C. 2011. Coping with environmental change: systemic responses and the roles of property and community in three fisheries. *In* World Fisheries: A Socio-Ecological Analysis. Wiley-Blackwell, Oxford, UK. 381-400 p.

McGilliard CR, Hilborn R, MacCall A, Punt AE, and Field JC. 2011. Can information from marine protected areas be used to inform control-rule-based management of small-scale, data-poor stocks? *ICES Journal of Marine Science* **68**: 201-211.

McGranahan G, Balk D, and Anderson B. 2007. The rising tide: assessing the risks of climate change and human settlements in low elevation coastal zones. *Environment & Urbanization* **19** (1): 17-37.

Mcleod E, Chmura GL, Bouillon S, Salm R, Björk M, Duarte CM, Lovelock CE, Schlesinger WH, and Silliman BR. 2011. A blueprint for blue carbon: toward an improved understanding of the role of vegetated coastal habitats in sequestering CO_2. *Frontiers in Ecology and the Environment* **9**: 552-560.

Miller AW, Reynolds AC, Sobrino C, and Riedel GF. 2009. Shellfish face uncertain future in high CO_2 world: influence of acidification on oyster larvae calcification and growth in estuaries. *PloS One* **4**: e5661.

Morris JT, Sundareshwar PV, Nietch CT, Kjerfve B, and Cahoon DR. 2002. Responses of coastal wetlands to rising sea level. *Ecology* **83**: 2869-2877.

Mote PW, Hamlet AF, Clark MP, and Lettenmaier DP. 2005. Declining mountain snowpack in western North America. *Bulletin of the American Meteorological Society* **86**: 39-49.

Mueter FJ, and Litzow MA. 2008. Sea ice retreat alters the biogeography of the Bering Sea continental shelf. *Ecological Applications* **18**: 309-320.

Murawski SA. 1993. Climate change and marine fish distributions: forecasting from historical analogy. *Transactions of the American Fisheries Society* **122**: 647-658.

Murray BC, Pendleton L, Jenkins WA, and Sifleet S. 2011. Green Payments for Blue Carbon: Economic Incentives for Protecting Threatened Coastal Habitats. Durham: Nicholas Institute for Environmental Policy Solutions. Available at: http://nicholasinstitute.duke.edu/economics/naturalresources/blue-carbon-report

Nelson E, Mendoza G, Regetz J, Polasky S, Tallis H, Cameron DR, Chan KMA, Daily GC, Goldstein J, Kareiva PM, Lonsdorf E, Naidoo R, Ricketts TH, and Shaw R. 2009. Modeling multiple ecosystem services, biodiversity conservation, commodity production, and tradeoffs at landscape scales. *Frontiers in Ecology and the Environment* **7**: 4–11.

Nicholls RJ, Hoozemans FMJ, and Marchand M. 1999. Increasing flood risk and wetland losses due to global sea-level rise: regional and global analyses. *Global Environmental Change* **9**: S69-S87.

NCA (National Climate Assessment). 2009. Global Climate Change Impacts in the United States. Karl TR, Melillo JM, and Peterson TC (eds). Cambridge University Press. Available at: http://www.globalchange.gov/publications/reports/scientific-assessments/us-impacts/full-report

NEFMC (New England Fishery Management Council). 2011. Framework Adjustment 45 to the Northeast Multispecies Fishery Management Plan. New England Fishery Management Council, 408p. Available at: http://www.nefmc.org/nemulti/frame/fw45/110120_Final_FW_45_Resubmit.pdf

NMFS (National Marine Fisheries Service). 2010. Fisheries economics of the United States, 2009. U.S. Department of Commerce, NOAA.

NOAA (National Oceanic and Atmospheric Administration). 2011. Regional mean sea-level trends. [cited 12/9/11. Available from http://tidesandcurrents.noaa.gov/sltrends/slrmap.html]

NOEP (National Ocean Economics Program). 2005. Data available at: http://www.oceaneconomics.org/

National Research Council. 2005. Valuing ecosystem services: Toward better environmental decision-making. National Academy Press, Washington DC.

Nolin AW, and Daly C. 2006. Mapping "at risk" snow in the Pacific Northwest. *Journal of Hydrometeorology* 7: 1164-1171.

Nye JA, Link JS, Hare JA, and Overholtz WJ. 2009. Changing spatial distribution of fish stocks in relation to climate and population size on the Northeast United States continental shelf. *Marine Ecology-Progress Series* **393**: 111-129.

Oswalt SN, Johnson TG, Howell M, and Bentley JW. 2009. Fluctuations in national forest timber harvest and removals: the southern regional perspective. Resource Bulletin - Southern Research Station, USDA Forest Service. Document No. SRS-148, 30 p.

Painter TH, Deems JS, Belnap J, Hamlet AF, Landry CC, and Udall B. 2010. Response of Colorado River runoff to dust radiative forcing in snow. *Proceedings of the National Academy of Sciences* **107**: 17125-17130.

PCAST (President's Council of Advisors on Science and Technology). 2011. Sustaining environmental capital: protecting society and the economy. 145 p. Available at: http://www.whitehouse.gov/administration/eop/ostp/pcast/docsreports

Pendleton LH, and Mendelsohn R. 1998. Estimating the economic impact of climate change on the freshwater sportsfisheries of the Northeastern U.S. *Land Economics* **74**: 483-496.

Pendleton L, King P, Mohn C, Webster DG, Vaughn R, and Adams PN. 2011. Estimating the potential economic impacts of climate change on Southern California beaches. *Climatic Change* **90**: 475-492.

Perry AL, Low PJ, Ellis JR, and Reynolds JD. 2005. Climate change and distribution shifts in marine fishes. *Science* **308**: 1912.

Perry RI, Cury P, Brander K, Jennings S, Möllmann C, and Planque B. 2010. Sensitivity of marine systems to climate and fishing: Concepts, issues and management responses. *Journal of Marine Systems* **79**: 427-435.

Pinnegar JK, Cheung WWL, and Heath MR. 2010. Marine Climate Change Impacts Science Review 2010-11: Fisheries. *In* Partnership Annual Report Card Science Review 2010–11. Available at: http://www.mccip.org.uk/arc

Post WM, and Kwon KC. 2008. Soil carbon sequestration and land-use change: processes and potential. *Global Change Biology* **6**(3): 317-327.

Powlson DS, Whitmore A, and Goulding KWT. 2011. Soil carbon sequestration to mitigate climate change: a critical re-examination to identify the true and the false. *European Journal of Soil Science* **62**: 42-55.

Radeloff VC, Nelson E, Plantinga AJ, Lewis DJ, Helmers D, Lawler J J, Withey JC, Beaudry F, Martinuzzi S, Butsic V, Lonsdorf E, White D, and Polasky S. 2012. Economic-based projections of future land use under alternative economic policy scenarios in the conterminous U.S. *Ecological Applications* **22**(3): 1036–1049.

Raymond PA, Zappa CJ, Butman D, Bott TL, Potter J, Mulholland P, Laursen AE, McDowell WH, and Newbold D. 2012. Scaling the gas transfer velocity and hydraulic geometry in streams and small rivers. *Limnology and Oceanography Fluids and Environments* **2**: 41-53.

Reyers B, Polasky S, Tallis H, Mooney H, and Larigauderie A. 2012. Finding a Common ground for biodiversity and ecosystem services. *BioScience* 62(5): 503-507.

Richardson RB, and Loomis JB. 2005. Climate change and recreation benefits in an alpine national park. *Journal of Leisure Research* **37**: 307-320.

Robbins J. 2008. Bark beetles kill millions of acres of trees in West. *In* The New York Times. Available at: http://www.nytimes.com/2008/11/18/science/18trees.html?pagewanted=all

Rodrigo A, and Perrings C. 2011. Paying for International Environmental Public Goods. *AMBIO* **40**: 798-806.

Rose JB, Epstein RPR, Lipp EK, Sherman BH, Bernard SM, and Patz JA. 2001. Climate variability and change in the United States: Potential impacts on water and foodborne diseases caused by microbiologic agents. *Environmental Health Perspectives* **109**(2): 211-221..

Roy SB, Chen L, Girvetz EH, Maurer EP, Mills WB, and Grieb TM. 2012. Projecting water withdrawal and supply for future decades in the U.S. under climate change scenarios. *Environmental Science and Technology* **46**(5): 2545–2556.

Rumpel C, and Kogel-Knabner I. 2010. Deep soil organic matter: key but poorly understood component of terrestrial C cycle. *Plant and Soil* **338**: 1-16.

Schindler D, Augerot X, Fleishman E, Mantua N, Riddell B, Ruckelshaus M, Seeb J, and Webster M. 2008. Climate change, ecosystem impacts, and management for Pacific salmon. *Fisheries* **33**: 502-506.

Schoennagel T, Sherriff RL, and Veblen TT. 2011. Fire history and tree recruitment in the Colorado Front Range upper montane zone: implications for forest restoration. *Ecological Applications* **21**(6): 2210-2222.

Scott D, McBoyle G, and Schwartzentruber M. 2004. Climate change and the distribution of climatic resources for tourism in North America. *Climate Research* **27**: 105-117.

Scott D, McBoyle G, Minogue A, and Mills B. 2006. Climate change and the sustainability of ski-based tourism in Eastern North America: a reassessment. *Journal of Sustainable Tourism* **14**: 376-398.

Scott D, Dawson J, and Jones B. 2008. Climate change vulnerability of the U.S. Northeast winter recreation-tourism sector. *Mitigation and Adaptation Strategies for Global Change* **13**: 577-596.

Semmens DJ, Briggs JS, and Martin DA. 2008. An ecosystem services framework for multi-disciplinary research in the Colorado River headwaters. Proceedings of the Third Interagency Conference on Research in the Watersheds, 8-11 September 2008, Estes Park, CO, *USGS Scientific Investigations Report* 2009-5049. http://pubs.usgs.gov/sir/2009/5049/pdf/Semmens.pdf

Shanks AL. 2009. Pelagic larval duration and dispersal distance revisited. *Biological Bulletin* **216**: 373-385.

Shaw WD, and Loomis JB. 2008. Frameworks for analyzing the economic effects of climate change on outdoor recreation. *Climate Research* **36**: 259-269.

Shepard C, Agostini VN, Gilmer B, Allen T, Stone J, Brooks W, and Beck MW. 2011. Assessing future risk: quantifying the effects of sea level rise on storm surge risk for the southern shores of Long Island, New York. *Natural Hazards* **60**: 1-19.

Sifleet S, Pendleton L, and Murray B. 2011. The state of the science on blue carbon: A summary for policy makers. Nicholas Institute for Environmental Policy Solutions, Duke University, Durham, North Carolina. Available at: http://nicholasinstitute.duke.edu/economics/naturalresources/state-of-science-coastal-blue-carbon/

Smith JE, Heath LS, and Nichols MC. 2007. U.S. forest carbon calculation tool: forest-land carbon stocks and net annual stock change. General Technical Report NRS-13. Newtown Square, PA: USDA Forest Service, Northern Research Station. 28 p. Software, User's Guide, and Example Data Sets: Available at: http://www.nrs.fs.fed.us/carbon/tools/

Stanton T, Echavarria M, Hamilton K, and Ott C. 2010. State of Watershed Payments: An Emerging Marketplace. Ecosystem Marketplace., Available online: http://www.forest-trends.org/documents/files/doc_2438.pdf.

Starbuck CM, Berrens RP, and Mckee M. 2006. Simulating changes in forest recreation demand and associated economic impacts due to fire and fuels management activities. *Forest Policy and Economics* **8**: 52-66.

Stevenson JC, Rooth JE, Sundberg KL, and Kearney MS. 2002. The Health and Long Term Stability of Natural and Restored Marshes in Chesapeake Bay. *Concepts and Controversies in Tidal Marsh Ecology* 9: 709-735.

Stevenson JC, and Kearney MS. 2009. Impacts of global climate change and sea level rise on tidal wetlands. *In* Silliman B, and Strong D (Ed), Anthropogenic Modification of North American Salt Marshes. University of California Press, Berkeley, CA.

Sumaila UR, Cheung WWL, Lam VWY, Pauly D, and Herrick S. 2011. Climate change impacts on the biophysics and economics of world fisheries. *Nature Climate Change* **1**: 449-456.

Syswerda S, Corbin A, Mokma D, Kravchenko A, and Robertson G. 2011. Agricultural management and soil carbon storage in surface vs. deep layers. *Soil Science Society of America Journal* **75**: 92-101.

Tallis H, Lester SE, Ruckelshaus M, Plummer M, McLeod K, Guerry A, Andelman S, Caldwell M, Conte M, Copps S, Fox D, Fujita R, Gaines SD, Gelfenbaum G, Gold B, Kareiva P, Kim C, Lee K, Papenfus M, Redman S, Silliman B, Wainger L, and White C. 2011. New Metrics for managing and sustaining the ocean's bounty. *Marine Policy* **36**(1): 303-306.

Temmerman S, Govers G, Wartel S, and Meire P. 2004. Modelling estuarine variations in tidal marsh sedimentation: response to changing sea level and suspended sediment concentrations. *Marine Geology* **212**: 1-19.

Tilman D, Balzer C, Hill J, and Befort BL. 2011. Global food demand and the sustainable intensification of agriculture. *Proceedings of the National Academy of Sciences* **108**(50): 20260-20264.

Titus JG, Anderson KE, Cahoon DR, Gesch DB, Gill SK, Gutierrez BT, Thieler ER, and Williams SJ. 2009. Coastal Sensitivity to Sea-level Rise: a Focus on the mid-Atlantic Region. U.S. Climate Change Science Program Syntheis and Assessment Product 4.1. Available at: www.globalchange.gov/publications/reports/scientific-assessments

Tol RSJ. 2009. The Economic Effects of Climate Change. *Journal of Economic Perspectives* **23**: 29–51.

Twilley RR. 2007. Gulf Coast Wetland Sustainability in a Changing Climate. Coastal Wetlands and Global Climate Change. Arlington, VA: Pew Center on Global Climate Change. Available at: www.c2es.org/docUploads/Regional-Impacts-Gulf.pdf

USACE (US Army Corps of Engineers). 2006. Louisiana Coastal Protection and Restoration (LACPR) Preliminary Technical Report to United States Congress. Available at: http://www.nap.edu/openbook.php?record_id=12215&page=1

USDA-NASS (U.S. Department of Agriculture, National Agricultural Statistics Service). 2011. Quick Stats: Agricultural Statistics Data Base. Available at: http://quickstats.nass.usda.gov/

USDA-NRCS (U.S. Department of Agriculture, Natural Resources Conservation Service). 2012. National Soil Survey Handbook, title 430-VI, Part 622. Available at: http://soils.usda.gov/technical/handbook/contents/part622.html

USEPA (U.S. Environmental Protection Agency). 2008. Inventory of US Greenhouse Gas Emissions and Sinks: 1990-2006. EPA 430-R-08-005. Forest sections of the Land use change and forestry chapter, and Annex by Smith JE, Heath LS, Skog KE, and EPA consultants. Available at: http://www.epa.gov/climatechange/emissions/usinventoryreport.htm

USGAO (US General Accounting Office). 2003. Alaska Native Villages: Most are affected by flooding and erosion, but few qualify for federal assistance. REport to Congressional Committees. GAO-04-142. Available at: www.gao.gov/new.items/d04142.pdf

Valiela I, Bowen JL, and York JK. 2001. Mangrove Forests: One of the World's Threatened Major Tropical Environments. *BioScience* **51**(10): 807-815.

van Beukering P, and Cesar HSJ. 2004. Ecological economic modeling of coral reefs: evaluating tourist overuse at Hanauma Bay and algae blooms at the Kīhei Coast, Hawaiʻi. *Pacific Science* **58**: 243-260.

Van Voorhees D, and Lowther A. 2011. Fisheries of the United States 2010. Fisheries Statistics Division, National Marine Fisheries Service, National Ocean and Atmospheric Administration, U.S. Department of Commerce, Silver Spring, Maryland.

Vitousek PM, Aber JD, Howarth RW, Likens GE, Matson PA, Schindler DW, Schlesinger WH, and Tilman DG. 1997. Human alteration of the global nitrogen cycle: sources and consequences. *Ecological Applications* 7:737–750.

Walker JF, Hay LE, Markstrom SL, and Dettinger MD. 2011. Characterizing climate-change impacts on the 1.5-yr flood flow in selected basins across the United States: a probabilistic approach. *Earth Interactions* **15**(18): 1-16.

Wamsley TV, Cialone MA, Smith JM, Atkinson JH, and Rosati JD. The Potential of Wetlands in Reducing Storm Surge. *Ocean Engineering* **37**(1): 59–68.

Waycott M, Duarte CM, Carruthers TJB, Orth RJ, Dennison WC, Olyarnik S, Calladine A, Fourqurean JW, Heck KL, Hughes AR, Kendrick GA, Kenworthy WJ, Short FT, and Williams SL. 2009. Accelerating loss of seagrasses across the globe threatens coastal ecosystems. *Proceedings of the National Academy of Sciences* **106**(30): 12377-12381.

Westerling AL, Hidalgo HG, Cayan DR, and Swetnam TW. 2006. Warming and earlier spring increases western U.S. Forest wildfire activity. *Science* **313**: 940-943.

Whitlock C. 2004. Land management: Forests, fires and climate. *Nature* **432**: 28-29.

Whitehead JC, Poulter B, Dumas CF, and Bin O. 2009. Measuring the economic effects of sea level rise on shore fishing. *Mitigation and Adaptation Strategies for Global Change* **14**: 777-792.

Williams, CN, Menne MJ, and Thorne PW. 2012. Benchmarking the performance of pairwise homogenization of surface temperatures in the United States. *Journal of Geophysical Research-Atmospheres* 117: D0511.

Impacts of Climate Change on Biodiversity, Ecosystems, and Ecosystem Services |
Technical Input to the 2013 National Climate Assessment

Chapter 4
Ecosystem Services

Woodbury PB, Smith JE, Heath LS. 2007. Carbon sequestration in the U.S. forest sector from 1990 to 2010. *Forest Ecology and Management* **241**(1-3): 14-27.

WB (World Bank). 2010. The economics of adaptation to climate change. Washington, D.C.: The World Bank. Available at: http://climatechange.worldbank.org/content/economics-adaptation-climate-change-study-homepage

Zeller D, Booth S, and Pauly D. 2007. Fisheries contributions to the gross domestic product: underestimating small-scale fisheries in the Pacific. *Marine Resource Economics* **21**: 355-374.

Zimmerman RJ, Minnillo TJ, and Rozas LP. 2000. Salt marsh linkages to productivity of Penaeid shrimps and blue crabs in the Northern Gulf of Mexico. *In* Weinstein MP (Ed), Concepts and Controversies in Marsh Ecology. Kluwer Academic, Amsterdam, Netherlands.

Impacts of Climate Change on Biodiversity, Ecosystems, and Ecosystem Services |
Technical Input to the 2013 National Climate Assessment

Chapter 4
Ecosystem Services

Table 4.1. Current status, and projected future impacts of climate on ecosystem services.

| Specific Service | Current Status | | Climate Change Impacts | Expected Future Climate Change Impacts |
	Ecosystem Service (ES)	Human Well-being	ES and Human Well-being	ES and Human Well-being
Coastal flood protection	Over the past 200 years, 0.4 million ha of salt marsh has been lost in North America (Sifleet and others, 2011). Globally, seagrasses have been disappearing at a rate of 110 km^2/yr since 1980 (Waycott and others, 2009); and mangrove systems worldwide have declined at 1.4 percent/yr (Valiela and others, 2001). Coastal protection services have been lost (Barbier and others, 2011). Decrease in sea ice extent and earlier breakup of sea ice (Gearhead and others, 2006; Jones and others, 2009), are contributing to erosion and flooding of coastal areas.	Coastal marshes provide $836/ha/yr in reduced hurricane damages (Costanza and others, 2008). Some Native Alaskan communities have had to relocate their villages due to loss of protective sea ice (U.S. General Accounting Office 2003).	Total impacts of sea-level rise are expected to put as many as 480,000 people at risk from a 100-yr flood event, causing ~$100 Billion in damages (Heberger and others, 2009). Modest and probable sea level rise in Long Island (0.5 m by 2080) increases the number of people (by 47 percent) and property loss (by 73 percent) impacted by storm surge (Shepard and others, 2011). Climate change contribution to losses in extent of coastal marsh, mangrove and seagrass habitats is uncertain.	Sea-level rise and warming temperatures may promote expansion of coastal habitats to higher latitudes or further inland, provided that space to migrate upslope is available. Climate change may also alter rainfall patterns, which would in turn change local salinity regimes and competitive interactions of coastal plant communities with other wetland species (USGS 2004). Ability of mangroves and coastal marshes to keep up with sea level rise is uncertain. There is greater certainty that coral reefs will suffer severe damage. As much as one-third of reef building corals worldwide are at risk of extinction from climate change (Carpenter and others, 2008) Sea ice will continue to decline in spatial extent (Doney and others, 2012).

Table 4.1.

Table 4.1. *Current status, and projected future impacts of climate on ecosystem services.*

Specific Service	Current Status		Climate Change Impacts	Expected Future Climate Change Impacts
	Ecosystem Service (ES)	Human Well-being	ES and Human Well-being	ES and Human Well-being
Coastal erosion protection	Coastal erosion and Hurricane Katrina's damages to the areas surrounding New Orleans have reduced the natural storm defenses around the city by more than 500 square miles (U.S. ACE 2006). (see Coastal flood protection)	Preventing beach erosion along an 8 km beach in Maine and New Hampshire was worth $4.45/household (Huang and others, 2007). Oceanfront property increases in value by $233 per meter of beach width in Tybee Island, Georgia (Landry and others, 2003). If erosion remains at current levels, the cost of allowing Delaware beaches to retreat inland is about $291 million (Parsons and Powell 2001).	Over the past 2-3 decades, wave heights have increased all along the coast of the U.S., causing higher rates of erosion (Komar and Allan, 2006, 2008). It is unclear whether this is due to climate change or environmental variability. Governments are incurring high costs to maintain their beaches. For example, from 1990-2000 Delaware paid $15-$20 million to replenish its 25 miles of beaches (Parsons and Powell 2001). (see Coastal flood protection)	(see *Coastal flood protection*)

Table 4.1.

Impacts of Climate Change on Biodiversity, Ecosystems, and Ecosystem Services | Technical Input to the 2013 National Climate Assessment

Chapter 4
Ecosystem Services

Table 4.1. *Current status, and projected future impacts of climate on ecosystem services.*

Specific Service	Current Status		Climate Change Impacts	Expected Future Climate Change Impacts
	Ecosystem Service (ES)	Human Well-being	ES and Human Well-being	ES and Human Well-being
Fire regulation	Average of 6.5M acres/yr burned in U.S. (NOAA 2011).	U.S. FS spent more than $1B/yr on fire suppression alone in 5 of the 7 years during 2003-2009 (Venn and Calkin 2011).	Increased evapotranspiration, earlier spring, and higher temperatures have lead to 4x increased incidence of wildfire and 6x increased area burned since mid 1980s.	Area burned in western U.S. forests would increase 3-6.5x with each 1°C increase. Plant communities expected to change w/ changing fire regimes (Westerling and others, 2011).

Table 4.1.

Impacts of Climate Change on Biodiversity, Ecosystems, and Ecosystem Services |
Technical Input to the 2013 National Climate Assessment

Table 4.1. *Current status, and projected future impacts of climate on ecosystem services.*

Specific Service	Current Status		Climate Change Impacts	Expected Future Climate Change Impacts
	Ecosystem Service (ES)	Human Well-being	ES and Human Well-being	ES and Human Well-being
Carbon storage and sequestration in forest biomass	Increased biomass sequestration and storage slows down rates of climate change. Forests in the lower 48 are sequestering approximately 191 Tg of C/year (Woodbury and others, 2007; EPA 2008); equivalent to 10 percent of the U.S.'s annual CO_2 emissions. Currently, forest biomass carbon stocks are highest in the Pacific Northwest (Washington, Oregon, northern California; Woodbury and others, 2007); moderate stocks occur along the Appalachian Mountains (Oswalt and others, 2009). Sequestration rates in managed forests are highest in the Northeast (E. Nelson analysis).	Increased biomass sequestration and storage slows down economic damages associated with climate change. Estimates of the value of every additional ton of C sequestered range from $25 to $675 (Tol 2009).	Climate change has induced perturbations in forest distribution, forest growth rates, and risk of degradation via fire, invasive species, and disease. These perturbations are reducing rates of sequestration and expectations for C storage periods (Allen and others, 2010).	Climate change is predicted to affect the rate of tree growth in managed forests, both positively and negatively (Latta and others, 2010). The types of trees and/or management practices in an area also may change. Further, the risks to forests from fire and disease will increase (Allen and others, 2010; Lata and others, 2010; Liu and others, 2010; Haim and others, 2011). Payment for C storage and sequestration services would generate private value and alter the distribution of wealth in the U.S. Biomass carbon payment programs could affect the 11.3 million private forest owners who own 171 million ha in the U.S. (Oswalt and others, 2009).

Table 4.1.

Impacts of Climate Change on Biodiversity, Ecosystems, and Ecosystem Services |
Technical Input to the 2013 National Climate Assessment

Chapter 4
Ecosystem Services

Table 4.1. *Current status, and projected future impacts of climate on ecosystem services.*

Specific Service	Current Status		Climate Change Impacts	Expected Future Climate Change Impacts
	Ecosystem Service (ES)	Human Well-being	ES and Human Well-being	ES and Human Well-being
Carbon storage and sequestration in soils	Transition from cropland to grassland and forest increases soil carbon (Post and Kwon 2000). However, how much additional soil is conserved in such transitions is open to debate (Dlugofl and others, 2010; Syswerda and others, 2011; Rumpel and Kögel-Knabner 2010; Powlson and others, 2011).	Increased sequestration and storage of carbon in the soil slows down rates of climate change and associated economic damages. Estimates of the value of every additional ton of C sequestered range from $25 to $675 (Tol 2009).	No known attribution of recent climate to changes in C storage and sequestration in soils.	Climate change is predicted to reduce the amount of carbon stored in soils world-wide (Parton and others, 1995). Payment for C storage and other ecosystem services would generate private value and alter the distribution of wealth in the U.S. The 2.2 million farms and 373 million hectares of farmland in the U.S. (40 percent of all U.S. land) could be impacted by such a payment program (U.S.DA-ERS 2012).
Carbon storage and sequestration in marine habitats	Coastal ecosystems sequester and store carbon in biomass in the short term and in sediments in the long term (Duarte and others, 2005; McLeod and others, 2011). Carbon sequestered by salt marshes, mangroves, and seagrass beds varies widely, from 0.003 to 17.13, 0.03 to 3.81, and –21 to 23.2 Mg of C per hectare, respectively (Sifleet and others, 2011).	Increased carbon sequestration and storage slows down rates of climate change and associated economic damages. Estimates of the value of every additional ton of C sequestered or not emitted range from $25 to $675 (Tol 2009).	The relationship between coastal habitat losses and climate change is unknown.	Changes in productivity of coastal habitats due to increasing temperature and changes in salinity are predicted to affect C storage and sequestration to an unknown degree. (see *Coastal flooding protection*)

Table 4.1.

Chapter 4
Ecosystem Services

Impacts of Climate Change on Biodiversity, Ecosystems, and Ecosystem Services |
Technical Input to the 2013 National Climate Assessment

Table 4.1. *Current status, and projected future impacts of climate on ecosystem services.*

| Specific Service | Current Status | | Climate Change Impacts | Expected Future Climate Change Impacts |
	Ecosystem Service (ES)	Human Well-being	ES and Human Well-being	ES and Human Well-being
Water quality regulation	Sediments and turbidity currently impair 25 percent of lakes and 17 percent of rivers. Phosphorus impairs 18 percent of lakes and 14 percent of rivers. Nitrogen impair10 percent of lakes and 4 percent of rivers (EPA 2011).	Studies estimating the costs of nitrogen pollution are rudimentary and range from less than $1.00/kg N exported to $56/kg N exported (Compton and others, 2011).	Precipitation and river discharge are negatively correlated with nitrogen retention (Howarth and others, 2012). Temperature is positively correlated with nitrogen retention (Howarth and others, 2012). Areas with expected increases in precipitation could lose this service, and areas with expected increases in temperature could gain it. Extreme precipitation and discharge events are positively correlated with waterborne disease outbreaks (Curriero and others, 2001).	Waterborne illness is predicted to increase with climate change because extreme precipitation events increase the loading of contaminants to waterways (Rose and others, 2001; Curriero and others, 2001; Ebi and others, 2006). Climate change predictions for the Mississippi Basin (under doubled CO_2) indicate a 20 percent increase in river discharge that can lead to higher nitrogen loads and a 50 percent increase in primary production in the Gulf, a 30-60 percent decrease in deep water dissolved oxygen concentration and an expansion of the dead zone (Justic and others, 1996).

Table 4.1.

Impacts of Climate Change on Biodiversity, Ecosystems, and Ecosystem Services |
Technical Input to the 2013 National Climate Assessment

Chapter 4
Ecosystem Services

Table 4.1. Current status, and projected future impacts of climate on ecosystem services.

Specific Service	Current Status		Climate Change Impacts	Expected Future Climate Change Impacts
	Ecosystem Service (ES)	Human Well-being	ES and Human Well-being	ES and Human Well-being
Timber yield	Since the 1950s, overall land devoted to timber production in the U.S. has stayed relatively constant, and the amount of reserved forests has increased 200 percent (Oswalt and others, 2009). Net growth of forest stock has consistently exceeded removals by approximately 3 percent. Timber mortality rates have remained well below 1 percent of inventory during the same time period; but mortality rates relative to inventory are currently at the highest level in 50 years (Oswalt and others, 2009).	Since the late 1980s the volume of the U.S. timber harvest has fallen slightly (approximately 450 million cubic meters of wood was produced in the U.S. in 2006 (Oswalt and others, 2009)), and imports are forming an increasingly larger portion of U.S. timber consumption (Oswalt and others, 2009).	Drought and warm temperatures across western North America in the last decade have led to extensive insect outbreaks and mortality throughout the region, affecting 20 million ha from Alaska to Mexico (Allen and others, 2010). It is uncertain whether current mortality rates are beyond the range of normal variability (USFS 2011). Wildland fire intensity and area burned have increased in recent decades (Running 2006; Westerling and others, 2006; Miller and others, 2008), and Federal agencies now spend more than $1 billion annually on fire suppression efforts (U.S. GAO 2006).	Overall increase in forest productivity is predicted to increase long-term timber inventory (Alig and others, 2002). "Timber harvests in most scenarios rise over the next 100 years, lowering timber prices, and reducing costs of wood and paper products to consumers and returns to owners. of timberland." (Alig and others, 2002, p. 9). How the increased risk to forests stands from fire and disease will affect these trends is unclear (Westerling and others, 2006, Oswalt and others, 2009; Allen and others, 2010; Liu and others, 2010). (see Carbon storage and sequestration in forest biomass)

Table 4.1.

Impacts of Climate Change on Biodiversity, Ecosystems, and Ecosystem Services |
Technical Input to the 2013 National Climate Assessment

Chapter 4
Ecosystem Services

Table 4.1. *Current status, and projected future impacts of climate on ecosystem services.*

Specific Service	Current Status		Climate Change Impacts	Expected Future Climate Change Impacts
	Ecosystem Service (ES)	Human Well-being	ES and Human Well-being	ES and Human Well-being
Agricultural yield	Currently, cropland accounts for 18 percent of land in the U.S. Pasture and rangeland account for another 27 percent of land in the U.S. (USDA ERS Datasets, "Major Land Uses"). In 2009, U.S. agriculture produced 31 percent of the world's coarse grains and 11 percent of the world's oilcrops (FAO STAT 2012).	The U.S. produced 10 percent of the globe's net production value in food in 2009 (FAOSTAT 2012).	Compared to the rest of the world, growing season weather has changed relatively little in the U.S. over the past 30 years (Lobell and others, 2011). This suggests yield trends in the U.S. over the past 30 years have been primarily driven by farm management, managed input use, technology, and cropland soil quality (Lobell and others, 2011).	Accelerated climate change may lead to greater yield impacts over the next 50 years. Temperature changes have had a more dramatic impact on corn, wheat, soybean, and rice yields around the world than changes in precipitation (Lobell and others, 2011). We estimate that in the Midwest U.S., climate change could reduce mid century maize yields by 2 to 14 percent compared to expected yields given no climate change; wheat yield could be reduced by 1 to 7 percent; and soybean yield could be reduced by 0.6 to 10 percent; (data and statistical model code can be found at http://www.bowdoin.edu/faculty/e/enelson/index.shtml).

Table 4.1.

Impacts of Climate Change on Biodiversity, Ecosystems, and Ecosystem Services |
Technical Input to the 2013 National Climate Assessment

Chapter 4
Ecosystem Services

Table 4.1. *Current status, and projected future impacts of climate on ecosystem services.*

Specific Service	Current Status		Climate Change Impacts	Expected Future Climate Change Impacts
	Ecosystem Service (ES)	Human Well-being	ES and Human Well-being	ES and Human Well-being
Water provision (all)	Water largely allocated, with some conflicts (Christian-Smith and others, 2012).	36 percent of U.S. counties have moderate or higher water-supply sustainability risk (Roy and others, 2012).	Observed changes in precipitation, increasing ET (Dai and others, 2011; Hamlet and others, 2007), increasing extremes (U.S. GCRP 2009), snow to rain events (Hamlet and others, 2005). Effects of climate-induced changes in water provision on human well being are not well documented.	Predictions indicate changes in precipitation patterns (esp. decreases in Southwest, increases in North), increasing ET (Hamlet and others, 2007; Diaz and others, 2011), increasing extremes (IPCC SREX, 2011), snow to rain (Adam and others, 2009).

Table 4.1.

Impacts of Climate Change on Biodiversity, Ecosystems, and Ecosystem Services |
Technical Input to the 2013 National Climate Assessment

Chapter 4
Ecosystem Services

Table 4.1. Current status, and projected future impacts of climate on ecosystem services.

Specific Service	Current Status		Climate Change Impacts	Expected Future Climate Change Impacts
	Ecosystem Service (ES)	Human Well-being	ES and Human Well-being	ES and Human Well-being
Marine fishery yields	In 2009, 7.9 billion pounds of fish and shellfish were landed in U.S. ports (NMFS 2010).	Fisheries added $48.3 billion and 1 million jobs to the U.S. economy in 2009 (NMFS 2010). Almost all communities within the Pacific Islands derive over 25 percent of their animal protein from fish, with some deriving up to 69 percent (NCA 2009).	Fish populations are shifting poleward and deeper (Nye and others, 2009; Murawski 1993; Mueter and Litzow 2008; Dulvy and others, 2008; Perry and others, 2005) and communities are transitioning from cold-water to warm-water species as local temperatures warm (Collie and others, 2008; Lucey and Nye 2010). Jobs, catch, and value for individual species are moving poleward as temperatures warm and as species shift poleward (McCay and others, 2011; Pinsky and Fogarty, written communication 2012.).	Globally, fish populations are predicted to shift 45-49 km/decade poleward (Cheung and others, 2009). Species like Atlantic croaker are predicted to increase in the northeastern U.S., while pollock, haddock and cod are predicted to decrease (Hare and others, 2010; Fogarty and others, 2007; Lenoir and others, 2010). Oceanic habitat for salmon is predicted to disappear from the Gulf of Alaska (Abdul-Aziz and others, 2011).

Table 4.1.

Impacts of Climate Change on Biodiversity, Ecosystems, and Ecosystem Services |
Technical Input to the 2013 National Climate Assessment

Chapter 4
Ecosystem Services

Table 4.1. *Current status, and projected future impacts of climate on ecosystem services.*

Specific Service	Current Status		Climate Change Impacts	Expected Future Climate Change Impacts
	Ecosystem Service (ES)	Human Well-being	ES and Human Well-being	ES and Human Well-being
Marine aquaculture yields	In 2009, 720 million pounds of marine aquaculture were produced in the U.S. (Van Voorhees and Lowther 2010).	Shellfish produced in the U.S. was worth $280 million in 2009 (Van Voorhees and Lowther 2010).	Ocean acidification impedes growth and reproduction, particularly in calcifying organisms such as shellfish (Kurihara 2008; Miller and others, 2009; Kroeker and others, 2010). New diseases have moved poleward as temperatures warmed (Hofmann and others, 2001).	Warm temperatures are predicted to increase aquaculture potential in poleward regions, but decrease it in the tropics (De Silva and Soto 2009). Acidification, will reduce growth and survival (Cooley and Doney 2009).

Table 4.1.

Impacts of Climate Change on Biodiversity, Ecosystems, and Ecosystem Services |
Technical Input to the 2013 National Climate Assessment

Chapter 4
Ecosystem Services

Table 4.1. Current status, and projected future impacts of climate on ecosystem services.

Specific Service	Current Status		Climate Change Impacts	Expected Future Climate Change Impacts
	Ecosystem Service (ES)	Human Well-being	ES and Human Well-being	ES and Human Well-being
Recreation-winter sports		26 percent of U.S. population participates in winter sports activities (NSRE 2000). The ski/snowboard/ snowshoe industry in U.S. is worth $66 billion and supports 556,000 jobs (Southwick Associates 2006). Snowmobiling adds $22 billion annually and 90,000 jobs (International Snowmobile Manufacturers Association).		Ski seasons are predicted to be shorter: the California season would be shorter by 49-103 days (Hayhoe and others, 2004); Michigan and Vermont shorter by 5-60 percent (Scott and others, 2006; Dawson and Scott 2007); 6-48 percent shorter ski season in Northeast (Scott and others, 2008). It is projected that the ski season will disappear in Arizona after 2050 (Bark and others, 2010). 12.5 percent of Cascades ski areas and 60 percent of Olympic ski areas at risk due to predicted warm winters (Nolin and Daly 2006). Severe losses of snowmobiling season (>50 percent) predicted in Northeast (Scott and others, 2008).

Table 4.1.

Chapter 4
Ecosystem Services

Impacts of Climate Change on Biodiversity, Ecosystems, and Ecosystem Services |
Technical Input to the 2013 National Climate Assessment

Table 4.1. *Current status, and projected future impacts of climate on ecosystem services.*

Specific Service	Current Status		Climate Change Impacts	Expected Future Climate Change Impacts
	Ecosystem Service (ES)	Human Well-being	ES and Human Well-being	ES and Human Well-being
Recreation - coral reefs	See Culver and others, 2012	Net benefits of $360 million annually to Hawaiian economy from 1660 square kilometers of reef area (Cesar and others, 2004); $50-60 million annual revenues from Hawaiian dive operations (Van Beukering and Cesar 2004).	See Culver and others, 2012	
Recreation- coastal	See Culver and others, 2012; Griffis and others, 2012.	Ocean-related tourism contributes $82 billion to GDP and supports 5 million jobs in leisure and hospitality in coastal states (NOEP 2011).	*(see Coastal erosion protection)*	Beach erosion projected to cost more than $1 billion annually in coastal state tourism losses; $63 million annually in southern California (Bin and others, 2007; Whitehead and others, 2009; Pendleton and others, 2011). Some economic gains may result from an increase in user days with better weather (Loomis and Crespi 1999).

Table 4.1.

Impacts of Climate Change on Biodiversity, Ecosystems, and Ecosystem Services |
Technical Input to the 2013 National Climate Assessment

Chapter 4
Ecosystem Services

Table 4.1. *Current status, and projected future impacts of climate on ecosystem services.*

Specific Service	Current Status		Climate Change Impacts	Expected Future Climate Change Impacts
	Ecosystem Service (ES)	Human Well-being	ES and Human Well-being	ES and Human Well-being
Recreation - angling	(see *Marine fishery yields*)	U.S. anglers take 74 million saltwater fishing trips annually, with combined saltwater and freshwater economic impact of more than $45 billion/year on trips and equipment (U.S. DOI, FWS, DOC, and U.S. CB 2006). 327,000 full- and part-time jobs are related to saltwater and freshwater recreational fisheries (NMFS 2010).	(see *Marine fishery yields*)	Predictions indicate a decrease in cold-water fishing (trout, salmon); may be offset by increase in warm-water fish catch, such as bass and perch (Pendleton and Mendelsohn 1998).

Table 4.1.

Impacts of Climate Change on Biodiversity, Ecosystems, and Ecosystem Services | Technical Input to the 2013 National Climate Assessment

Chapter 4
Ecosystem Services

Table 4.1. Current status, and projected future impacts of climate on ecosystem services.

Specific Service	Current Status		Climate Change Impacts	Expected Future Climate Change Impacts
	Ecosystem Service (ES)	Human Well-being	ES and Human Well-being	ES and Human Well-being
Recreation-other			Campground closures are projected due to hazard trees and fire risk (Robbins 2008; Starbuck and others, 2006); decreases in visitation to parks suffering catastrophic fires (Starbuck and others, 2006); decreased reliability of whale-watching opportunities (Lambert and others, 2010); increase in visitation to Rocky Mountain NP with increased temperatures (Richardson and Loomis 2005); redistribution of "winter sun" and "summer cool" destinations in North America (Scott and others, 2004); increase in golfing, boating, and other activities promoted by warmer, drier weather (Loomis and Crepsi 1999; Shaw and Loomis 2008). The net effect is predicted to be a redistribution of the industry and its economic impact, with visitors and tourism dollars shifting away from some communities in favor of others.	

Table 4.1.

Impacts of Climate Change on Biodiversity, Ecosystems, and Ecosystem Services | Technical Input to the 2013 National Climate Assessment

Chapter 4
Ecosystem Services

Table 4.1. *Current status, and projected future impacts of climate on ecosystem services.*

Specific Service	Current Status		Climate Change Impacts	Expected Future Climate Change Impacts
	Ecosystem Service (ES)	Human Well-being	ES and Human Well-being	ES and Human Well-being
Subsistence hunting and foraging		For indigenous Alaskans, wildlife hunting provides a large component of the diet, contributes to cash income, and serves as an important cultural touchstone. Subsistence hunting of wildlife (whales, seals, walrus, caribou, fish, and birds) is greater than 100kg per capita among coastal Inupiat (Gearhead and others, 2006); hunters also earn cash income from seal, narwhal, and polar bear hunts.	Wildlife migratory patterns and abundance are changing, and weather conditions becoming more hazardous and unpredictable, leading to decreased reliability of traditional ecological knowledge and fewer days spent hunting. Predictions are for decreases in sea ice extent and earlier breakup of sea ice (Gearhead and others, 2006); changes to abundance and migratory patterns of wildlife, including bowhead whales and geese; decline in Porcupine caribou herd of up to 85 percent over 40 years (Kruse and others, 2004); less predictable weather (Ford and others, 2006); increased windiness/ storminess leading to fewer boatable days (Ford and others, 2006; Hinzman and others, 2005).	

Table 4.1.

Impacts of Climate Change on Biodiversity, Ecosystems, and Ecosystem Services |
Technical Input to the 2013 National Climate Assessment

Chapter 4
Ecosystem Services

Table 4.2. *Factors affecting adaptation responses to climate change impacts*

Specific Service	Ecosystem Effects on Human well-being	Interacting Stressors	Most vulnerable geographic region	Most vulnerable sector or part of society	Human Response (list of actions that may be taken)
Coastal flood protection	Sea-level rise would increase risk of storm related coastal hazards for many coastal communities. Currently, no published studies quantify the marginal change in human well-being due to impacts on hazard reduction due to storm surge dampening.	Coastal development, sediment and nutrient runoff, nearshore management.	Southeast; the Atlantic coast of North America may experience one of the world's largest losses in wetlands (Nichollas and others, 1999). Losses in extent of coastal marshes have already impaired human well-being. This is especially evident in the Gulf coast with respect to hurricane damage.	Recreation, residential, insurance	Sea walls, restoration and protection of habitats, relocation of people or infrastructure.

Table 4.2.

Table 4.2. *Factors affecting adaptation responses to climate change impacts*

Specific Service	Ecosystem Effects on Human well-being	Interacting Stressors	Most vulnerable geographic region	Most vulnerable sector or part of society	Human Response (list of actions that may be taken)
Coastal erosion protection	14,000 people currently live in the 41 sq. miles of coastline that is predicted to be lost to sea-level rise and coastal hazards by 2100 (Heberger and others, 2009).	Wave heights (which lead to higher erosion) have increased all along the coast of the US with greater increases occurring in higher latitudes (Komar and Allan, 2006, 2008), but it is unclear whether due to climate change or variability.	Pacific coast (Boruff and others, 2005); especially Alaska in places where protective sea ice is disappearing (Jones and others, 2009).	see *Coastal flood protection*	see *Coastal flood protection*
Fire regulation	Where warmer, drier temperatures occur and where fuel build-up due to fire suppression has taken place, fires will be more frequent and/or more intense.	Fuel loads, invasive species, disease, forest management.	Western U.S.	Forest products, rural residential, carbon emissions	Forest/fuels management

Table 4.2.

Impacts of Climate Change on Biodiversity, Ecosystems, and Ecosystem Services |
Technical Input to the 2013 National Climate Assessment

Chapter 4
Ecosystem Services

Table 4.2. *Factors affecting adaptation responses to climate change impacts*

Specific Service	Ecosystem Effects on Human well-being	Interacting Stressors	Most vulnerable geographic region	Most vulnerable sector or part of society	Human Response (list of actions that may be taken)
Carbon storage and sequestration in forest biomass	Climate change is projected to continue perturbations in forest distribution, forest growth rates, and risk of degradation via fire, invasive species, and disease. These perturbations will continue to reduce rates of sequestration and expectations for C storage periods (Allen and others, 2010).	Economic drivers of land use change (for example, 220 million hectares of forest are expected in the U.S. by 2051, due in part to cropland and pasture abandonment Radeloff and others, 2012), forest management, invasive species, disease.	Western U.S.	Global impact; local recipients of C sequestration projects	Markets for forest carbon sequestration projects exist and are expanding (Canadell and Raupach 2008; Rodrigo and Perrings 2011); forest management.
Carbon storage and sequestration in soils	Land-use change will have a large impact on carbon soil sequestration and storage, with transitions from cropland to forest and urban areas having a positive impact on soil carbon storage (E. Nelson analysis).	Economic drivers of land-use change; agricultural and timber management practices affecting erosion.	Soils in Minnesota, Iowa, Vermont, New York, and Maine have the potential to store the most carbon (E. Nelson analysis).	Global impact; local recipients of C sequestration projects.	Programs that pay landowners to increase their soil carbon (IBRD/WB 2011; Glenk and Colombo 2011); agricultural and timber management practices affecting erosion.

Table 4.2.

Impacts of Climate Change on Biodiversity, Ecosystems, and Ecosystem Services |
Technical Input to the 2013 National Climate Assessment

Chapter 4
Ecosystem Services

Table 4.2. *Factors affecting adaptation responses to climate change impacts*

Specific Service	Ecosystem Effects on Human well-being	Interacting Stressors	Most vulnerable geographic region	Most vulnerable sector or part of society	Human Response (list of actions that may be taken)
Carbon storage and sequestration in marine habitats	If climate change reduces the extent of marine features that have positive sequestration rates, or reduces their capacity to sequester and store carbon, all else equal, global economic damages could increase (Westerling and others, 2006).	Coastal development, sediment and nutrient runoff, nearshore management.	Atlantic coast of North America may experience one of the world's largest losses in wetlands (Nichollas and others, 1999, 2004); SE U.S. where mangroves occur.	Global impact; local recipients of C sequestration projects.	Programs that pay landowners to increase their marine habitat-based carbon; restoration and protection of habitats.
Water quality regulation	Not aware of estimates of current climate impacts on water quality regulation.	Nitrogen and phosphorus application rates will strongly interact with climate change (NCA 2009).	See *Water Resources Chapter,* NCA 2012	Households, industries reliant on natural water supplies.	Increased water treatment; increased health care to counteract health impacts; altered land use practices (fertilizer application, tillage practices, buffers, feed and livestock management, manure management). (NCA 2009, 2012).

Table 4.2.

Impacts of Climate Change on Biodiversity, Ecosystems, and Ecosystem Services |
Technical Input to the 2013 National Climate Assessment

Chapter 4
Ecosystem Services

Table 4.2. Factors affecting adaptation responses to climate change impacts

Specific Service	Ecosystem Effects on Human well-being	Interacting Stressors	Most vulnerable geographic region	Most vulnerable sector or part of society	Human Response (list of actions that may be taken)
Timber yield	The effects of climate change on forestry remains somewhat uncertain. Changes in weather patterns could lead to more rapid tree growth and greater harvest volumes and profits, or to less rapid tree growth and smaller harvest volumes and profits. It is thought that both dynamics will occur in the Pacific Northwest (Latta and others, 2010). Greater risk of forest destruction due to fire and/or disease could lower the profits of timber firms, resulting in job losses.	Economic drivers of land-use change (for example, Radeloff and others, 2012), forest management, invasive species, disease.	Pacific Northwest, Southeast	Logging and mill workers, construction industry	Forest management

Table 4.2.

Impacts of Climate Change on Biodiversity, Ecosystems, and Ecosystem Services |
Technical Input to the 2013 National Climate Assessment

Chapter 4
Ecosystem Services

Table 4.2. *Factors affecting adaptation responses to climate change impacts*

Specific Service	Ecosystem Effects on Human well-being	Interacting Stressors	Most vulnerable geographic region	Most vulnerable sector or part of society	Human Response (list of actions that may be taken)
Agricultural yield	Yields impacts expected over the next 50 years. Temperature changes have had a more dramatic impact on corn, wheat, soybean, and rice yields around the world than changes in precipitation (Lobell and others, 2011).	Drivers of agricultural land conversion (Radeloff and others, 2012).	Agriculture in the areas of the U.S. that will experience the most dramatic climate change will have the greatest transition costs.	Agriculture and fertilizer, pesticide, food processing.	Agricultural management, subsidies.
Water provision (all)	U.S. counties with water-supply sustainability risk would double to 70 percent (Roy and others, In Press).	Changing demands from households, industry, agriculture.	Southwest, Great Plains, Southeast U.S.	Agriculture, municipal, and wetland/aquatic ecosystems.	Increase water-use efficiency, price adjustments, recycling.

Table 4.2.

Chapter 4
Ecosystem Services

Impacts of Climate Change on Biodiversity, Ecosystems, and Ecosystem Services |
Technical Input to the 2013 National Climate Assessment

Table 4.2. Factors affecting adaptation responses to climate change impacts

Specific Service	Ecosystem Effects on Human well-being	Interacting Stressors	Most vulnerable geographic region	Most vulnerable sector or part of society	Human Response (list of actions that may be taken)
Marine fisheries	Fisheries are predicted to decline in the lower 48 states, but increase in parts of Alaska (Cheung and others, 2010). Costs of fishing are predicted to increase as fisheries transition to new species and as processing plants and fishing jobs shift poleward (NCA 2009; Sumaila and others, 2011).	Fishing (Hare and others, 2010), habitat destruction (Beck and others, 2001), eutrophication and coastal water quality, and invasive species (NCA 2009).	The continental U.S. and Hawaii (Cheung and others, 2010).	Coastal states and communities (Coulthard 2009; McCay and others, 2011).	Switch to warm-water species (Sumaila and others, 2011); adjust fisheries quotas or subsidies (Hare and others, 2010); conduct international negotiations over transboundary species.
Marine aquaculture	U.S. mollusk fisheries may have economic losses of $0.3-5.1 billion in Net Present Value by 2060 (Cooley and Doney, 2009); aquaculture operations face increased costs and less predictability (De Silva and Soto 2009).	Coastal water quality; eutrophication.	West Coast U.S. in areas of upwelling (Feely and others, 2008); areas of land runoff, hypoxia, sulfur dioxide precipitation, and eutrophication (Kelly and others, 2011).	Aquaculture industry, coastal states and communities (Da Silva and Soto 2009).	Switch to less sensitive species (Da Silva and Soto 2009); mitigate sources of local acidification (Kelly and others, 2011).

Table 4.2.

Impacts of Climate Change on Biodiversity, Ecosystems, and Ecosystem Services |
Technical Input to the 2013 National Climate Assessment

Chapter 4
Ecosystem Services

Table 4.2. *Factors affecting adaptation responses to climate change impacts*

Specific Service	Ecosystem Effects on Human well-being	Interacting Stressors	Most vulnerable geographic region	Most vulnerable sector or part of society	Human Response (list of actions that may be taken)
Recreation- winter sports	Doubling of cost of snowmaking (+5 percent total operating costs to ski areas) under high emissions scenario (Scott and others, 2008; Dawson and Scott 2007); lower home prices in ski areas where snow reliability is low (Bustic and others, 2011).		Ski areas located at lower elevation or lower latitude (Bark and others, 2010); snowmobiling operations where snowmaking is not an option (Scott and others, 2008).	Winter sport industry and tourism.	Snowmaking (Scott and others, 2006; Scott and others, 2008; Bark and others, 2010).
Recreation - coral reefs	Loss of coral cover due to lowering of ocean pH, warm temps (Culver and others, 2012; Griffis and others, 2012).	UV stress, coastal development, recreational impacts, invasive species.	Areas with coral	Tourism, recreational and commercial fishing on coral-dependent species.	Protection and restoration, reduction of pollution and habitat-destroying activities.

Table 4.2.

Impacts of Climate Change on Biodiversity, Ecosystems, and Ecosystem Services |
Technical Input to the 2013 National Climate Assessment

Chapter 4
Ecosystem Services

Table 4.2. Factors affecting adaptation responses to climate change impacts

Specific Service	Ecosystem Effects on Human well-being	Interacting Stressors	Most vulnerable geographic region	Most vulnerable sector or part of society	Human Response (list of actions that may be taken)
Recreation-coastal	Losses due to beach erosion (Bin and others, 2007; Whitehead and others, 2009; Pendleton and others, 2011); potential increase in user days with better weather, resulting in economic gains (Loomis and Crespi 1999).	coastal development, sediment impoverishment from upstream changes to hydrology.	Gulf and Pacific coasts (Culver and others, 2012)		Sand replenishment on beaches
Recreation - angling	Decrease in cold-water fishing (trout, salmon); may be offset by increase in warm-water fish catch, such as bass and perch (Pendleton and Mendelsohn 1998).	Overfishing, pollution	Atlantic coast	Recreation & tourism	Stocking with warm-water species; fishery management

Table 4.2.

Impacts of Climate Change on Biodiversity, Ecosystems, and Ecosystem Services |
Technical Input to the 2013 National Climate Assessment

Chapter 4
Ecosystem Services

Table 4.2. *Factors affecting adaptation responses to climate change impacts*

Specific Service	Ecosystem Effects on Human well-being	Interacting Stressors	Most vulnerable geographic region	Most vulnerable sector or part of society	Human Response (list of actions that may be taken)
Subsistence hunting & foraging	Increased hazards to hunters and travelers (Ford and others, 2008; Ford and others, 2006); less time spent hunting (Ford and others, 2006; Berman and Kofinas 2004); obsolescence of traditional ecological knowledge about weather prediction and risk assessment (Ford and others, 2008; Ford and others, 2006); decreased harvest of wildlife or switch to lower-value wildlife species (Ford and others, 2006; Hinzman and others, 2005; Kruse and others, 2004).		Alaska	Indigenous people	Hunters may get improved access to weather prediction and safety technology (Ford and others, 2006). Hunters may switch to different prey with less associated risk (Ford and others, 2006).

Table 4.2.

Chapter 5. Impacts of Climate Change on Already Stressed Biodiversity, Ecosystems, and Ecosystem Services

Convening Lead Authors: Amanda Staudt, Allison K. Leidner
Lead Authors: Jennifer Howard, Kate A. Brauman, Jeff Dukes, Lara Hansen, Craig Paukert, John Sabo, Luis A. Solórzano
Contributing Author: Kurt Johnson

Key Findings

- Biodiversity, ecosystems, and ecosystem services are already under stress from a variety of sources (for example, land use and land cover change, extraction of natural resources, biological disturbances, and pollution); in most cases, these interacting stressors have had a greater effect on the overall health of these systems than climate change. However, climate change effects are projected to be an increasingly important source of stress in the future.
- Climate change has been shown to exacerbate the effects of other stressors. Ecosystems that are already being affected by other stressors are likely to have faster and more acute reactions to climate change.
- Interactions between climate change and other stressors must be considered in climate adaptation strategies. In many cases, managers can draw upon existing strategies to address the interaction of climate and other stressors.
- Climate change responses employed by other sectors (for example, energy, agriculture, transportation) may create new ecosystem stresses and interact with existing climate and non-climate stresses.

5.1. INTRODUCTION

Climate change, in conjunction with other human activities, causes a variety of interacting stresses to biodiversity, ecosystems, and ecosystem services (MA, 2005; Mooney and others, 2009). A stress is an activity that induces an adverse effect and therefore degrades the condition and viability of a natural system (Groves and others, 2000; EPA, 2008a). In most environments, the stressors that have most impacted natural systems fall into four general categories: land use and land cover change (including habitat fragmentation, urbanization, and infrastructure development), biological disruptions (such as the introduction of non-native invasive species, disease, and pests), extractive activities (such as fishing, forestry, and water withdrawals), and pollution (including chemicals, heavy metals, and nutrients). Climate change is a stress in its own right, and it interacts with these other stressors in complex ways.

Environmental stresses have caused extensive transformation and deterioration of terrestrial, freshwater, and marine environments (MA, 2005; Brook and others, 2008; Butchart and others, 2010). Recent reports estimate that more than 75 percent of Earth's ice-free land shows anthropogenic alterations (Ellis and Ramankutty, 2008). In the United States, about 43 percent of the native ecosystems, including forests, grasslands, deserts, and wetlands, have been converted for agriculture, urban growth, and other economic activities (Lubowski and others, 2006). Nearly two thirds of global river discharge is moderately to highly threatened, which in turn threatens aquatic habitats (Vorosmarty and others, 2010). In the United States, humans appropriate the equivalent of more than 40 percent of renewable supplies of freshwater in more than 25 percent of all watersheds. The numbers are even higher in the arid Southwest, where the equivalent of 76 percent of all renewable freshwater is appropriated (Sabo and others, 2010a). In addition, there are virtually no areas of the oceans that are not impacted by some anthropogenic driver of ecological change (Halpern, 2008a). Just over 40 percent of the world's oceans are considered to have an anthropogenic impact rating of at least "medium high" (Halpern, 2008a), and within the United States Exclusive Economic Zone about two thirds of ocean areas fall into the "medium high" to "very high" category (Kappel and others, 2009).

Consideration of observed and projected impacts of climate change in the context of other environmental stressors is essential for effective planning and management, especially given the existing impact of these other stressors. Although climate change is not currently the biggest threat to most natural systems, its impact will likely increase (Mooney and others, 2009). Other chapters of this technical report present the current understanding of how climate change is impacting biodiversity, ecosystems, and ecosystem services. In this chapter, we focus on how climate change interacts with other anthropogenic environmental stresses and the implications for developing effective response strategies.

5.2. CONCEPTUAL FRAMEWORK OF CLIMATE CHANGE INTERACTIONS WITH OTHER STRESSORS

Climate change, interacting with environmental stressors and human response strategies to climate change, affects biodiversity, ecosystems, and ecosystem services through a variety of pathways and with complex, interacting effects (Mora and others, 2007; Brook and others, 2008; Halpern and others, 2008ab). The interactions of climate change and other stressors typically result in increased stress on natural systems, though individual stresses can have counteracting effects. These interactions can affect the timing, distribution, and severity of the stresses experienced by ecosystems. For natural systems that are relatively undisturbed by human activities, climate change may increase their susceptibility to other environmental stresses. Human responses to climate change may further complicate these relationships, presenting additional and novel sources of stress.

A clear conceptual framework that helps identify the pathways, types, and character of climate interactions with other stressors can clarify natural resource management strategies, including potential adaptation responses, and inform decision-making (Didham and others, 2007; Crain and others, 2008; Darling and Cote, 2008; Halpern and others, 2008b). We present such a framework in this section and then apply it to the discussion of a case study in Section 5.4.

Pathway of interaction

Climate change can stress ecosystems *directly* or *indirectly*. Often both pathways occur simultaneously. For example, a forest could experience both climate-induced drought stress (a direct effect) and stress from invasive pests whose life cycle or range has been altered by climate change (an indirect effect) (Dukes and others, 2009; Carnicer and others, 2011; Luedeling, 2011). In addition, climate change may alter the *interaction of different stressors* with one another. For example, changes in precipitation from climate change may affect water levels in rivers; reduced water volume can result in increased pollutant concentrations that reduce the fitness of an organism, making it more susceptible to stressors such as disease or warmer water (Johnson and others, 2010; Johnson and others, 2011). Finally, *human responses to climate change*, particularly adaptation strategies undertaken by other sectors, may put biodiversity, ecosystems, or ecosystem services under additional stress. For example, sea walls put in place to counteract sea level rise and erosion can negatively affect habitat structure and species movement patterns (Bulleri and Chapman, 2010).

Figure 5.1 illustrates how the interaction between climate change and a single environmental stressor might be represented. With just this interaction, there are four potential pathways for climate change to affect an ecosystem, and multiple pathways can occur simultaneously. Climate change can (a) directly affect an ecosystem or (b) indirectly affect an ecosystem by affecting a stressor. Climate change can also induce a climate mitigation or adaptation response that has (c) direct effects on the ecosystem or (d) indirectly affects an ecosystem by affecting a stressor.

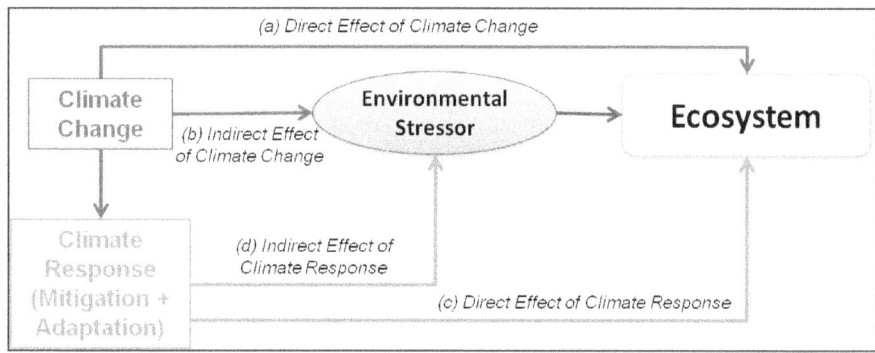

Figure 5.1. Conceptual diagram of climate change interaction with a single other environmental stress. Climate change can interact with other environmental stressors to (a) directly or (b) indirectly affect ecosystems. Climate change can also induce a climate mitigation or adaptation response that (c) directly or (d) indirectly affects an ecosystem by affecting an environmental stressor.

Most ecosystems are subjected to multiple environmental stressors and multiple climate stressors, opening many more pathways by which climate change can affect them. Thus, a more complicated conceptual diagram is necessary to represent the numerous potential stressor interactions, as shown in **Figure 5.2.** We present a generic diagram here and a version illustrating specific stressors for the California Central Valley case study in section 5.4. Despite the multiple pathways represented, this figure does not show the full potential for interactions because each environmental stressor can interact with any of the other stressors and there may be interactions that involve multiple stressors.

Type of interactions

Climate change can alter the direction of interacting stressors' impacts, most often *exacerbating* them but potentially *ameliorating* them (Folt and others, 1999; Breitburg and Riedel, 2005). This interaction may be *linear* or *additive*, such that the effect of multiple stressors is equal to the sum when each acts alone. Alternatively, the impacts of climate change may be *non-linear* or *synergistic*, whereby the combined impacts have a greater effect than the

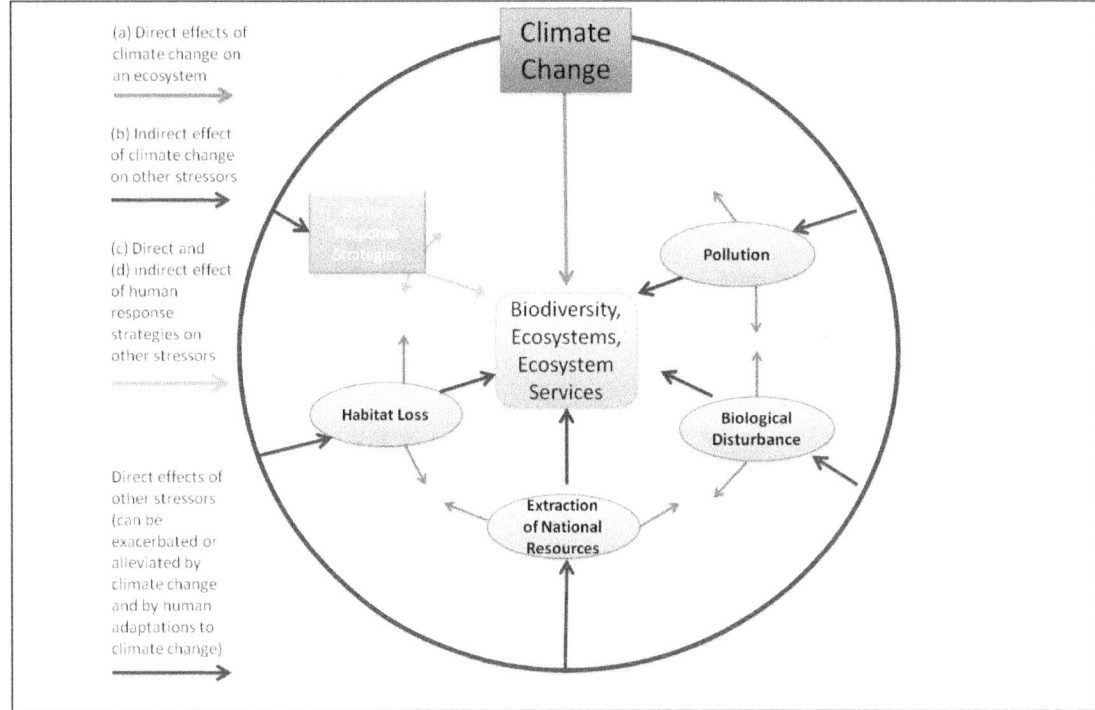

Figure 5.2. Conceptual diagram illustrating the multiple interacting environmental stresses that can affect natural systems, including climate change. The figure does not completely represent the full possibility of interactions. For example, each environmental stressor can interact with climate change and with any of the other stressors, and there may be interactions that involve multiple stressors. Although most of the interactions will magnify the stress on the ecosystem, there are some situations where the interaction will diminish the total stress.

sum of the individual contributions of multiple stresses (Brook and others, 2008). Often, observed examples of interacting stresses in natural systems cannot be explicitly classified as linear or synergistic because we lack baseline information on the condition of natural systems and the independent impacts of individual stressors (Mora and others, 2007; Halpern and others, 2008ab).

Character of interactions

The interaction of climate change with another stressor will likely change the character of that stressor, altering its timing (for example, onset, duration, and frequency), its spatial extent, or its intensity. For example increased incidence of drought may impact the *timing* of water stress by causing water withdrawals for irrigation earlier in the season, for a longer period of time, or at higher frequency (Fischer and others, 2007). The water hyacinth, on the other hand, has expanded its *spatial* extent into higher latitudes as temperatures rise; it is now considered invasive in freshwater systems in over 50 countries on five continents (Villamagna and Murphy, 2010). Finally, climate change may alter the *intensity* of another stressor. In one river system in the southern United States, regional drought and high summer air temperatures in combination with altered water withdrawal management practices resulted in increased stream temperatures, which in turn raised mortality rates of thermally-sensitive mussels (Galbraith and others, 2010).

5.3. INTERACTIONS BETWEEN STRESSORS AND CLIMATE CHANGE ARE ALREADY BEING OBSERVED

Climate change is already interacting with multiple environmental stressors to affect biodiversity, ecosystems, and ecosystem services. Understanding and quantifying these effects is an active area of research. Here we survey current knowledge of the impacts of climate change interactions with four major categories of environmental stressors: land use and land cover change, extraction of natural resources, biological disturbances, and pollution. Illustrative examples of these interactions are offered in the text and in **Table 5.1**.

5.3.1. Land use and land cover change

Land use and land cover change in the United States is widespread (Lubowski and others, 2006). In addition to causing habitat loss and fragmentation, these changes alter hydrological and climatic regimes through complex processes associated with alterations to solar radiation absorption, surface aerodynamic roughness, rooting depth, and carbon cycling (Kvalevåg and others, 2010; Pongratz and others, 2010; Bonan, 2011).

Habitat loss, degradation, and fragmentation is a leading cause of terrestrial biodiversity loss and the impairment of ecosystem functioning, resulting in a loss of ecosystem services (IPCC, 2007; Fahrig, 2003; Krauss and others, 2010; La Sorte, 2006). Habitat loss occurs when natural areas are actively converted to other land cover types, such as agriculture or urban development, or when land cover changes due to climate change, natural disturbances, or indirect influences of environmental stress. Habitat fragmentation results when natural areas that were once continuous become isolated due to changes in land cover or land use between fragments; this can limit dispersal of species between natural areas. Isolated populations are frequently at higher risk for extinction and have reduced genetic diversity that may reduce their adaptive potential (Leimu and others, 2010). Species that are able to disperse across fragmented landscapes may be stressed by translocation, reestablishment of territory, and increased competition from other species (Vögeli and others, 2011).

Land use change need not be extensive to have substantial impacts. Linear features, such as roads, oil and gas seismic exploration grids, shipping lanes, transmission lines, and drainage ditches, can disrupt ecological functions of adjacent lands (Dale and others, 2011, Kareiva and others, 2007). Local land cover change can have regional impacts. For example, urban sprawl in northern latitudes appears related to declines in abundances in some migratory birds in southern latitudes (Valiela and Martinetto, 2007). Not all land cover changes pose the same threats: some open passageways for species invasion, some create new habitats with new ecosystems (seen often in urban environments), some completely change the regional landscape, and some have very little impact.

Climate change is likely to interact with land cover change in ways that impact biodiversity, ecosystems, and ecosystem services, often in a way that exacerbates the detrimental effects of land cover change. For example, habitat fragmentation may limit the pathways and increase the distance that species would need to disperse in response to climate change (Coristine and Kerr, 2011). Assessing the magnitude of the effects of climate and land use change interactions is difficult. A major challenge is that different species and taxonomic groups respond differently to habitat loss and climate change: insects, fish, birds, and primates have strong correlations between habitat, climate and success, but the correlations for plants, small vertebrates, and birds are not as strong (Hockey and Curtis, 2009).

A recent meta-analysis of empirical studies based on 1,779 observations from 168 studies worldwide found that biodiversity was more likely to be negatively impacted by habitat loss in areas that show evidence of precipitation changes (Mantyka-Pringle and others, 2012). One empirical study based on 35 years of surveys showed that locations in the Sierra Nevada mountains that underwent the greatest declines in butterfly species richness were at lower elevations, where habitat loss was more severe (Forister and others, 2010). An experimental manipulation of rotifers, a microorganism, showed that habitat fragmentation and overfishing combined with increasing temperatures could lead to a decline in rotifer populations up to 50 times faster than when either threat acted alone (Mora and others, 2007).

In addition to impacting biodiversity, ecosystems, and ecosystem services directly and in conjunction with climate change, land use and land cover change can have indirect effects as a driver of climate change (IPCC, 2007). Land use and land cover change in both rural and urban environments has been a major contributor to greenhouse gas emissions (Grimm and others, 2008; Satterthwaite, 2008). Land use and land cover change such as clear-cutting of forests can also affect local weather patterns, for example by reducing rainfall and increasing temperature, and potentially reduce biomass storage of carbon (Dale and others, 2011).

5.3.2. Extraction of Natural Resources

Climate change can complicate the management of natural resources, often by exacerbating the stress that extraction of that resource puts on biodiversity, ecosystems, and ecosystem services. In some cases, climate change can have positive effects on systems that are already affected by extraction. Here we identify a variety of interacting impacts of climate change with fisheries, forest harvest, water withdrawals, and acid mine drainage.

Climate change has had both negative and positive effects on fisheries. Overexploitation of fish stocks is a concern in many freshwater and marine environments, and any exploited fishery needs to have sufficient replacement of the stock being exploited. However, water temperature, which is affected by climate change, can substantially affect vital rates—such as individual growth rates, survival, and reproduction—and thereby rates of population replacement. In some cases, climate change may make new resources available. In the Arctic, for example, potential new fishing grounds have recently been made more available by melting sea ice (NPFMC, 2009).

Overharvest of forests has the dual effect of causing local environmental damage that can decrease the resilience of an ecosystem to climate change, as well as potentially compounding the magnitude of climate change itself by releasing stored carbon into the atmosphere. For example, in the Hoh temperate rainforest in Washington State, the climatic conditions that support the Hoh are enabled by other nearby forest types. As those forest types have been harvested by commercial logging, the health of the Hoh has diminished. In some cases forest harvest can result in localized cooling (Gibbard and others, 2005), and in some cases increased harvest has been presented as a tool for addressing a climate change-caused problem, such as pest infestation (Nelson, 2007).

The extraction of water for human uses can be another stress on biodiversity and ecosystems, even while providing an important ecosystem service. There are over 75,000 large dams in the continental United States (Graf, 1999). In the West, reservoirs store up to 6 times the annual runoff, compared to only about 2 times the runoff in the East (Sabo and others, 2010a). Yet, watersheds in the East are more fragmented by smaller dams (Sabo and others, 2010a). Controlled water releases from large dams often affect the flow and thermal regimes of rivers, downstream sediment inputs, and hence habitat for the biota (Olden and Naiman, 2010; Sabo and others, In Press) and the composition of river food webs (Sabo and others, 2010a). These effects can either exacerbate or alleviate impacts of climate change. In the Colorado River, for example, coldwater releases from dams have been implicated in the decline of native fishes such as the Federally endangered humpback chub (Petersen and Paukert, 2005). Climate-induced declines in reservoir levels lead to discharge of warmer water, which has helped the humpback chub and other native species, though the warmer water has reduced the growth of the economically important sport fishes such as rainbow trout (Paukert and Petersen, 2007). In the Kiamichi River (Oklahoma), drought-induced changes in river flow and temperature patterns in conjunction with climate-induced temperature increases and changes in reservoir release rates resulted in changes to mussel communities and higher mortality rates for certain species (Galbraith and others, 2010).

Water withdrawals also impact terrestrial systems. For example, lowered water tables caused by increased aridity, groundwater pumping or stream flow diversions can trigger state changes in riparian plant communities characterized by shifts from tall forests to shrublands with deep roots and short canopies (Stromberg and others, 2007; Stromberg and others, 2010). In some cases, the combined effects of water withdrawals and climate change can have positive effects on biodiversity. For example, one study found higher levels of butterfly biodiversity adjacent to irrigated fields, suggesting that irrigation may mitigate water-limitation effects of climate change in ecosystems adjacent to agricultural fields (González-Estébanez and others, 2011).

5.3.3. Biological Disturbance

Biological disturbances can occur naturally—as in the case of native pest species and pathogens—or as the consequence of human activities when new species, pests, or diseases are introduced into an ecosystem. Climate change is already exacerbating the impacts of these stresses.

Invasive species have responded to recent changes in climate (Walther and others, 2009). Most notably, warming, an increase in the frequency of extreme events, and increasing carbon dioxide concentrations are thought to have facilitated invasive species' spread (Driscoll and others, 2011). For instance, the expansion of buffelgrass in the southwestern United States is associated with warmer winters experienced since the 1980s (Archer and Predick, 2008). Invasive species can also benefit when extreme climatic events stress or kill native species, increasing native communities' susceptibility to invasion (Diez and others, In Press; Brown and others, 2011). Changing climate patterns may also create new niches that can be exploited by invasive species and used by those species to expand their distribution. For example, climate change can increase the frequency and intensity of fires, which will impact the distribution of fire-adapted species such as cheatgrass, an invasive species considered particularly noxious in rangelands in the western United States. Climate-driven changes in fire regimes will affect native species as well, likely reducing population viability for some species as they cope with rising temperatures and shifting habitats caused by changes in fire frequency (Keith and others, 2008).

Climate change is also affecting the geographic ranges, ability to spread, and virulence of many pests, pathogens, parasites, and disease vectors. In turn, the ability to manage or control outbreaks has been impaired. Ranges of many pests and diseases have recently expanded, including the near epidemic spread of pine and bark beetles in the American West (Bentz and others, 2010) and the northward expansion of the oyster diseases MSX and dermo to Nova Scotia (Ford and Smolowitz, 2007). Climate change impacts on pests may have cascading effects: projected increases in temperature in boreal forests in Alaska may increase the frequency and severity of insect outbreaks and the wildfires resulting from associated tree mortality (Wolken and others, 2011).

It is well established that wildlife and zoonotic disease emergence are linked to environmental variables such as climate (NRC, 2001). For example, climate-warmed habitats for vectors are believed to have facilitated the outbreak of hantavirus (Clement and others, 2009), and climate-induced habitat change has expanded the range of amphibian mortality due to fungal pathogens (Pounds and others, 2006). However, climate change may not result in a net disease increase. Areas that become climatically suitable for a new pest, pathogen, or host may contain other barriers to the spread of disease. For example, the newly introduced pest, pathogen, or host may be limited by competition with other species, physical barriers, or predation. Furthermore, climate change may induce other habitat changes that are less favorable to the spread of disease (Slenning, 2010). Some have argued that it is possible that climate change will be associated with the decline of pathogens, vectors, and hosts (Lafferty, 2009).

5.3.4. Pollution

The additional stress of climate change may magnify the adverse effects of pollutants—including metals, pesticides, persistent organic pollutants, excessive nutrients, endocrine disruptors, and atmospheric ozone—on humans, wildlife, and the environment (Hansen and Hoffman, 2011). Climate change can alter temperature, pH, dilution rates, salinity, and other environmental conditions that in turn modify the availability of pollutants, the exposure and sensitivity of species to pollutants, transport patterns, and the uptake and toxicity of pollutants (Noyes and others, 2009). For example, increasingly humid conditions could result in the increased use of fungicides, whereas altered pH can change the availability of metals. In the first case, a greater quantity of a contaminant is introduced into the environment; in the second, the contaminant is simply more biologically available to cause damage. In cases where climate change affects transport patterns of environmental pollutants, pollutants may reach and accumulate in new places, exposing biota to risk in different habitats. Climate change effects on uptake and toxicity can be the result of direct increases in the toxicity of some chemicals or increased sensitivity in the target species. Sensitivity can be increased due to general metabolic stress due to environmental changes or inhibition of physiological processes that govern detoxification. Some contaminants that were thought to be diminishing in concern, such as PCBs, are being remobilized in the environment by climate change. In recent years it has been shown that persistent organic pollutants, deposited in glaciers during the period of heavy use in the mid-twentieth century, are being released due to climate-induced melt (Blais and others, 2001).

Heavy metals are a widely dispersed class of pollutants with both lethal and sublethal effects on organisms. Metals can be deposited from anthropogenic sources (smelting, fossil fuel combustion, waste incineration) and are particularly dangerous because they do not break down in the environment, and can accumulate in soil, water, sediments, and biological tissues. They also have significant adverse biological impacts (Sorvari and others, 2007; Ayeni and others, 2010; Brumbaugh and others, 2010, Campbell and others, 2010; Kouba and others, 2010; Stephansen and others, 2012). Various manifestations of climate change are very likely to increase the availability and toxicity of heavy metals. For example, altered pH can make metals more biologically available in aquatic systems, thereby increasing their adverse impact on the environment. Similarly, increasing temperature can increase exposure to metals by increasing respiration rates of many ectotherms such as fish (Ficke and others, 2007). Heavy metal pollution is a product of fossil fuel combustion rates as well as other industries such as mining (Renberg, 1986), so future release of these pollutants will, in part, be determined by mitigation responses to climate change.

Excess nitrogen and phosphorus can leach into soil and waterways through the manufacturing and application of fertilizer, and nitrogen is also an atmospheric pollutant produced in the combustion of fossil fuels (Rabalais and others, 2009). Excess nitrogen can hamper or boost the growth of nitrogen sensitive plants, including algae (Filippino and others, 2011; Howarth and others, 2006). While increased primary production can lead to an increase in fish abundance, when the increase in nutrients exceeds the capacity of a system to absorb the increased phytoplankton production, algal blooms and oxygen depletion will occur, reducing the abundance of fish (Glibert and others, 2005). Climate change and increased anthropogenic nutrient loading will make coastal ecosystems more susceptible to the development of hypoxia through enhanced stratification, decreased oxygen solubility, increased metabolism and remineralization rates, and increased production of organic matter (Boesch and others, 2007).

Table 5.1. Interacting stressors and non-speculative examples of their effects on biodiversity, ecosystems and ecosystem services when combined with climate change.

Stress	Climate Change Interaction Example
Habitat Loss	
Fragmentation	The dramatic decline of the green salamander, *Aneides aeneus*, a species with a highly fragmented habitat, in the southern Appalachians of the United States, is due in some part to the increase in the July temperature and greater fluctuations in the January temperature since 1970, coupled with its limited ability to disperse in landscapes modified by logging, resort development, and dams (Corser, 2001).
Urban development	Urban areas create their own microclimate; particularly well documented is the urban "heat island" effect where mean temperatures can be several degrees (6-9°C) higher than in surrounding rural areas. The heat island effect interacts with climate change and can contribute to stress in plants, animals and humans (Imhoff, 2010).

Land use changes	In an attempt to lower carbon emissions, land is being set aside for wind farms. However, studies show high levels of bird and bat fatalities related to wind turbines. Facilities in West Virginia and Pennsylvania had over 2000 bat fatalities in a 6-week period and facilities in California report over 1000 raptor fatalities a year. Wind farms currently do not pose a major threat to overall population levels but proposed development of wind power in migratory flyways containing high numbers of species are causing concern. Bats and raptors will potentially be the most affected due to their low birth rates and thus slow population growth (GAO, 2005).

Extraction of Natural Resources

Exploitation (Forest, Fisheries)	The Atlantic cod is an important food fish found along both the western and eastern parts of the North Atlantic Ocean and many of the historically large populations have been severely depleted by overharvesting. Cod species recruitment is strengthened during cold years and weakened during warm years. Therefore, it is predicted that a full recovery of the North Sea cod stock might not be expected until the environment becomes more favorable (Olsen and others, 2011).
Mining	Mines are point sources for metal and acid pollution, which can accumulate in the topsoil during dry spells and then get washed into receiving streams during heavy rains poising a danger to aquatic life. Climate change is lengthening dry summers in the western United States and rainstorms are further apart and more intense when they happen, leading to increased risk of pollution from mines. To be prepared for more extreme conditions, remediation efforts would need to increase the capacity of engineered designs (Kirk, 2009).
Irrigation	California's Central Valley is one of the most productive agricultural regions in the world producing about 250 different crops with an estimated value of $17 billion per year. This irrigated agriculture relies heavily on surface-water diversions and groundwater pumping. The Central Valley also is rapidly becoming an important area for California's expanding urban population. This surge in population has increased the competition for water resources within the Central Valley and Statewide. Projected climate changes include less snowpack, which would mean less natural springtime replenishment of water storage in the surface-water reservoirs. More variability in rainfall could place more stress on the reliability of existing flood management and water-storage systems (Faunt, 2009).

Biological Disruptors

Invasive species	Climate change is projected to increase the incidence of drought in some areas, thus expanding arid habitats. Cheatgrass (*Bromus tectorum*) is invasive in arid and semi-arid shrublands and grasslands of the Intermountain West. As do several other invasive annual grasses, cheatgrass promotes fire, creating a positive feedback cycle favoring further invasion, the exclusion of

	native plants, and loss of carbon (Crowl and others, 2008). Although cheatgrass abundance is likely to increase in some regions, climate envelope models suggest that other areas in which the species is currently abundant may become wetter and thus less hospitable, potentially providing opportunities for ecological restoration (Bradley, 2009).
Disease	Disease outbreaks caused by the bacteria *Vibrio* have been shown to correspond with increased precipitation, and thus decreased salinity, as well as increases in ocean temperature. In 2005, an outbreak of *Vibrio parahaemolyticus* caused the deaths of otters in Puget Sound, and seasonal expansion of *V. vulnificus* is responsible for illnesses associated with oysters harvested from the Gulf of Mexico. Recent data from the Centers for Disease Control indicate *Vibrio* infections in the United States have increased since 2000, corresponding to the frequency and severity of extremes in temperature and precipitation. These observations suggest that climate change in the United States may expand the risk of illness in wildlife and humans from *Vibrio* (Martinez-Urtaza and others, 2010).
Pests	The hemlock woolly adelgid (*Adelges tsugae*), an invasive insect introduced to eastern North America from Japan, has decimated stands of Eastern and Carolina hemlock from Georgia to Connecticut. However, its spread across central and northern New England has been slowed substantially by its inability to tolerate cold winter temperatures. In the future, rising winter temperatures due to climate change (IPCC, 2007; USGCRP, 2009) are likely to remove the conditions currently limiting adelgid spread, and facilitate northward expansion into new habitat (Paradis and others, 2008).
Disturbance Regimes	Recent studies have documented an increase in the occurrence of large fires during the last few decades in mid-elevation regions of the northern Rocky Mountains due to increases in temperature and drier conditions. If fire frequency increases forests will re-burn before they have re-accumulated the carbon lost in the previous fire. As a result, Rocky Mountain forests could become carbon sources instead of carbon sinks, which could worsen global climate change. More frequent fires would also mean that mature and old-growth forests, in the central and northern Rockies, will be increasingly replaced by young forests or even by non-forest vegetation during this century (Smithwick and others, 2010).
Pollution	
Persistent Organic Pollutants	As glaciers melt as a result of climate change, persistent organic pollutants (POPs), such as PCBs and DDT that were incorporated into the ice through atmospheric deposition prior to being banned years ago, are released. POPs make their way into the ecosystem and food chain increasing the level and type of POP exposure to wildlife. This has been recently demonstrated for Western Hudson Bay polar bears, where temporal changes in sea ice conditions (1991–2007) were found to be linked to polar bear diet and an increase in bioaccumulated pollutants (Letcher and others, 2010).
Heavy Metal	Interactions between environmental temperature and metal pollution strongly affect physiological tolerance to both stressors and can limit the survival and

Impacts of Climate Change on Biodiversity, Ecosystems, and Ecosystem Services |
Technical Input to the 2013 National Climate Assessment

Chapter 5
Multiple Stressors

	distribution of ectotherm populations in the face of global climate change and increasing anthropogenic pollution of aquatic environments. An increase in metabolic rates at elevated temperatures may contribute to metal accumulation in ectotherms due to a higher energy demand. Studies show that an increase in environmental temperature results in elevated mortality rates in metal-exposed ectotherms in 80 percent of the cases (Sokolova and Lanning, 2008)
Nutrient Loading / Eutrophication	Eutrophication increases production of phytoplankton, including harmful algal blooms; decreased water clarity, resulting in loss of seagrasses; altered food chains; and severe depletion of dissolved oxygen in the water column (hypoxia). Despite substantial expenditures to reduce nutrient pollution in the Chesapeake Bay, reports of record-sized hypoxic zones in 2003 and 2005 raised concerns about whether progress is really being made. Hypoxia in the Chesapeake Bay, and in most other regions experiencing this phenomenon, is greatly affected by climate (river inflows, warm temperatures, and relatively calm summer winds), as well as by nutrient inputs from human activities (Boesch and others, 2007).

5.4. ANTICIPATED INTERACTIONS OF CLIMATE CHANGE WITH OTHER STRESSORS

At present, land use and land cover change is the main driver of degradation for terrestrial systems and over-exploitation is the main driver for marine systems. As climate change persists, it is projected to become a dominant stress on biodiversity and ecosystems, and the interaction of climate change with existing and future environmental stressors is expected to exacerbate losses (MA, 2005). Understanding how climate change and these other stressors will interact to affect biodiversity, ecosystems, and ecosystem services is a daunting challenge (MA, 2005). It requires information not only on future scenarios of climate change and other global changes that cause environmental stresses, but also models of how climate change and these others stressors interact to affect biological systems. Consequently, only a few studies have explicitly considered the relative importance of climate change and other stressors in the future and even fewer studies specifically quantify whether there is a linear or non-linear effect. Furthermore, there is a notable lack of research on aquatic systems and little information that connects estimates of biodiversity loss to impacts on ecosystem services (Pereira and others, 2010).

Our general understanding about the combined effects of climate change and other environmental stressors on the future of biodiversity is informed by research on global biodiversity scenarios (Sala and others, 2000; MA, 2005; Sala and others, 2005). Scenarios, or storylines, present alternative futures given assumptions about indirect drivers of global change, such as human demography and economic development. These indirect drivers are then used to project changes in direct drivers, such as land use change and climate change, which can then be used as input to biodiversity and ecosystem services models (Pereira and others, 2010). Scenario exercises can thus provide an integrated view about the combined impacts of multiple global environmental changes, and can elucidate the general importance of various stressors on biodiversity and associated ecosystem services.

Sala and others (2000) developed global scenarios of biodiversity for terrestrial and freshwater biomes in 2100 using five drivers of change: land use, atmospheric carbon dioxide concentration, nitrogen deposition and acid rain, climate, and biotic exchanges. Averaged across biomes, land use change was found to be the most important driver, followed by climate change. In freshwater systems, biotic exchange was also important. The Millennium Ecosystem Assessment (MA) developed four global change scenarios that were used to evaluate impacts on biodiversity and ecosystem services in the terrestrial, marine, and freshwater realms (Sala and others, 2005). Overall, the scenarios project biodiversity losses, with associated deterioration of ecosystem services. For terrestrial systems, land cover change had the greatest impacts on biodiversity, followed by climate change and nitrogen deposition.

Although findings from the scenario analyses provide insight about the role of climate change and other environmental stressors on biodiversity and ecosystem services, they provide only a general overview of potential impact. Other studies, some of which make use of the MA scenarios, provide a more detailed perspective on the interaction of climate change and other stressors, either by examining combined effects or by comparing the future effects of climate change to historical effects of other stressors. In this section, we review key studies of the potential interactions between climate change and other stressors in the future that focus on the same grouping of environmental stresses discussed in section 5.3.

5.4.1. Projections for Climate Change Interactions with Land Use and Land Cover Change

The combined effects of climate change and land cover change, including habitat loss, are projected to contribute to species extinctions. Estimates of projected extinctions, and even the number of species at risk for extinction, vary based on taxonomy and are influenced by the climate change and land use scenarios evaluated as well as the modeling methodology employed. Resolution differences for modeling land use changes, biodiversity, and climate change make it challenging to evaluate species' response under future climate scenarios (de Chazal and Rounsevell, 2009). Nonetheless, most studies indicate that biodiversity will be negatively impacted by climate change and habitat loss.

Studies that examine the effects of climate change on biodiversity without explicit treatment of other environmental stressors can provide initial insight into the combined effects of climate change and habitat loss because they often include assumptions about species dispersal, a variable influenced by habitat availability. Across a wide range of terrestrial organisms analyzed under several climate change scenarios and species dispersal models, extinction rates ranged from 11 percent to 34 percent for a 0.8°C-1.7°C increase in temperature and from 33 percent to 58 percent for a greater than 2°C change in temperature (Thomas and others, 2004). Within global biodiversity hotspots, extinctions among endemics were projected to range from less than 1 percent to 43 percent, depending on the models and scenarios analyzed (Malcolm and others, 2006). Of 25 global hotspots analyzed, the California Floristic Province was identified to be one of the areas most vulnerable to climate change (Malcolm and others, 2006). In both studies, extinction risks were higher when species were modeled under limited dispersal scenarios, situations that reflect both the inherent biological constraints of many organisms and the ways in which habitat fragmentation may impede species dispersal.

Only a few modeling studies have explicitly examined the combined effects of land cover change and climate change, most notably two studies that examine possible impacts on bird populations. Jetz and others (2007) use the four MA scenarios, which incorporate information on climate change and land use change, to estimate impacts on the breeding range of 8,750 land bird species. Based on scenarios of habitat change, between about 4.5-10 percent of species were projected to have more than half of their current range transformed to different habitats by 2050, and about 10-20 percent were projected to have their range transformed by 2100. Climate change was the dominating effect driving range contractions in temperate regions, whereas land cover and land use change was the dominating effect in the tropics. With a different set of assumptions that allowed species to change their elevational limits in response to climate change, Sekercioglu et al (2008) used an intermediate estimate of warming of 2.8°C combined with the MA land cover change scenarios to project that by 2100, 4.5-6 percent of species would go extinct and an additional 20-30 percent of species would be at risk of extinction.

Quantitative projections also indicate that the combined effects of land use change and climate change will result in the loss of vascular plant diversity. Based on the MA scenarios, by 2050 there could be a loss of 7-24 percent of plant diversity relative to 1995 (Van Vuuren and others, 2006). Between 2000-2050, land use change was a bigger contributor to the loss of species diversity relative to climate change, but the impacts of climate change after 2050 were expected to become increasingly important (Van Vuuren and others, 2006).

5.4.2. Projections for Climate Change Interactions with Water Extraction

Water flow regimes shape the biodiversity of freshwater systems and increasing interference of flow regimes has been associated with greater ecological changes (Poff and Zimmerman, 2010). Climate change can affect riverine ecosystems both through changes in temperature and through changes in flow regime, which can be further altered by human modifications such as water withdrawals and dams. Xenopoulos and others (2005) examined the combined effects of increased water withdrawal and climate change projected that 25 percent of rivers were forecasted to lose more than 22 percent of fish species by 2070. For three out of the four rivers in the United States for which data were presented, the combined effects of climate change and water withdrawal were notably greater than the effect of climate change alone. Spooner and others (2011) also found significant declines in mussels in scenarios where climate change and water withdrawals were considered.

The impact of climate change on freshwater systems is expected to be greater in extent and intensity than the current impact of dams and water withdrawals. Using two global climate models and two emissions scenarios (A2 and B2), Döll and Zhang (2010) found that climate change was forecasted by 2050 to significantly change seasonal water flow regimes across 90 percent of the global land area (excluding Greenland and Antarctica), which is a notably higher impact than the 25 percent of the global land area that has been significantly affected by water withdrawals and dams as of 2002. Furthermore, river discharge was projected to change on about 33 percent of global land area due to climate change, compared to less than 5 percent of the land area that has already been affected by water withdrawals and dams (Döll and Zhang, 2010). Since this analysis used 2002 data for dams and water withdrawals in the future projections, the expansion of dams and increased demand for water could further worsen the impacts of climate change in many, but perhaps not all, locations. However, in areas where there is currently extensive irrigation, such as the High Plains Aquifer in the western U.S., future climate change may not have a greater impact than current dams and water withdrawals.

5.4.3. Projections for Climate Change Interactions with Biological Disturbances

Although many recent studies have projected how the distributions of invasive species and disease vectors could be affected by different climate change scenarios (Crowl and others, 2008; Bradley and others, 2010), few of these studies have quantified how changes in distributions could lead to specific losses in biodiversity, ecosystems, or ecosystem services. Forecasting changes in impacts from pests, pathogens, or invasive plant species is often fraught with uncertainties beyond those in forecasting climate change alone; often too little is known about the climatic tolerances or responses of the species of concern to make confident projections under a given scenario (Dukes and others, 2009). However, general principles and several case studies suggest that, in some areas, some pests, pathogens, and invasive species will respond strongly to future conditions, likely increasing their impacts. In other cases or areas, the climate is likely to become less favorable for certain unwanted species, reducing their impacts (Bradley and others, 2009).

Warming is likely to increase the ranges of several invasive plant species in the southern and western United States (Bradley and others, 2009b, 2010), potentially increasing their impacts. In addition, projected increases in temperature in boreal forests in south-central and Kenai Peninsula, Alaska will likely increase the probability of establishment of invasive plant species (Wolken and others, 2011). Hemlock wooly adelgids, which have killed many eastern hemlocks in recent years, are likely to expand their ranges to the north as climates warm, spreading into portions of the hemlocks' range that were previously too cold for the insect pests (Paradis and others, 2008; Dukes and others, 2009). Although warming may have strong effects on future ranges, changes in the abundance of habitat types within a region have the potential to be equally or more important in controlling abundance of some invasive plant species (Ibáñez and others, 2009). Changes in precipitation will also affect invasive species, although the nature of the effects will depend on the type and timing of precipitation change (Blumenthal and others, 2008; Suttle and others, 2007).

The lack of research on disease is potentially troublesome in freshwater environments, where emerging diseases may pose a notable threat (Okamura and Feist, 2011). Given that climate change may exacerbate or ameliorate the impacts of disease, for example through disease range expansions or contractions (Harvell and others, 2002; Lafferty, 2009), a better understanding of future projections can help prioritize monitoring and adaptation efforts. For example, based on an assumption of a 2°C warming and changes in precipitation, Benning and others (2002) found that extant Hawaiian honeycreepers may be driven to extinction through the combined effects of climate change, avian malaria, and historical land use changes.

5.4.4. Projections for Climate Change Interactions with Pollutants

Few studies have examined the projected impacts on biodiversity, ecosystems, or ecosystem services of future interactions between climate change and pollutants. However, one area that has been explored is how elevated levels of atmospheric carbon dioxide may interact with increases in nitrogen availability, resulting from human activities, to affect ecosystems and their ability to help regulate climate. Increased abundances of atmospheric carbon dioxide often stimulate plant growth, if other nutrients are not limiting. With increased nitrogen deposition, plants may sequester more carbon and therefore help mitigate climate change (for example, Reich and others, 2006ab; McCarthy and others, 2010; Norby and others, 2010). Yet, the interactions of climate change with nitrogen deposition are not always this straightforward. In a study conducted in a brackish marsh, a short-term increase in plant productivity in response to carbon dioxide and nitrogen additions reversed in subsequent years because the additions ultimately led to a change in plant community composition, and the new plant community was less productive (Langley and Megonigal, 2010).

Atmospheric ozone is created in sunlight-dependent reactions involving nitrogen oxides and volatile organic compounds. Atmospheric ozone concentrations are determined by emission rates of these precursor compounds, as well as atmospheric circulation patterns, air temperatures, and other factors. Over the course of this century, climate change is projected to cause increases in ozone concentrations in many regions of the world (Sitch and others, 2007; Ebi and McGregor, 2008; Selin and others, 2009), including biodiversity hotspots (Royal Society, 2008). Exposure to ozone damages plants and animals, with major economic consequences; damage to global crops already has an estimated annual fiscal impact of $14-26 billion (van Dingenen and others, 2009). By 2050, annual costs to human health from global ozone pollution could reach $580 billion (relative to ozone levels under preindustrial conditions), with more than 2 million deaths due to acute ozone exposure over that time period (Selin and others, 2009). Although effects on natural ecosystems are less well studied, ground-level ozone reduces growth of many wild plant species (Hayes and others, 2007; Wittig and others, 2009). Ozone damage will likely offset some productivity gains due to rising atmospheric CO_2 levels, reducing carbon storage on land, and thus contributing to climate change (Sitch and others, 2007). At the same time, increases in atmospheric CO_2 will cause plants to open their stomata less, limiting the damage from ozone (Mattysek and others, 2010). Similarly, drought conditions that are severe enough to cause stomatal closure limit plants' exposure to ozone (Löw and others, 2006), and these conditions are likely to become more frequent in some regions. Very few studies have examined the potential consequences of increasing ozone concentrations for biodiversity, and predictions are complicated by the possibility of interactions in responses across trophic levels. However, the documented reductions in wild plant productivity in response to ozone exposure suggest that these pressures on biodiversity be taken into account alongside those from climate change.

5.5. CASE STUDY: WATER USE IN CALIFORNIA'S CENTRAL VALLEY

Water use in California has profound effects on regional biodiversity and ecosystem processes. Future water demand, combined with climate change and other environmental stresses, is likely to have even greater impacts on the biota, their ecosystems and services. Agriculture is the biggest user of water in California (Kenny and others, 2009), providing crops for local consumption and export to other States and countries (USDA, 2009; Hoekstra and Mekonnen, 2012). Climate change is likely to both increase agricultural water demand and decrease available water (Center for Irrigation Technology, 2011). Human consumption places the next greatest demand on water; this demand is expected to grow with California's expanding population. These uses have resulted in decreased water for in-stream flows, creating a major stress to biodiversity, ecosystems, and ecosystem services. For example, this has implications for salmon population viability (Ligon and others, 1995), the health of nearshore coastal and delta fisheries (for example, shellfish) that are dependent on freshwater inputs (Drinkwater and Frank, 1994), and water quality in rivers and deltas (Sabo and others, 2010b).

Salmon, once abundant in California's coastal and Central Valley rivers, are now imperiled; several stocks are listed as threatened or endangered under the U.S. Endangered Species Act. The decline in the California Chinook salmon fishery resulted in economic impacts on the order of $22.7 million between 1993-2005 (UOP, 2010). As illustrated in **Figure 5.3**, a changing climate will reduce the quality and quantity of habitat for native freshwater species like salmon directly by increasing water temperatures and the likelihood and severity of droughts (Battin and others, 2007) Salmon habitat is also likely to experience indirect effects of climate change, for example by inducing more fertilizer and pesticide use, and a subsequent increase in water pollution, or by creating greater demand for water to irrigate drought-stricken crops.

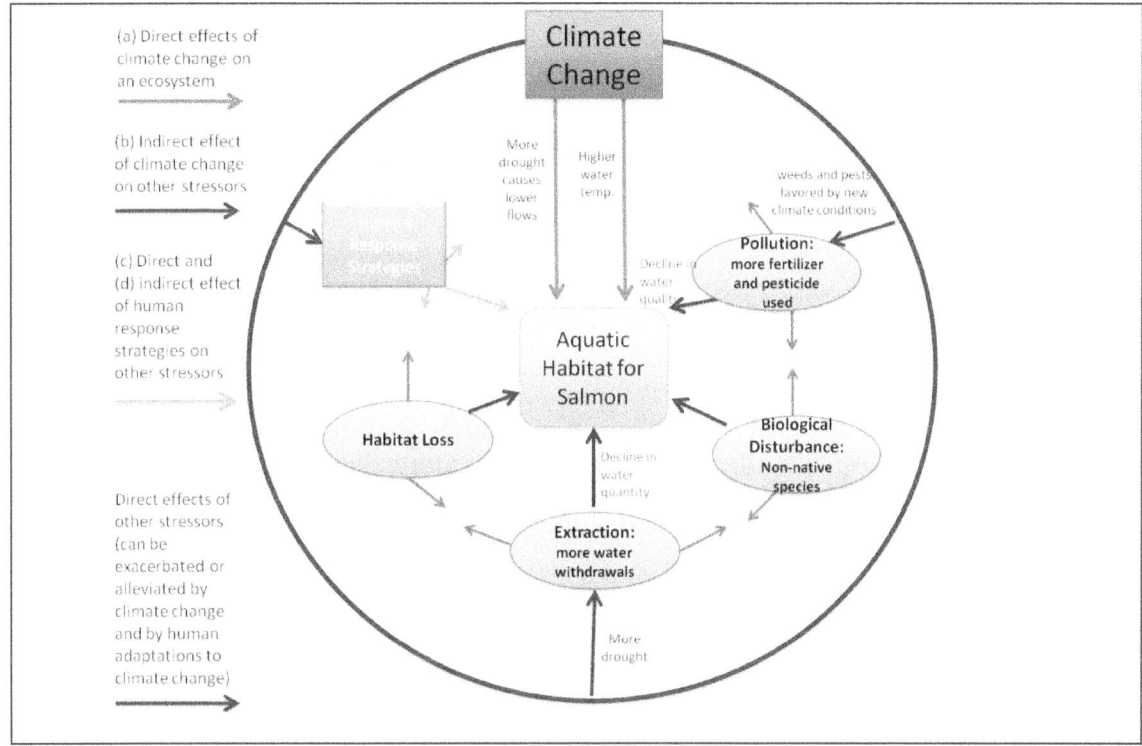

Figure 5.3 *Conceptual diagram of the impacts of climate change interaction with multiple environmental stressors on salmon and their aquatic habitat in California. Human response strategies to climate change may interact with many of these stressors to exacerbate or alleviate the effects of stressors to salmon.*

Climate change is also likely to interact with other stressors to impact many other fish species. Fishes in freshwater and in deltas are thought to be declining due to reduction in the quantity, quality (temperature), and variability in surface water, including imperiled species such as green sturgeon and the iconic delta smelt (CADFG, 2011). Projected increases in water temperatures with climate change will increase the competitive edge of non-native salmonids (Wegner and others, 2010), causing range-wide declines in available habit for native trout, some of which are also threatened. The encroachment of non-native fishes will also likely accelerate declines of native frogs (Knapp and others, 2007).

Many potential interactions between climate change and other stressors hinge on the way in which climate change adaptation policies for water management are implemented over the next century, and may even challenge us to do conservation triage. For example, winter flooding of rice fields in California's northern Central Valley can provide habitat for migratory waterbirds

(Elphick and Oring, 1998) and winter floods provide juvenile salmon access to productive rearing habitats on floodplains (Sommer and others, 2001). Climate change induced reductions in stream flow may limit water releases such that adequate habitat for both fish and birds cannot be maintained; water managers would have to find a way to optimize water release regimes to balance competing biodiversity needs as well as address demands for agricultural production and human consumption.

5.6. INTEGRATING CLIMATE ADAPTATION AND CONSERVATION STRATEGIES FOR OTHER STRESSORS

Reducing the extent that other stressors affect species, biodiversity, ecosystems, and ecosystem services can be an important component of adapting to climate change. Climate adaptation strategies (for example, USFWS and NOAA, 2012) and literature reviews (for example, Glick and others, 2009, Heller and Zavaleta, 2009, Lawler, 2009, Mawdsley and others, 2009, Hansen and Hoffman, 2010) have highlighted the importance of tackling other stressors. For example, the National Fish, Wildlife and Plant Climate Adaptation Strategy draft (USFWS and NOAA, 2012) identifies reducing "non-climate stressors to help fish, wildlife, plants, and ecosystems adapt to a changing climate" as one of its goals and identifies several specific management objectives under that goal (see **Box 5.1**). It is important that these strategies consider the complexities of interactions between climate change and other stressors in order to achieve the most effective solutions.

Box 5.1. National Fish, Wildlife and Plant Climate Adaptation Strategy draft: Goal 7 (USFWS and NOAA 2012)

Goal 7: Reduce non-climate stressors to help fish, wildlife, plants, and ecosystems adapt to a changing climate. Reducing existing threats such as habitat degradation and fragmentation, invasive species, pollution, and over-use can help fish, wildlife, plants, and ecosystems better cope with the additional stresses caused by a changing climate.

1. Slow and reverse habitat loss and fragmentation.
2. Slow, mitigate, and reverse where feasible ecosystem degradation from anthropogenic sources through land/ocean-use planning, water resource planning, pollution abatement, and the implementation of best management practices.
3. Use, evaluate, and as necessary, improve existing programs to prevent, control, and eradicate invasive species and manage pathogens.

Natural resource managers often have substantial training and experience addressing other stressors and have management strategies that can serve as a framework for integrating climate considerations (Glick and others, 2009). Therefore resource management strategies developed to help natural systems adapt to climate change need not be novel, as existing practices aim to reduce the impacts of various environmental stresses. For example, improved management practices have been shown to reduce erosion, excessive nutrients, and other possible pollutants into the landscape. Harvest strategies for fish, wildlife, and forests are often in place to ensure these systems remain healthy and intact. Regulatory limits have been set for emissions of contaminants from point and nonpoint sources into water and the air. Although developing and implementing these strategies can be challenging because of the diverse stakeholder groups and interests involved, particularly at larger spatial scales that cross political boundaries, the know-how to develop them is often available.

However, in the context of changing climate, existing actions may not be sufficient to achieve desired outcomes in an efficient manner (Hansen and Hoffman, 2011). Current formulations of natural resource management strategies often omit the compounding effects that climate change may cause, which can reduce the efficacy of what may be best practices from past environmental conditions. For example, **Figure 5.4** shows how an environmental stressor, coupled with climate change, may have a larger consequence for the condition of biodiversity than an individual stressor alone (Hansen and Hoffman, 2011).

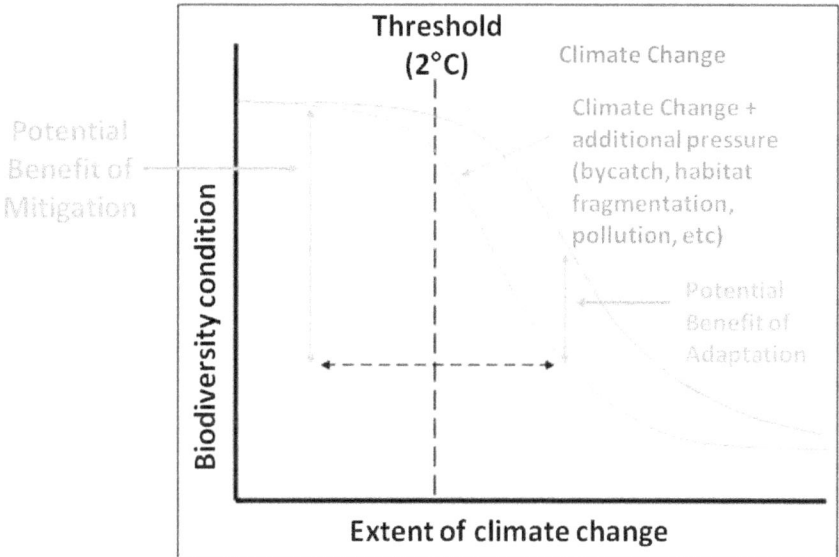

Figure 5.4 *An environmental stressor, coupled with climate change, may have a larger consequence for the condition of biodiversity than an individual stressor alone. Adaptation actions can be implemented to ameliorate the impact of these combined stressors (Modified from Hansen and Hoffman, 2011).*

As discussed in *Chapter 6* of this report, many current conservation plans do not consistently include consideration of interactions of existing environmental stressors with climate change. As of 2008, only 10 percent of U.S. threatened and endangered species recovery plans indicated climate change was a threat (Povilitis and Suckling, 2010). While more plans have considered climate change in recent years, the number of them that include actions to address this threat is still quite low.

The potential for maladaptive management strategies that address one stressor but exacerbate another is an additional concern for effective stewardship of biodiversity, ecosystems, and ecosystem services. As portrayed in our conceptual framework (**Figures 5.1** and **5.2**), efforts to respond to climate change—either for the benefit of human systems or natural systems—can become new stressors themselves. For example, some have suggested that efforts to improve habitat connectivity could also facilitate the expansion of invasive species (for example, Proches and others, 2005; although, see Damschen and others, 2006 for a contrasting example). Research is still needed to clarify how best to identify possible unintended consequences and reconcile different objectives.

In the remainder of this section, we highlight some examples of important challenges that climate change poses for conservation strategies that address a range of environmental stressors, as well as opportunities to assist climate adaptation by using existing management strategies. **Table 5.2** provides some illustrative examples of conservation strategies that are being reconsidered to account for climate change. Some of these strategies are more developed, while others are still in a conceptual phase. A few of these strategies have been initiated in limited ways, but on-the-ground examples are still relatively limited.

5.6.1. Habitat connectivity imperative for addressing habitat loss

A recent survey of published literature indicated that increasing connectivity of protected areas is one of the most common recommendations for managing biodiversity in the face of climate change (Heller and Zavaleta, 2009). The patchwork of existing protected lands and waters are an important tool for protecting biodiversity and ecosystems; however, climate change may render the current collection of protected areas insufficient as habitats and species shift their locations. Thus, many of the climate adaptation strategies that are now being designed or implemented focus on how best to enhance habitat connectivity.

Many suggestions have been made for how to locate, design, and connect terrestrial reserves to accommodate new climate conditions (Lawler, 2009). Many strategies for improving habitat connectivity have focused on expanding protected areas to provide more space for species to shift, cover broader climatic conditions, or to span multiple biomes. One such strategy that could be considered is for existing programs (for example, the National Wildlife Refuge system) to increase the number of functionally connected units, which may increase resilience to climate change for migratory species (Griffith and others, 2009). Recently, efforts have been made to locate new or expand existing protected areas based on projected shifts in species distributions under different climate scenarios (for example, Game and others, 2011; Hannah and others, 2007; Vos and others, 2008; Ackerly and others, 2010; Williams and others, 2005; Rose and Burton, 2009). Another option for increasing habitat connectivity might be to increase the permeability of the landscape, thereby facilitating the ability of wildlife to travel from one protected area to another (Franklin and Lindenmayer, 2009).

Fragmentation is also a concern for coastal habitats, where losses from development would be compounded by sea level rise, projected to inundate beaches, wetlands, and agricultural areas critical to many species (Griffith and others, 2009). One response option is to use existing conservation programs available through the Farm Bill (for example, Wildlife Habitat Incentives Program, Wetland Reserve program) to provide compensation to willing landowners to protect and enhance areas likely to be flooded under future sea level rise (USFWS and NOAA, 2012). However, existing conservation tools may be insufficient to maintain healthy coastal habitats, especially in locations where they can not migrate due to existing development (Titus and others, 2009).

5.6.2. Managing harvest must contend with new situations

Harvest of plants and animals are frequently managed through regulations or policies that limit how much or which sorts of specimens can be removed. These strategies will likely need to be adjusted to accommodate impacts of climate change on the structure and functioning of ecosystems.

One good example is the need to modify size and catch limits of fish to reflect the impacts that changing climate conditions, such as ocean temperatures, will have on the distribution and abundance of economically important species. This issue was recently confronted in the Arctic when a potential new fishing ground was made more available by melting sea ice. The North Pacific Fisheries Management Council (NPFMC) determined that they would close this area to commercial fishing until "adequate scientific information on fish stocks and how commercial fisheries might affect the Arctic environment are available" (NPFMC, 2009). This is a case where an adaptation strategy (closing the fishery) was developed due to a high degree of uncertainty (virtually no scientific information about fish harvest in a region that has never had large-scale commercial fish harvest) (Stram and Evans, 2009).

Similarly, forest harvest strategies may need to be adapted in the face of a changing climate. Harvesting trees to create a diverse array of age structure and communities (for example, selective clear cutting, harvesting specific trees), target select ages of trees to harvest (harvest only trees that are sexually mature), or target species of trees to harvest (for example, harvest tree species that are not resilient to climate change) are all strategies than can be used to selectively harvest forest to increase the resiliency to climate change (Steenberg and others, 2011).

5.6.3. Best practices for managing invasive species, pest, and disease outbreaks need to be reconsidered

Invasive species management strategies depend on the extent and magnitude to which the invasion has progressed. Responses include preventing the introduction of invasive species in the first place, early detection combined with rapid response, containing or minimizing the harmful effects of invasive species that are already established, and ecosystem restoration once the invasive species have been removed (EPA, 2008b). These approaches will remain the foundation for dealing with invasive species, but the specific strategies may need to be revisited to consider the impacts of climate change. In particular, more intensive monitoring would help to detect invasive species as climate change will make it possible for species to establish in new locations, expand their range into new territories, and even become invasive in response to new conditions (EPA, 2008b). A recent review found that most State plans for addressing invasive species have not considered how to modify their strategies in the face of climate change (EPA, 2008b).

Many of the tools used to control pests and disease outbreaks, such as pesticides, can have adverse effects that can be compounded by climate change. However in some cases the ability to take action appropriate for the increased challenge caused by climate change, such as in the case of the pine beetle, requires considering new conditions. For example, past practices typically allowed pine beetle infestations to run their course, a historically one or two year outbreak. However, this management strategy was proven ineffective given the new environmental conditions that allowed the beetle to propagate unchecked when the cold weather conditions that typically regulate outbreaks failed to occur owing to warming in the region (Bentz and others, 2010).

5.6.4. Pollution regulations can be undermined by climate change effects

Regulations to protect the environment and human populations from the adverse effects of pollutants (metals, pesticides, POPs, excessive nutrients, ground-level ozone) often do not account for the additional stresses of climate change (altered temperature, pH, dilution rates,). Awareness of the inadequacy of these regulations in the context of a changing climate is only now surfacing. Traditional regulatory tools, such as the Total Maximum Daily Load (TMDL), have been found to be vulnerable to the effects of climate change. Developing approaches to contaminant regulations to prepare for and respond to the interactive effects will be challenging. For one thing, the permitting process may not be flexible enough to accommodate altered sensitivity to pollutants or the fluctuations in perceived baseline conditions of an ecosystem being perturbed by climate change. In addition, it may be necessary to refine monitoring protocols to assess interactions and to develop new regulatory mechanisms in ways that cross jurisdictional boundaries. There is an additional need to focus increased attention on pollutants that are likely to increase vulnerability to climate change and to consider tools for incorporating uncertainty into regulatory mechanisms (Hansen and Hoffman, 2011).

Table 5.2. Example strategies for conserving and managing natural resources and ways that the strategies have been modified to integrate climate change adaptation.

Example Conservation and Management Strategies	How Strategies Could Be Modified to Integrate Climate Change Adaptation
Habitat Loss	
Vernal pools protected to maintain important habitat for amphibians	Many vernal pools in New Jersey coastal areas have been lost due to urbanization and development. In addition, projected sea level rise would fragment habitat by inundating the migratory routes used by eastern tiger salamander and Cope's gray treefrog (USFWS and NOAA, 2012). To create new corridors for amphibians, New Jersey is identifying areas to create new vernal pools at elevations above the projected sea level rise and in places adjacent to existing protected lands.
Wildlife refuges established to protect key habitat in the face of development	As climate conditions shift, increasing connectivity between habitats will be an important strategy for facilitating species' movements. One way to increase connectivity would be to increase the permeability of the landscape (Franklin and Lindenmayer, 2009), through selective harvest or retention cuts, tree-planting, alternative zoning, and rotational grazing (Kohm and Franklin 1997, Manning and others, 2009).
Farmers compensated for setting aside their lands for conservation purposes	Farm Bill programs, such as Wetland Reserve program, can provide compensation to landowners to protect areas projected to be flooded under future sea level rise, providing important habitat for wetland birds (USFWS and NOAA, 2012).

Extraction of Natural Resources	
Regulations on fish size and catch limits established to maintain healthy fish populations	The melting of Arctic sea ice will open new fishing grounds. One such fishing area was closed until adequate scientific information is obtained on how commercial fishing may impact the Arctic (NPFMC, 2009).
Forest harvest and management practices	As climate change brings an increased risk of forest fires due to longer, hotter, and drier summers (Westerling and others, 2006), forest managers will need to adjust management schemes. This may include focusing on ecological process and ecosystem service as opposed to forest structure and composition. Reducing tree density and supporting seed banks for post-fire regeneration may also be helpful (Peterson and others, 2011).
Water quality and quantity managed to provide adequate habitat for cold-water fish species	Water managers may need to alter reservoir release patterns or purchase or lease water rights in order to sustain flows necessary for habitat protection, especially during drought conditions (Palmer and others, 2009). For example, strategic releases of cold water from the Shasta reservoir may provide relief to Chinook salmon (*Oncorhynchus tshawytscha*) in the Sacramento River, where in-stream temperatures are expected to exceed temperature thresholds for the species (Yates and others, 2008).
Biological Disruptors	
Monitoring to detect and subsequently eradicate invasive species before they become well established	The State of Washington's Aquatic Invasive Species plan identifies species such as the giant salvinia (*Salvinia molesta*) and the water hyacinth (*Eichonria crassipes*) as species that should be monitored, even though temperatures in the State are currently too cold for these species (Bierwagen and others, 2008).
Allow naturally occurring pest outbreaks (for example, pine bark beetles) to run their course	Warmer winters have allowed the beetle to propagate unchecked by cold, causing widespread tree deaths (Bentz and others 2010). Adaptive management approaches may be necessary to contain pest outbreaks (Chmura and others, 2011).
Pollution	
Total maximum daily load (TMDL) levels set to minimize nutrient pollution	The 2002 phosphorus TMDL for Lake Champlain was challenged recently for failing to include the implications of climate change, such as altered precipitation patterns and flow in the watershed (Zamudio, 2011). Consequently, the EPA is now working to update the TMDL to include relevant climate information, an approach that is being applied to other TMDL processes around the country.

5.7. CONCLUDING THOUGHTS

The increasing climate change impacts on biodiversity, ecosystems, and ecosystem services that are already affected by other environmental stresses adds significant complexity to the types of research questions and management strategies relevant to safeguarding these resources into the future. Relatively little research attention has been devoted to interactions between climate change and multiple environmental stressors. Consequently, there is only a nascent understanding about the precise pathways, types, and character of interactions, and such elements can only rarely be quantified for observed or projected impacts. Yet, combinations of stressors will shape the ecosystems of the future, likely leading to thresholds and tipping points. One future research priority could be to develop analytical frameworks and tools to screen ecosystems for vulnerability, and to model and identify critical thresholds.

A critical barrier to investigating how multiple stressors interact is the lack of national data networks that combine climate, biological, and stressor information, including explicit data on population structure and abundance for invasive, rare, threatened, endangered, and other key species. Such data networks would allow researchers to combine information on projected climate changes with species biological data to understand possible future range shifts, and also to consider how other environmental stressors can influence future species distributions. It would also allow better attribution of impacts to climate change and other stressors, which can in turn, inform decision makers about how to prioritize natural resource management strategies.

Considering the multifaceted context in which biodiversity, ecosystems, and ecosystem services are being stressed will be critical for informing and prioritizing responses strategies. A failure to appreciate interactions may result in the implementation of climate adaptation responses that are at best inefficient and at worst harmful to the biodiversity and ecosystems that form the life support for Earth and are the foundation for ecosystem services. Fortunately, natural resource managers already possess a toolkit that can begin to address these complex challenges, and there are already examples where the implementation of such actions has ameliorated the interacting effects of climate change and other environmental stressors.

5.8. LITERATURE CITED

Ackerly DD, Loarie SR, Cornwell WK, Weiss SB, Hamilton H, Branciforte R, and Kraft NJB. 2010. The geography of climate change: implications for conservation biogeography. *Diversity and Distributions* **16**: 476-487.

Archer SR and Predick KI. 2008. Climate change and ecosystems of the southwestern United States. *Rangelands* **30**: 23–28.

Ayeni OO, Ndakidemi PA, Snyman RG, and Odendaal JP. 2010. Chemical, biological and physiological indicators of metal pollution in wetlands. *Science Research Essays* **5**: 1938-1949.

Battin J, Wiley MW, Ruckelshaus MH, Palmer RN, Korb E, Bartz KK, and Imaki H. 2007. Projected impacts of climate change on salmon habitat restoration. *Proceedings of the National Academy of Sciences* **104**: 6720–6725.

Benning TL, LaPointe D, Atkinson CT, and Vitousek PM. 2002. Interactions of climate change with biological invasions and land use in the Hawaiian Islands: modeling the fate of endemic birds using a geographic information system. *Proceedings of the National Academy of Sciences* **99**: 14246–14249.

Bentz BJ, Regniere J, Fettig CJ, Hansen EM, Hayes JL, Hickey JA, Kelsey RA, Negron JF, and Seybold SJ. 2010. Climate change and bark beetles of the western United States and Canada: direct and indirect effects. *BioScience* **60**: 602-613.

Bierwagen BG, Thomas R, Kane A. 2008. Capacity of management plans for aquatic invasive species to integrate climate change. *Conservation Biology* **22**:568-74.

Blais JM, Schindler DW, Muir DCG, Sharp M, Donald D, Lafreniere M, Braekevelt E and Stachan MMJ. 2001. Melting glaciers: A major source of persistent organochlorines to subalpine Bow Lake in Banff National Park, Canada. *Ambio* **30**: 410-415.

Blumenthal D, Chimner RA, Welker JM, and Morgan JA. 2008. Increased snow facilitates plant invasion in mixedgrass prairie. *New Phytologist* **179**: 440-448.

Bonan GB. 2011. Forests and global change. *In* Levia DF, Carlyle-Moses D, and Tanaka T (Eds), Forest Hydrology and Biogeochemistry. Springer, Netherlands. 711-728 p.

Boesch DF, Coles VJ, Kimmel DG, and Miller WD. 2007. Ramifications of climate change for Chesapeake Bay hypoxia. *In* Ebi KL, Meehl GA, Blanchet D, Twilley RR, and Boesch DF (Eds), Regional impacts of climate change, four case studies in the United States. Pew Center on Global Climate Change, Arlington, Virginia.

Bradley BA. 2009. Regional analysis of the impacts of climate change on cheatgrass invasion shows potential risk and opportunity. *Global Change Biology* **15**: 196-208.

Bradley BA, Oppenheimer M, and Wilcove DS. 2009a. Climate change and plant invasion: restoration opportunities ahead? *Global Change Biology* **15**: 1511-1521.

Bradley BA, Oppenheimer M, and Wilcove DS. 2009b. Climate change increases risk of plant invasion in the eastern United States. *Biological Invasions* **12**: 1855-1872.

Bradley BA, Blumenthal DM, Wilcove DS, and Ziska LH. 2010. Predicting plant invasions in an era of global change. *Trends in Ecology and Evolution* **25**: 310-318.

Breitburg D and Riedel G. 2005. Multiple stressors in marine systems. *In* Norse E and Crowder L (Eds), Marine conservation biology. Island Press, Washington, D.C. 167-182 p.

Brook BW, Sodhi N, and Bradshaw CJA. 2008. Synergies among extinction drivers under global change. *Trends in Ecology and Evolution* **23**: 453-460.

Brown D R, Sherry TW, and Harris J. 2011. Hurricane Katrina impacts the breeding bird community in a bottomland hardwood forest of the Pearl River basin, Louisiana. *Forest Ecology and Management* **261**: 111-119.

Brumbaugh W, Mora M, May TW, and Phalen DN. 2010. Metal exposure and effects in voles and small birds near a mining haul road in Cape Krusenstern National Monument, Alaska. *Environmental Monitoring and Assessment* **170**: 73-86.

Bulleri F and Chapman MG. 2010. The introduction of coastal infrastructure as a driver of change in marine environments. *Journal of Applied Ecology* **47**: 26–35.

Butchart SHM, Walpole M, Collen B, van Strien A, Scharlemann JPW, Almond REA, Baillie JEM, Bomhard B, Brown C, Bruno J, Carpenter KE, Carr GM, Chanson J, Chenery AM, Csirke J, Davidson NC, Dentener F, Foster M, Galli A, Galloway JN, Genovesi P, D Gregory R, Hockings M, Kapos V, Lamarque J-F, Leverington F, Loh J, McGeoch MA, McRae L, Minasyan A, Hernández Morcillo M, Oldfield TEE, Pauly D, Quader S, Revenga C, Sauer JR, Skolnik B, Spear D, Stanwell-Smith D, Stuart SN, Symes A, Tierney M, Tyrrell TD, Vié J-C, and Watson R. 2010. Global biodiversity: indicators of recent declines. *Science* **328**: 1164-1168.

CADFG (California Department of Fish and Game). 2011. State and Federally listed endangered and threatened animals of California. http://www.dfg.ca.gov/biogeodata/cnddb/pdfs/TEAnimals.pdf

Campbell JW, Waters MN, Tarter A, and Jackson J. 2010. Heavy metal and selenium concentrations in liver tissue from wild American alligator (*Alligator mississippiensis*) livers near Charleston, South Carolina. *Journal of Wildlife Disease* **46**:1234–1241.

Carnicer J, Coll M, Ninyerolac M, Ponsd X, Sáncheze G, and Peñuelasa J. 2011. Widespread crown condition decline, food web disruption, and amplified tree mortality with increased climate change-type drought. *Proceedings of the National Academy of Sciences* **108**: 1474-1478.

Center for Irrigation Technology Staff Report. 2011. Agricultural water use in California: A 2011 update. The Center for Irrigation Technology. Fresno: California State University. Available at: http://www.californiawater.org/docs/CIT_AWU_Report_v2.pdf

Chmura DJ, Anderson PD, Howe GT, Harrington CA, Halofsky JE, Peterson DL, Shaw DC and St. Clair JB. 2011. Forest responses to climate change in the northwestern United States: Ecophysiological foundations for adaptive management. *Forest Ecology and Management* **261**:1121–1142.

Clement J, Vercauteren J, Verstraeten W, Ducoffre G, Barrios J, Vandamme A-M, Maes P, and Van Ranst M. 2009. Relating increasing hantavirus incidences to the changing climate: the mast connection. *International Journal of Health Geographics* **8**: 1-11.

Coristine LE and Kerr JT. 2011. Habitat loss, climate change, and emerging conservation challenges in Canada. *Canadian Journal of Zoology* **89**: 435-451.

Corser JD. 2001. Decline of disjunct green salamander (*Aneides aeneus*) populations in the southern Appalachians. *Biological Conservation* **97**: 119-126.

Crain CM, Kroeker K., and Halpern BS. 2008. Interactive and cumulative effects of multiple human stressors in marine systems. *Ecological Letters* **11**: 1304–1315.

Crowl T, Parmenter R, and Crist T. 2008. The spread of invasive species and infectious disease as drivers and responders of ecosystem change at regional to continental scales. *Frontiers in Ecology and the Environment* **6**: 238–246.

Dale VH, Efroymson RA, and Kline KL. 2011. The land use–climate change–energy nexus. *Landscape Ecology* **26**(6): 755-773.

Damschen EI, Haddad NM, Orrock JL, Tewksbury JJ, and Levey DJ. 2006. Corridors increase plant species richness at large scales. *Science* **313**: 1284-1286.

Darling ES and Coté IM. 2008. Quantifying the evidence for ecological synergies. *Ecological Letters* **11**, 1278–1286.

de Chazal J and Rounsevell MDA. 2009. Land-use and climate change within assessments of biodiversity change: a review. *Global Environmental Change-Human and Policy Dimensions* **19**: 306-315.

Didham R, Tykianakis J, Gemmell N, Rand T, Ewers R. 2007. Interactive effects of habitat modification and species invasion on native species decline. *Trends in Ecology and Evolution* **22**: 489–496.

Diez JM, D'Antonio CM, Dukes JS, Grosholz ED, Olden JD, Sorte CJB, Blumenthal DM, Bradley BA, Early R, Ibáñez I, Jones SJ, Lawler JJ, and Miller LP. 2012. Will extreme climatic events facilitate biological invasions? *Frontiers in Ecology and the Environment*. Accepted pending minor revision.

Döll P and Zhang J. 2010. Impact of climate change on freshwater ecosystems: a global-scale analysis of ecologically relevant river flow alterations. *Hydrology and Earth System Sciences* **14**(5), 783–799.

Drinkwater KF and Frank KT. 1994. Effects of river regulation and diversion on marine fish and invertebrates. *Aquatic Conservation- Marine and Freshwater Ecosystems* **4**: 135-151.

Driscoll DA, Felton A, Gibbons P, Felton AM, Munro NT and Lindenmaye DB. 2011. Priorities in policy and management when existing biodiversity stressors interact with climate-change. *Climatic Change*. Online early 29 July 2011, pp. 1-25. http://www.springerlink.com/content/l07j35gk61074m41/

Dukes JS, Pontius J, Orwig DA, Garnas JR, Rodgers VL, Brazee NJ, Cooke BJ, Theoharides KA, Stange EE, Harrington RA, Ehrenfeld JG, Gurevitch J, Lerdau M, Stinson K, Wick R, and Ayres MP. 2009. Responses of insect pests, pathogens, and invasive plant species to climate change in the forests of northeastern North America: What can we predict? In NE Forests 2100: A Synthesis of Climate Change Impacts on Forests of the NortheasternU.S. and Eastern Canada. *Canadian Journal of Forest Research* **39**: 231-248.

Ebi KL and McGregor G. 2008. Climate change, tropospheric ozone and particulate matter, and health impacts. *Environmental Health Perspectives* **116**: 1449-1455.

Ellis EC and Ramankutty N. 2008. Putting people in the map: anthropogenic biomes of the world. *Frontiers in Ecology and the Environment* **6**:439-447.

Elphick CS and Oring LW. 1998. Winter management of Californian rice fields for waterbirds. *Journal of Applied Ecology* **35**: 95-108.

EPA (U.S. Environmental Protection Agency). 2008a. U.S. EPA's 2008 Report on the Environment (Final Report). U.S. Environmental Protection Agency, Washington, D.C., EPA/600/R-07/045F (NTIS PB2008-112484).

EPA (U.S. Environmental Protection Agency). 2008b. Effects of climate change for aquatic invasive species and implications for management and research. National Center for Environmental Assessment, Washington, DC; EPA/600/R-08/014. Available from the National Technical Information Service, Springfield, VA, and online at http://www.epa.gov/ncea.

Fahrig L. 2003. Effects of habitat fragmentation on biodiversity. *Annual Review of Ecology, Evolution and Systematics* **34**: 487–515.

Faunt CC (ed). 2009. Groundwater availability of the Central Valley Aquifer, California: U.S. *Geological Survey Professional Paper* 1766. 225 pp.

Ficke AD, Myrick CA and Hansen LJ. 2007. Potential impacts of global climate change on freshwater fisheries. *Reviews in Fish Biology and Fisheries* **17**: 581-613.

Filippino KC, Mulholland MR, and Bernhardt PW. 2011. Nitrogen uptake and primary productivity rates in the Mid-Atlantic Bight (MAB). *Estuarine, Coastal and Shelf Science* **91**(1): 13-23.

Fischer G, Tubiello FN, van Velthuizen H, and Wiberg DA. 2007. Climate change impacts on irrigation water requirements: Effects of mitigation, 1990. *Technological Forecasting and Social Change* **74**:1083-1107.

Folt CL, Chen CY, Moore MV, and Burnaford J. 1999. Synergism and antagonism among multiple stressors. *Limnology and Oceanography* **44**: 864–877.

Ford SE and Smolowitz R. 2007. Infection dynamics of an oyster parasite in its newly expanded range. *Marine Biology* **151**:119-33.

Forister ML, McCall AC, Sanders NJ, Fordyce JA, Thorne JH, O'Brien J, Waetjen DP, and Shapiro AM. 2010. Compounded effects of climate change and habitat alteration shift patterns of butterfly diversity. *Proceedings of the National Academy of Sciences* **107**: 2088-2092.

Franklin JF and Lindenmayer D. 2009. Importance of matrix habitats in maintaining biological diversity. *Proceedings of the National Academy of Science* **106**: 349-350.

Galbraith HS, Spooner DE, and Vaughn CC. 2010. Synergistic effects of regional climate patterns and local water management on freshwater mussel communities. *Biological Conservation* **143**: 1175-1183.

Game ET, Lipsett-Moore G, Saxon E, Peterson N, and Sheppard S. 2011. Incorporating climate change adaptation into national conservation assessments. *Global Change Biology* **17**: 3150-3160.

GAO (U.S. Government Accountability Office). 2005. Wind Power: Impacts on Wildlife and Government Responsibilities for Regulating Development and Protecting Wildlife. GAO-05-906.

Gibbard S, Caldeira K, Bala G, Phillips TJ, and Wickett M. 2005. Climate effects of global land cover change. *Geophysical Research Letters* **32**: L23705.

Glibert PM, Seitzinger S, Heil CA, Burkholder JM, Parrow MW, Codispoti LA, and Kelly V. 2005. The role of eutrophication in the global proliferation of harmful algal blooms. *Oceanography* **18**: 198–209.

Glick P, Staudt A, and Stein B. 2009. A New Era for Conservation: Review of Climate Change Adaptation Literature. National Wildlife Federation, Washington DC. Available at: http://www.nwf.org/News-and-Magazines/Media-Center/Reports/Archive/2009/New-Era-for-Conservation.aspx

González-Estébanez FJ, García-Tejero S, Mateo-Tomás P, and Olea PP. 2011. Effects of irrigation and landscape heterogeneity on butterfly diversity in Mediterranean farmlands. Agriculture. *Ecosystems and Environment* **144**: 262–270.

Graf WL. 1999. Dam nation: A geographic census of American dams and their large-scale hydrologic impacts. *Water Resources Research* **35**: 1305-1311.

Griffith BD, Scott JM, Adamcik R, Ashe D, Czech B, Fishman R, Gonzales B, Lawler G, McGuire AD, and Pidgorna A. 2009. Climate change adaptation for the U.S. National Wildlife Refuge system. *Environmental Management* **44**:1043-1052.

Grimm NB, Faeth SH, Golubiewski NE, Redman CL, Wu J, Bai X, and Briggs JM. 2008. Global change and the ecology of cities. *Science* **319**: 756-760.

Groves C, Valutis L, Vosick D, Neely B, Wheaton K, Touval J and Runnels B. 2000. Designing a geography of hope: a practitioner's handbook for ecoregional conservation planning 2nd edition. The Nature Conservancy, Arlington VA.

Halpern BS, Walbridge S, Selkoe KA, Kappel CV, Micheli F, D'Agrosa C, Bruno JF, Casey KS, Ebert C, Fox HE, Fujita R, Heinemann D, Lenihan HS, Madin EMP, Perry MT, Selig ER, Spalding M, Steneck R, and Watson R. 2008a. A global map of human impact on marine ecosystems. *Science* **319**: 948–952.

Halpern BS, McLeod KL, Rosenberg AA, and Crowder LB. 2008b. Managing for cumulative impacts in ecosystem-based management through ocean zoning. *Ocean Coast Management* **51**: 203–211.

Hannah L, Midgley GF, Andelman S, Araujo MB, Hughes G, Martinez-Meyer E, Pearson RG, and Williams P. 2007. Protected area needs in a changing climate. *Frontiers in Ecology and the Environment* **5**: 131-138.

Hansen LJ and Hoffman JR. 2011. Climate Savvy: Adapting Conservation and Resource Management to a Changing World. Island Press, Washington DC.

Harvell CD, Mitchell CE, Ward JR, Altizer S, Dobson AP, Ostfeld RS, and Samuel MD. 2002. Climate warming and disease risks for terrestrial and marine biota. *Science* **296**: 2158–2162.

Hayes F, Jones MLM, Mills G, and Ashmore M. 2007. Meta-analysis of the relative sensitivity of semi-natural vegetation species to ozone. *Environmental Pollution* **146**: 754-762.

Heller NE and Zavaleta ES. 2009. Biodiversity management in the face of climate change: a review of 22 years of recommendations. *Biological Conservation* **142**: 14-32.

Hockey PAR and Curtis OE. 2009. Use of basic biological information for rapid prediction of the response of species to habitat loss. *Conservation Biology* **23**: 64-71.

Hoekstra AY and Mekonnen MM. 2012. The water footprint of humanity. PNAS online early. http://www.pnas.org/content/early/2012/02/06/1109936109.full.pdf+html

Howarth RW, and Marino R. 2006. Nitrogen as the limiting nutrient for eutrophication in coastal marine ecosystems: evolving views over three decades. *Limnology and Oceanography* **51**: 364–376.

Ibáñez I, Sinlander Jr. JA, Wilson AM, LaFleur N, Tanaka N, and Tsuyama I. 2009. Multivariate forecasts of potential distributions of invasive plant species. *Ecological Applications* **19**: 359-375.

Imhoff ML, Zhang P, Wolfe RE, and Bounoua L. 2010. Remote sensing of the urban heat island effect across biomes in the continental USA. *Remote Sensing of Environment* **114**: 504-513.

IPCC (Intergovernmental Panel on Climate Change). 2007 Climate Change 2007: Synthesis Report. Contributions of Working Groups I, II, and III to the Fourth Assessment Report of the Intergovernmental Panel on Climate Change (AR4). Core writing team, Pachauri, R.K. and Reisinger, A. (eds.). IPCC, Geneva, Switzerland.

Jetz W, Wilcove DS, and Dobson AP. 2007. Projected impacts of climate and land-use change on the global diversity of birds. *PLoS Biol* **5**: e157.

Johnson PTJ, Townsend AR, Cleveland CC, Glibert PM, Howarth RW, McKenzie VJ, Rejmankova E, and Ward MH. 2010. Linking environmental nutrient enrichment and disease emergence in humans and wildlife. *Ecological Applications* **20**: 16-29.

Johnson PTJ, McKenzie VJ, Peterson AC, Kerby JL, Brown J, Blaustein AR, and Jackson T. 2011. Regional decline of an iconic amphibian associated with elevation, land-use change, and invasive species. *Conservation Biology* **25**(3): 556-566.

Kappel CV, Halpern BS, Martone RG, Micheli F, and Selkoe KA. 2009. In the zone comprehensive ocean protection. *Issues in Science and Technology* **25**: 33–44.

Kareiva P, Watts S, McDonald R, and Boucher T. 2007. Domesticated Nature: Shaping Landscapes and Ecosystems for Human Welfare. *Science* **316**: 1866-1869.

Keith DA, Akçakaya HR, Thuiller W, Midgley GF, Pearson RG, Phillips SJ, Regan HM, Araújo MB, and Rebelo TG. 2008. Predicting extinction risks under climate change: coupling stochastic population models with dynamic bioclimatic habitat models. *Biology Letters* **4**: 560-563.

Kenny JF, Barber NL, Hutson SS, Linsey KS, Lovelace JK, and Maupin MA. 2009. Estimated use of water in the United States in 2005: *U.S. Geological Survey Circular* 1344, 52 p.

Kirk D. 2009. Acid rock drainage and climate change. *Journal of Geochemical Exploration* **100**: 97-104.

Knapp RA, Boiano DM, and Vredenburg VT. 2007. Removal of nonnative fish results in population expansion of a declining amphibian (mountain yellow-legged frog, *Rana muscosa*). *Biological Conservation* **135**:11-20.

Kohm KA and Franklin JF (eds). 1997. Creating a forestry for the 21st Century: the science of ecosystem management. Washington, DC: Island Press.

Kouba A, Buřič M, and Kozák P. 2010. Bioaccumulation and effects of heavy metals in crayfish: A review. *Water, Air, & Soil Pollution* **211**: 5-16.

Krauss J, Bommarco R, Guardiola M, Heikkinen RK, Helm A, Kuussaari M, Lindborg R, Öckinger E, Pärtel M, Pino J, Pöyry J, Raatikainen KM, Sang A, Stefanescu C, Teder T, Zobel M, and Steffan-Dewenter I. 2010. Habitat fragmentation causes immediate and time-delayed biodiversity loss at different trophic levels. *Ecology Letters* **13**: 597-605.

Kvalevåg MM, Myhre G, Bonan G, and Levis S. 2010. Anthropogenic land cover changes in a GCM with surface albedo changes based on MODIS data. *International Journal of Climatology* **30**: 2105-2117.

La Sorte FA. 2006. Geographical expansion and increased prevalence of common species in avian assemblages: implications for large-scale patterns of species richness. *Journal of Biogeography* **33**: 1183–1191.

Lafferty KD. 2009. The ecology of climate change and infectious diseases. *Ecology* 90: 888-900.

Langley JA and Megonigal JP. 2010. Ecosystem response to elevated CO_2 levels limited by nitrogen-induced plant species shift. *Nature* **466**: 96–99.

Lawler JJ. 2009. Climate change adaptation strategies for resource management and conservation planning. *Annals of the New York Academy of Sciences* **1162**: 79-98.

Leimu R, Vergeer P, Angeloni F, and Ouborg NJ. 2010. Habitat fragmentation, climate change, and inbreeding in plants. *Annals of the New York Academy of Sciences* **1195**: 84-98.

Letcher RJ, Bustnes JO, Dietzc R, Jenssend BM, Jørgensene EH, Sonnec C, Verreaulta J, Vijayanf MM, Gabrielsen GW. 2010. Exposure and effects assessment of persistent organohalogen contaminants in Arctic wildlife and fish. *Science of The Total Environment* **408**: 2995-3043.

Ligon FK, Dietrich, WE, Trush WJ. 1995. Downstream ecological effects of dams. *Bioscience* **45**: 183-192.

Löw, M, Herbinger, K. 2006. Extraordinary drought of 2003 overrules ozone impact on adult beech trees (*Fagus sylvatica*). *Trees - Structure and Function* **20**: 539-548.

Lubowski RN, Vesterby M, Bucholtz S, Roberts MJ. 2006. Major uses of land in the United States, 2002; Economic Research Service, USDA.

Luedeling E, Steinmann KP, Zhang M, Brown PH, Grant J, and Girvetz EH. 2011. Climate change effects on walnut pests in California. *Global Change Biology* **17**: 228-238.

MA (Millennium Ecosystem Assessment). 2005. Ecosystems and Human Well-being: Biodiversity Synthesis. World Resources Institute, Washington, DC.

Manning AD, Gibbons P, and Lindenmayer DB. 2009. Scattered trees: a complementary strategy for facilitating adaptive responses to climate change in modified landscapes? *Journal of Applied Ecology* **46**: 915-919.

Malcom JR, Liu C, Neilson RP, Hansen L, and Hannah L. 2006. Global warming and extinctions of endemic species from biodiversity hotspots. *Conservation Biology* **20**: 538–548.

Mantyka-Pringle CS, Martin TG, and Rhodes JR. 2012. Interactions between climate and habitat loss effects on biodiversity: a systematic review and meta-analysis. *Global Change Biology* **18**: 1239–1252.

Martinez-Urtaza J, Bowers JC, Trinanesc J, and DePaola A. 2010. Climate anomalies and the increasing risk of *Vibrio parahaemolyticus* and *Vibrio vulnificus* illnesses. *Food Research International* **43**: 1780-1790.

Matyssek R, Karnosky DF, Wieser G, Percy K, Oksanen E, Grams TEE, Kubiske M, Hanke D, and Pretzsch. 2010. Advances in understanding ozone impact on forest trees: Messages from novel phytotron and free-air fumigation studies. *Environmental Pollution* **158**: 1990-2006.

Mawdsley JR, O'Malley R, and Ojima DS. 2009. A review of climate-change adaptation strategies for wildlife management and biodiversity conservation. *Conservation Biology* **23**: 1080-1089.

McCarthy HR, Oren R, Johnsen KH, Gallet-Budynek A, Pritchard SG, Cook CW, LaDeau SL, Jackson RB, and Finzi AC. 2010. Re-assessment of plant carbon dynamics at the Duke free-air CO_2 enrichment site: interactions of atmospheric $[CO_2]$ with nitrogen and water availability over stand development. *New Phytologist* **185**: 514-528.

Mooney H, Larigauderie A, Cesario M, Elmquist T, Hoegh-Guldberg O, Lavorel S, Mace GM, Palmer M, Scholes R, and Yahara T. 2009. Biodiversity, climate change, and ecosystem services. *Current Opinion in Environmental Sustainability* **1**: 46-54.

Mora C, Metzger R, Rollo A, and Myers RA. 2007. Experimental simulations about the effects of overexploitation and habitat fragmentation on populations facing environmental warming. *Proceedings of the Royal Society of London, Series B: Biological Sciences* **274**:1023–1028.

NRC (National Research Council). 2001. Under the weather: climate, ecosystems and infectious diseases. Washington DC, United States of America: National Academy Press. 146 pp.

Nelson H. 2007. Does a crisis matter? Forest policy responses to the mountain pine beetle epidemic in British Columbia. *Canadian Journal of Agricultural Economics* **55**: 459-70.

Norby RJ, Warren JM, Iversen CM, Medlyn BE, and McMurtrie RE. 2010. CO_2 enhancement of forest productivity constrained by limited nitrogen availability. *Proceedings of the National Academy of Sciences* **107**: 19368-19373.

NPFMC (North Pacific Fisheries Management Council). 2009. Fisheries Management Plan for
    Fish Resources of the Arctic Management Area: Public Review Draft.
    www.fakr.noaa.gov/npfmc/current_issues/Arctic/ArcticFMP109.pdf.

Noyes PD, McElwee MK, Miller HD, Clark BW, Van Tiem LA, Walcott KC, Erwin KN and
    Levin ED. 2009. The toxicology of climate change: Environmental contaminant in a
    warming world. *Environmental International* **35**: 971-986.

Okamura B and Feist SW. 2011. Emerging diseases in freshwater systems. *Freshwater Biology*
    56: 627–637.

Olden JD and Naiman RJ. 2010. Incorporating thermal regimes into environmental flows
    assessments: modifying dam operations to restore freshwater ecosystem integrity.
    Freshwater Biology **55**: 86-107.

Olsen EM, Ottersen G, Llope M, Chan K-S, Beaugrand GG, Stenseth NC. 2011. Spawning stock
    and recruitment in North Sea cod shaped by food and climate. *Proceedings of the Royal
    Society B: Biological Sciences* **278**, 504-510.

Palmer MA, Liermann CAR, Nilsson C, Flörke M, Alcamo J, Lake PS, and Bond N. 2009.
    Climate change and river ecosystems: Protection and adaptation options. *Environmental
    Management* **44**:1053–1068.

Paradis A, Elkinton J, Hayhoe K, Buonaccorsi J. 2008. Role of winter temperature and climate
    change on the survival and future range expansion of the hemlock woolly adelgid
    (*Adelges tsugae*) in eastern North America. *Mitigation and Adaptation Strategies for
    Global Change* **13**: 541-554.

Paukert CP and Petersen JH. 2007. Simulated effects of temperature warming on rainbow trout
    and humpback chub in the Colorado River, Grand Canyon. *Southwestern Naturalist* **52**:
    232-242.

Pereira HM, Leadley PW, Proenca V, Alkemade R, Scharlemann JPW, Fernandez-Manjarres JF,
    Araujo MB, Balvanera P, Biggs R, Cheung WWL, Chini L, Cooper HD, Gilman EL,
    Guenette S, Hurtt GC, Huntington HP, Mace GM, Oberdorff T, Revenga C, Rodrigues P,
    Scholes RJ, Sumaila UR, and Walpole M. 2010. Scenarios for global biodiversity in the
    21st century. *Science* **330**: 1496-1501.

Peterson DL, Millar CI, Joyce LA, Furniss MJ, Halofsky JE, Neilson RP, and Morelli TL. 2011.
    Responding to climate change in national forests: a guidebook for developing adaptation
    options. U.S. Department of Agriculture, Forest Service, Pacific Northwest Research
    Station, Portland, OR.

Petersen JH and Paukert CP. 2005. Development of a bioenergetics model for humpback chub
    and evaluation of water temperature changes in Grand Canyon, Colorado River.
    Transactions of the American Fisheries Society **134**: 960-974.

Poff NL and Zimmerman JKH. 2010. Ecological responses to altered flow regimes: a literature
    review to inform environmental flows science and management. *Freshwater Biology*
    55:194-20.

Pongratz J, Reick CH, Raddatz T, and Claussen M. 2010. Biogeophysical versus biogeochemical
    climate response to historical anthropogenic land cover change. *Geophysical Research
    Lett*ers **37**: L08702.

Pounds AJ, Bustamante MR, Coloma LA, Consuegra JA, Fogden MPL, Foster PN, La Marca E,
    Masters KL, Merino-Viteri A, and Puschendorf R. 2006. Widespread amphibian
    extinctions from epidemic disease driven by global warming. *Nature* **439**: 161-167.

Povilitis A and Suckling K. 2010. Addressing climate change threats to endangered species in U.S. recovery plans. *Conservation Biology* **24**(2):372-376.

Proches S, Wilson JRU, Veltman R, Kalwij JM, Richardson DM, and Chown SL. 2005. Landscape corridors: possible dangers? *Science* **310**: 778-779.

Rabalais NN, Turner RE, Diaz R, and Justić D. 2009. Global change and eutrophication of coastal waters. *ICES Journal of Marine Science*: *Journal du Conseil* **66**: 1528-1537.

Reich PB, Hobbie SE, Lee T, Ellsworth DS, West JB, Tilman D, Knops JMH, Naeem S, and Trost J. 2006a. Nitrogen limitation constrains sustainability of ecosystem response to CO_2. *Nature* **440**: 922-925.

Reich PB, Hungate BA, and Luo YQ. 2006b. Carbon-nitrogen interactions in terrestrial ecosystems in response to rising atmospheric carbon dioxide. *Annual Review of Ecology Evolution and Systematics* **37**: 611-636.

Renberg I. 1986. Concentration and annual accumulation values of heavy metals in lake sediments: Their significance in studies of the history of heavy metal pollution. *Hydrobiologia* **143**: 379-385.

Rose NA and Burton PJ. 2009. Using bioclimatic envelopes to identify temporal corridors in support of conservation planning in a changing climate. *Forest Ecology and Management* **258**: S64-S74.

Royal Society. 2008. Ground-level ozone in the 21st century: future trends, impacts and policy implications. Science Report, Royal Society. Available at: http://royalsociety.org/policy/publications/2008/ground-level-ozone/

Sabo JL, Bestgen K, Graf W, Sinha T, and Wohl EE. 2012. Dams in the Cadillac Desert-- downstream effects in a geomorphic context. The Year in Ecology and Conservation Biology (Proceedings of the New York Academy of Sciences). In Press.

Sabo JL, Sinha T, Bowling LC, Schoups GHW, Wallender WW, Campana ME, Cherkauer KA, Fuller PS, Graf WL, Hopmans JW, Kominoski JS, Taylor C, Trimble SW, Webb RH, and. Wohl EE. 2010a. Reclaiming freshwater sustainability in the Cadillac Desert. *Proceedings of the National Academy of Sciences* **107**:21263-21270.

Sabo JL, Finlay JC, Post DM, and Kennedy T. 2010b. The role of discharge variation in scaling between drainage area and food chain length in rivers. *Science* **330**: 965-967.

Sala OE, van Vuuren D, Pereira H, Lodge D, Alder J, Cumming GS, Dobson A, Wolters V, and Xenopoulos M. 2005. Biodiversity across Scenarios. *In* Carpenter SR, Pingali PL, Bennett EM, and Zurek M (Eds), Ecosystems and Human Well-Being: Scenarios. Island Press, Washington DC.

Sala OE, Chapin FS, III, Armesto JJ, Berlow E, Bloomfield J, Dirzo R, Huber-Sanwald E, Huenneke LF, Jackson RB, Kinzig A, Leemans R, Lodge DM, Mooney HA, Oesterheld M, Poff NL, Sykes MT, Walker BH, Walker M, and Wall DH. 2000. Global biodiversity scenarios for the year 2100. *Science* **287**: 1770-1774.

Satterthwaite D. 2008. Cities' contribution to global warming: notes on the allocation of greenhouse gas emissions. *Environment and Urbanization* **20**(2): 539-550.

Sekercioglu CH, Schneider SH, Fay JP, and Loarie SR. 2008. Climate change, elevational range shifts, and bird extinctions *Conservation Biology* **22**: 140-150.

Selin NE, Wu S, Nam KM, Reilly JM, Paltsev S, Prinn RG, and Webster MD. 2009. Global health and economic impacts of future ozone pollution. *Environmental Research Letters* **4**: 044014.

Sitch S, Cox PM, Collins WJ, and Huntingford C. 2007. Indirect radiative forcing of climate change through ozone effects on the land-carbon sink. *Nature* **448**:791-795.

Slenning BD. 2010. Global Climate Change and Implications for Disease Emergence. *Veterinary Pathology Online* **47**: 28-33.

Smithwick EAH, Westerling AL, Romme WH, Turner MG, and Ryan MG. 2010. Climate, fire and carbon: tipping points and landscape vulnerability in the Greater Yellowstone Ecosystem. Joint Fire Sciences Program Final Report, project number 09-3-01-47.

Sokolova IM and Lannig G. 2008. Interactive effects of metal pollution and temperature on metabolism in aquatic ectotherms: implications of global climate change. *Climate Research* **37**: 181-201.

Sommer TR, Nobriga ML, Harrell WC, Batham W, and Kimmerer WJ. 2001. Floodplain rearing of juvenile chinook salmon: evidence of enhanced growth and survival. *Canadian Journal of Fisheries and Aquatic Sciences* **58**:325-333.

Sorvari J, Rantala LM, Rantala ML, Hakkarainen H, and Eeva T. 2007. Heavy metal pollution disturbs immune response in wild ant populations. *Environmental Pollution* **145**: 324-328.

Spooner DE, Xenopoulos MA, Schneider C, and Woolnough DA. 2011. Coextirpation of host–affiliate relationships in rivers: the role of climate change, water withdrawal, and host-specificity. *Global Change Biology* **17**: 1720–1732.

Steenberg JWN, Duinker PN, and Bush PG. 2011. Exploring adaptation to climate change in the forests of central Nova Scotia, Canada. *Forest Ecology and Management* **262**:2316-2327.

Stephansen DA, Nielsen AH, Hvitved-Jacobsen T, and Vollertsen J. 2012. Bioaccumulation of heavy metals in fauna from wet detention ponds for stormwater runoff. *In* Rauch S, and Morrison GM (Eds), Urban Environment: Proceedings of the 10th Urban Environment Symposium, Springer Netherlands. *Alliance for Global Sustainability Bookseries* **19**(3): 329-338.

Stram DL and Evans DCK. 2009. Fishery management responses to climate change in the North Pacific. – *ICES Journal of Marine Science* **66**: 1633–1639.

Stromberg JC, Lite SJ, Marler R, Paradzick C, Shafroth PB, Shorrock D, White J, and White M. 2007. Altered stream flow regimes and invasive plant species: the *Tamarix* case. *Global Ecology and Biogeography* **16**:381-393.

Stromberg JC, Lite SJ, and Dixon MD. 2010. Effects of stream flow patterns on riparian vegetation of a semiarid river: implications for a changing climate. *River Research and Applications* **26**:712-729.

Suttle KB, Thomsen MA, and Power ME. 2007. Species interactions reverse grassland responses to changing climate. *Science* **315**:640-642.

Thomas CD, Cameron A, Green RE, Bakkenes M, Beaumont LJ, Collingham YC, Erasmus BFN, de Siqueira MF, Grainger A, Hannah L, Hughes L, Huntley B, van Jaarsveld AS, Midgley GF, Miles L, Ortega-Huerta MA, Peterson AT, Phillips OL, and Williams SE. 2004. Extinction risk from climate change. *Nature* **427**: 145-148.

Titus JG, Anderson KE, Cahoon DR, Gesch DB, Gill SK, Gutierrez BT, Thieler ER, and Williams SJ. 2009. Coastal Sensitivity to Sea-level Rise: a Focus on the mid-Atlantic Region. U.S. Climate Change Science Program Syntheis and Assessment Product 4.1. Available at: www.globalchange.gov/publications/reports/scientific-assessments

UOP, U. o. t. P. 2010. Employment Impacts of California Salmon Fishery Closures in 2008 and 2009. Business Forecasting Center, UOP.

USDA (U.S. Department of Agriculture). 2009. 2007 Census of Agriculture. National Agricultural Statistics Service.

USGCRP (U. S. Global Change Research Program). 2009. Global Climate Change Impacts in the United States. Cambridge, UK: Cambridge University Press.

USFWS (U.S. Fish and Wildlife Service) and NOAA (National Oceanic and Atmospheric Administration). 2012. National Fish, Wildlife, and Plants Climate Adaptation Strategy: Public Review Draft. USFWS and NOAA, Washington, DC.

Valiela I and Martinetto P. 2007. Bird Migratory Status and Habitat. *BioScience* **57**: 645.

Van Dingenen R, Dentener FJ, Raes F, Maarten C. Krol MC, Emberson L, and Cofala J. 2009. The global impact of ozone on agricultural crop yields under current and future air quality legislation. *Atmospheric Environment* **43**: 604–618.

Van Vuuren DP, Sala OE, and Pereira HM. 2006. The future of vascular plant diversity under four global scenarios. *Ecology and Society* **11**: 25. Available at: http://www.ecologyandsociety.org/vol11/iss2/art25/

Villamagna AM and Murphy BR. 2010. Ecological and socio-economic impacts of invasive water hyacinth (*Eichhornia crassipes*): a review. *Freshwater Biology* **55**: 282-298.

Vögeli M, Lemus JA, Serrano D, Blanco G, and Tella JL. 2011. An island paradigm on the mainland: host population fragmentation impairs the community of avian pathogens. *Proceedings of the Royal Society B: Biological Sciences* **278**: 2668-2676.

Vorosmarty VJ, McIntyre PG, Gessner MO, Dudgeon D, Prusevich A, Green P, Glidden S, Bunn SE, Sullivan CA, Reidy Liermann C, and Davies PM. 2010. Global threats to human water security and river biodiversity. *Nature* **467**: 555-561.

Vos CC, Berry P, Opdam P, Baveco H, Nijhof B, O'Hanley J, Bell C, and Kuipers H. 2008. Adapting landscapes to climate change: examples of climate-proof ecosystem networks and priority adaptation zones. *Journal of Applied Ecology* **45**: 1722-1731.

Walther GR, Roques A, Hulme PE, Sykes MT, Pysek P, Kuhn I, Zobel M, Bacher S, Botta-Dukat Z, Bugmann H, Czucz B, Dauber J, Hickler T, Jarosik V, Kenis M, Klotz S, Minchin D, Moora M, Nentwig W, Ott J, Panov VE, Reineking B, Robinet C, Semenchenko V, Solarz W, Thuiller W, Vila M, Vohland K, and Settele J. 2009. Alien species in a warmer world: risks and opportunities. *Trends in Ecology & Evolution* **24**: 686-693.

Wenger SJ, Isaak DJ, Luce CH, Neville HM, Fausch KD, Dunham JD, Dauwalter DC, Young MK, Elsner MM, Rieman BE, Hamlet AF, and Williams JE. 2011. Flow regime, temperature, and biotic interactions drive differential declines of trout species under climate change. *Proceedings of the National Academy of Sciences* **108**: 14175–14180.

Westerling AL, Hidalgo HG, Cayan DR, and Swetnam TW. 2006. Warming and earlier spring increase western U.S. forest wildfire activity. *Science* **313**(5789): 940-943.

Williams PH, Hannah L, Andelman SJ, Midgley GF, Araújo MB, Hughes G, Manne L, Martinez-Meyer E, and Pearson RG. 2005. Planning for climate change: identifying minimum-dispersal corridors for the Cape Proteaceae. *Conservation Biology* **19**: 1063-1074.

Wittig VE, Ainsworth EA, Naidu SL, Karnosky DF, Long SP. 2009. Quantifying the impact of current and future tropospheric ozone on tree biomass, growth, physiology and biochemistry: a quantitative meta-analysis. *Global Change Biology* **15**:396-424.

Wolken JM, Hollingsworth TN, Rupp TS, Chapin FS III, Trainor SF, Barrett TM, Sullivan PF, Mcguire AD, Euskirchen ES, Hennon PE, Beever EA, Conn JS, Crone LK, D'amore DV, Fresco N, Hanley TA, Kielland K, Kruse JJ, Patterson T, Schuur EAG, Verbyla DL, and Yarie J. 2011. Evidence and implications of recent and projected climate change in Alaska's forest ecosystems. *Ecosphere* **2**(11): 124.

Xenopoulos MA, Lodge DM, Alcamo J, Märker M, Schulze K and Van Vuuren DP. 2005. Scenarios of freshwater fish extinctions from climate change and water withdrawal. *Global Change Biology* **11**: 1557–1564.

Yates D, Galbraith H, Purkey D, Huber-Lee A, J. Sieber J, West J, Herrod-Julius S, and Joyce B. 2008. Climate warming, water storage, and Chinook salmon in California's Sacramento Valley. *Climatic Change* **91** (3-4): 335-350.

Zamudio HM. 2011. Predicting the future and acting now: Climate change, the Clean Water Act and the Lake Champlain phosphorus TMDL. *Vermont Law Review* **37**: 975-1022.

Chapter 6. Adaptation to impacts of climate change on biodiversity, ecosystems, and ecosystem services

Convening Lead Authors: Bruce A. Stein and Amanda Staudt
Lead Authors: Molly S. Cross, Natalie Dubois, Carolyn Enquist, Roger Griffis, Lara Hansen, Jessica Hellman, Josh Lawler, Eric Nelson, Amber Pairis
Contributing Authors: Doug Beard, Rosina Bierbaum, Evan Girvetz, Patrick Gonzalez, Susan Ruffo, Joel Smith

Key findings

- Climate adaptation has experienced a dramatic increase in attention since the last National Climate Assessment (NCA) and become a major emphasis in biodiversity conservation and natural resource policy and management.

- Adaptation can range from efforts to retain status quo conditions to actively managing system transitions; however, even the most aggressive adaptation strategies may be unable to prevent irreversible losses of biodiversity or serious degradation of ecosystems and their services.

- Static protected areas will not be sufficient to conserve biodiversity in a changing climate, requiring an emphasis on landscape-scale conservation, connectivity among protected habitats, and sustaining ecological functioning of working lands and waters.

- Mainstreaming biodiversity and ecosystem considerations into adaptation decisions in other societal sectors will be important to reduce the likelihood that human responses to climate impacts will harm biodiversity, ecosystems, and the services they provide.

- Ecosystem-based adaptation has emerged as a framework for understanding and promoting the role of ecosystem services in moderating climate impacts on people, although this concept currently is used more internationally than in the United States.

- Climate change can magnify the effects of existing stressors, and effective adaptation strategies will require an understanding of how to reduce their interacting and cumulative effects.

- Approaches to adaptation planning and implementation have advanced considerably since the last NCA and highlight the need to reassess conservation and management goals to ensure they are forward-looking and climate-informed.

- Agile and adaptive management approaches will be increasingly important for land and water managers given the pace and magnitude of climate change, and should incorporate monitoring, experimentation, and a capacity to evaluate and modify management actions.

6.1. RESPONDING TO CLIMATE CHANGE

Rapid climate change already is having significant effects on the nation's biodiversity, ecosystems, and ecosystem services, and these impacts are projected to become increasingly severe in the future. As documented in other chapters of this technical input document, these impacts are expected to lead to significant loss of genetic, population, and species diversity, alteration or disruption in ecosystem composition and functioning, and deterioration or collapse of key ecosystem services that support human welfare and the nation's economy. The scale and potential consequences of this problem require that we identify appropriate response strategies to address the most harmful impacts of climate change on biodiversity, ecosystems, and ecosystem services, and to capitalize on potential benefits.

Over the past two decades climate change research and policy largely has focused on: 1) an enhanced understanding of current and future climatic conditions and their associated effects on human societies and natural systems; and 2) efforts to stabilize or reduce the concentration of greenhouse gases in the atmosphere that constitute the climate's principal anthropogenic forcing agent, or what is known as climate *mitigation* (NRC, 2011). As the pace of climate change has become better understood, along with the magnitude and inevitability of potential impacts on everything from urban infrastructure and agricultural systems to natural ecosystems, significant attention has begun focusing as well on how to prepare for and cope with the impacts from climate change—what is known as climate *adaptation*.

This chapter addresses climate adaptation from a biodiversity, ecosystems, and ecosystem services perspective, emphasizing changes and advances in the field since the last National Climate Assessment.

6.1.1. What is Adaptation?

Climate change adaptation is an emerging field that focuses on how to prepare for and respond to the impacts of current and future climate change. As a relatively new field, definitions of adaptation are still somewhat in flux. The Intergovernmental Panel on Climate Change (IPCC) Fourth Assessment, for example, defined **adaptation** as "initiatives and measures to reduce the vulnerability of natural and human systems against actual or expected climate change effects" (IPCC, 2007a) and as "adjustment in natural or human systems in response to actual or expected climatic stimuli or their effects, which moderates harm or exploits beneficial opportunities" (IPCC, 2007b). The recent IPCC special report on extreme events (IPCC, 2012) defines adaptation as:

> In human systems, the process of adjustment to actual or expected climate and its effects, in order to moderate harm or exploit beneficial opportunities. In natural systems, the process of adjustment to actual climate and its effects; human intervention may facilitate adjustment to expected climate.

Adaptation stems from a structured process that considers the effects of climate change on valued resources such that appropriate management responses can be identified and carried out. Indeed, because adaptation fundamentally is about managing change (see section 6.1.2), it can best be thought of as a process, rather than a fixed outcome. Although specific adaptation goals and objectives may be set (see section 6.3), ongoing environmental and climatic change, and the ecological and human responses to these changes, will require continual reevaluation and adjustment of adaptation approaches (Fazey and others, 2010).

Actions undertaken to prepare for anticipated climate change impacts are often referred to as proactive or anticipatory adaptation, while actions in response to climate-driven impacts (or natural disasters) are often referred to as reactive adaptation (Adger and others, 2005). Using the example of adaptation to increasingly severe drought and forest fires, anticipatory actions might include targeted application of prescribed burns or selective forest thinning, while reactive adaptation might include a broadening the genetic composition of plant materials used for post-fire restoration to ensure they are better suited for future climatic conditions.

Several other adaptation-relevant terms and concepts are important to define in the context of biodiversity and ecosystem conservation and management. These include adaptive management, vulnerability, adaptive capacity, resilience, and maladaptation.

Impacts of Climate Change on Biodiversity, Ecosystems, and Ecosystem Services |
Technical Input to the 2013 National Climate Assessment

Chapter 6
Adaptation

Adaptive management is a concept that has been applied to resource management for many years (Williams and others, 2009), but recently has received renewed attention as a tool for helping resource managers make decisions in response to climate change. Adaptive management seeks to improve and inform decisions in the face of uncertainty by learning from management outcomes and incorporating that information into a structured process of flexible decision making. Specifically, this approach encourages management actions to be framed as hypotheses that can be tested and evaluated against expected results. Adaptive management frequently is invoked within the context of climate adaptation as a way to address and respond to the inherent uncertainties associated with predicting human and biological responses to climate change. Because of the semantic similarity between adaptation and adaptive management, these two concepts are sometimes confused with one another. In short, adaptive management may be used in the implementation of an adaptation strategy, but adaptation does not require adaptive management, nor does adaptive management necessarily lead to adaptation.

As noted in one of the IPCC definition of adaptation above, reducing the vulnerability of systems to climate change is central to adaptation. **Vulnerability** refers to the degree to which an ecological system or individual species is likely to experience harm as a result of changes in climate (IPCC, 2007b). Vulnerability to climate change is a function of exposure to climate change (i.e., the magnitude, intensity and duration of the climate changes experienced), the sensitivity of the species or community to these changes, and the adaptive capacity of the system (Williams and others, 2008; Glick and others, 2011a). Species and ecosystems that are more vulnerable are likely to experience greater impacts from climate change, whereas those that are less vulnerable may be more likely to persist or even benefit from changes in climate.

Among the three components of vulnerability, **adaptive capacity** is perhaps the most challenging to put into practice. Adaptive capacity refers to the ability of a natural system to accommodate climate change impacts with minimal disruption (Smith and Wandel 2006; Williams and others, 2008; Glick and others, 2011a). One aspect of adaptive capacity is the 'intrinsic adaptation potential' of a natural system to climate change (Glick and others, 2011a), resulting from factors such as dispersal ability, genetic diversity, and plasticity at the population or species level, or factors such as functional redundancy and patch size at the ecosystem level. However, the ability of a system to realize its intrinsic adaptive capacity is also influenced by extrinsic factors, such as barriers to dispersal and permeability of the landscape (Glick and others, 2011a). In assessing the vulnerability of human communities and institutions, the term adaptive capacity is also used to refer to the potential to implement planned adaptation measures (Metzger and others, 2005).

Resilience is a concept that frequently is invoked when describing the potential responses of biological systems to the impacts of climate change. In the ecological literature, resilience has been defined as a measure of the persistence of systems and their ability to absorb change (Holling, 1973, 1996). Similarly, for social-ecological systems, resilience has been characterized as the capacity of a system to absorb disturbance and reorganize while still maintaining the same relationships among its components (Walker and others, 2004). In practice, resilience must be defined contextually, in terms of the specific disturbance and the desired attributes or functions of the system that will be maintained (Carpenter and others, 2001). However, within the climate change and adaptation literature, the concept of resilience has been applied more broadly, often in varied and inconsistent ways. Some authors treat resilience as an index of the non-exposure components of vulnerability (for example, Magness and others, 2011), whereas other authors limit resilience to the adaptive capacity component of vulnerability (Gallopín, 2006). In its broadest applications, the concept of resilience is sometimes expanded to encompass properties of the system that promote resistance (that is, the ease or difficulty of changing the system) (Walker and others, 2004), allow systems to function "differently" (Lawler 2009), or even facilitate the emergence of new system trajectories (Folke, 2006).

Finally, **maladaptation**, is a concept that refers to responses to climate change that may actually be detrimental to the system of interest. Maladaptation has been formally defined as "any changes in natural or human systems that inadvertently increase vulnerability to climatic stimuli; an adaptation that does not succeed in reducing vulnerability but instead increases it" (IPCC, 2001). In terms of human activities in response to climate change, maladaptation is more likely to occur when climate impacts are considered on particular system components in isolation, without assessing the net benefit within and across sectors. For example, implementing hard engineering options, such as sea walls to protect infrastructure from sea level rise, may limit the ability of coastal marshes to adapt to climate change and reduce the storm protection these natural systems can provide.

6.1.2. Managing Change

Notwithstanding the poetic notion of "the balance of nature" (Egerton, 1973), ecological systems always have been dynamic, characterized by variability at daily, annual, and decadal scales (Holling, 1973, Landres and others, 1999). Nonetheless, stationarity—the idea that natural systems fluctuate within an unchanging envelope of variability—has been a foundational concept in many fields of conservation and natural resource management (Milly and others, 2008). Although the assumption of stationarity has long been compromised in many natural systems by anthropogenic disturbances, directional changes in climatic variables have now made clear that, in the words of Milly and others (2008) "stationarity is dead."

Biodiversity conservation efforts long have relied on strategies that seek either to preserve current ecological conditions or restore them to some historical state (Cole and Yung, 2010). Because conservation plans traditionally assumed a stationary climate, they have resulted in static configurations of protected areas that largely fail to incorporate large-scale dynamic processes (Stein and others, written communication 2012). Conservation and protected area management strategies typically are designed to address a suite of existing environmental stressors, such as direct habitat destruction, spread of invasive species, disruption of key ecological processes, and overharvesting of biological resources (Wilcove and others, 1998). Climate change will complicate existing conservation efforts in several ways. First, climate change will magnify the effect of many existing stresses on ecosystems and species; second, as individual species track shifts in climate (Chen and others, 2011; Nye and others, 2009; Kelly and Goulden, 2008), they will begin to shift out of reserves in which they are currently protected (Monzón and others, 2011), with the result that ecological communities may begin to disassemble, and ecosystem services degrade. Finally, human responses to climate change (for example, climate-driven population shifts, infrastructure-based adaptation to protect people and property, implementation of climate mitigation efforts) will place increasing pressure on remaining natural areas and the connections among them.

Adaptation to climate change in the context of biodiversity conservation and natural resource management will therefore largely be about managing change (Millar and others, 2007; Link and others, 2010; West and others, 2009). Approaches to change management can range from resisting changes, to protect high value and climate-sensitive assets, to actively facilitating changes, so that inevitable system transitions might retain desirable ecological attributes, rather than result in complete collapse of ecosystem functions and services.

One commonly used framework for adaptation responses to climate change consists of the continuum of: 1) resistance; 2) resilience; and 3) transformation (Millar, 2007[4]; Glick and others, 2011a). Under this framework, resistance actions are intended to assist the species or system to forestall impacts, thus maintaining status quo conditions. The term "resilience" has multiple meanings (see section 6.1.1), but in this context typically refers to actions designed to improve the capacity of a system to return to desired conditions after disturbance, or as a means to maintain some level of functionality in an altered state. Transformation in this context refers to efforts that enable or facilitate the transition of ecosystems to new functional states.

Peterson and others (2011) recently offered an expanded version of this continuum, charting out the following four steps: 1) resistance; 2) resilience; 3) response; and 4) realignment. They define response as proactive strategies that work directly with the changes climate is provoking, and assist system transitions in ways that minimize undesired outcomes (comparable to the "transformation" stage above). Realignment, in contrast, focuses on systems that already have been disturbed beyond historical ranges of natural variability, and focuses on restoration of the system, although not necessarily to its historic or predisturbance condition.

Moser and Ekstrom (2010) consider a continuum of change from the perspective of the relationship between time and effort required to achieve a desired outcome, and the time scale of adaptation goals (see **Figure 6.1**). They describe three classes of actions: 1) "coping measures," defined as short-term responses to deal with projected climate impacts and a return to status quo conditions; 2) "more substantial adjustments," defined as change in some aspects of a system without complete transformation; and 3) system transformation, which may require more profound shifts in institutional and ecological paradigms. To more explicitly distinguish among the attributes embedded in these change frameworks, Stein and others (written communication, 2012) differentiate between level of change, strategic response, and desired outcome (**Table 6.1**).

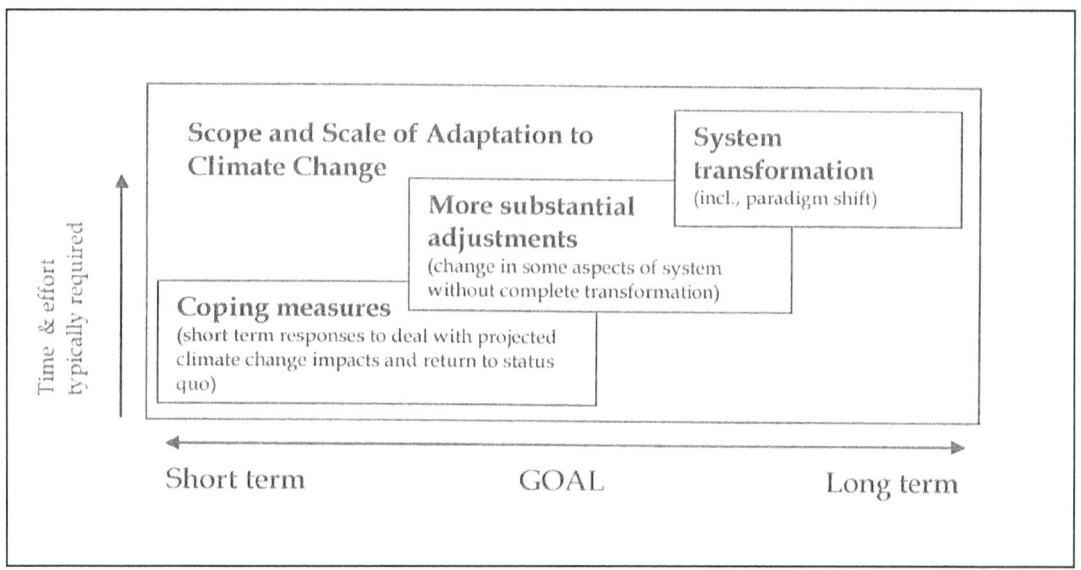

Figure 6.1. Scope and scale of adaptation efforts (from Moser and Ekstrom, 2010).

---

[4] In Millar and others (2007) this continuum is characterized as resistance, resilience, and response.

Impacts of Climate Change on Biodiversity, Ecosystems, and Ecosystem Services |
Technical Input to the 2013 National Climate Assessment

Chapter 6
Adaptation

Table 6.1. Change Continuum and Strategic Responses (Stein and others, written communication 2012).

Level of Change	Strategic Response	Desired Outcome
System Transformation	Facilitate Change/ Remediate Impacts	Maintain/Restore System Functions
Moderate Change	Accommodation/Coping Strategies	Buffer Impacts
Little/No Change	Resistance Strategies	Maintain Status Quo

Early adaptation thinking within the biodiversity and ecosystem conservation community has focused mostly on strategies to promote resistance and enhance resilience, often with the intent of "buying time" for significant biological features (Hansen and others, 2003). Enhancing resilience, in particular, has become a common catch-phrase among planners and practitioners, although it is most commonly invoked with a focus on the "rebound" oriented definitions as a means to sustain status quo conditions. In the past few years, however, more scientists and conservationists have begun seriously focusing not just on retaining existing ecological conditions, but also on managing or facilitating what many now see as inevitable system transformations.

6.1.3. Targets of adaptation: From species to services

Biodiversity and ecosystem-oriented adaptation can be applied at a variety of biological levels (for example, genes, species, ecosystems) and, therefore, can benefit different attributes of natural systems, such as the components of biodiversity (for example, species diversity and ecological patterns), particular ecosystem processes (for example, disturbance regimes, nutrient cycles, hydrological cycles), or the services that ecosystems provide (for example, water production, carbon sequestration). Adaptation strategies that focus on different biological levels or attributes can create trade-offs or provide co-benefits, depending on the goals and approaches. Simply put, what is viewed as adaptive for one conservation purpose or target might be maladaptive for another.

Many adaptation strategies are focused at the level of species because many Federal and State mandates for conservation focus on this level, and because key goods and services, such as timber and fisheries, are provided by species. Adaptation strategies could also focus on the maintenance of genetic diversity, a common conservation goal, particularly for endangered species. Genetic diversity is important to prevent inbreeding depression and risk of catastrophic species loss from random events, but it also can be important in fostering evolution under climate change (Hoffman and Sgrò, 2011). Evolution under climate change often will rely on genetic diversity already present in a species of interest and thus necessitates preservation of existing diversity (Reusch and Wood, 2007).

Adaptation strategies can also be applied at the level of species assemblages and interactions. For example, conservation of some specialist butterflies, like the endangered Karner blue butterfly (*Lycaeides melissa samuelis*) will depend on existence of its host plant, which in turn depends on the maintenance of certain ecosystem processes (for example, periodic disturbances). Facultative associations, such as between soil mycorrhizae and plants, may enable manipulation of one species to foster another under climate change (Carey and others, 1992; Wilson and Hartnett, 1997).

Entire ecosystems can also be the focus of adaptation, with ecosystem processes or services as the targets of conservation attention. Adaptation at the ecosystem level can address either biotic components of the system (that is, species) or abiotic components. Fluvial systems, for instance, depend on the quantity and timing of water flows, and aquatic adaptation efforts often focus on managing river flows (Richter and others, 2003; Palmer and others, 2008). Forest carbon dynamics is a topic of considerable interest to the climate change community, and forest adaptation efforts may focus on managing abiotic disturbance regimes (for example, fire) in order to maintain or enhance carbon stocks (Drever and others, 2006).

Adaptation pursued at different levels of biological organization, or for different purposes, can be competing or complementary. Because different adaptation strategies are likely to be motivated by different goals among different stakeholders, it is possible that controversy will emerge about which adaptation actions are desirable. For example, adaptation aimed at biodiversity conservation might focus on maximizing the population size and habitat area of an existing endangered species, but adaptation aimed at an ecosystem service (for example, water production or storm attenuation) might actively promote ecosystem change so that a new service can be delivered in the area. In other cases, different goals could be met with coordination of adaptation activities. For example, changing seed zones for reforestation to maximize forest productivity and timber production under changing climate could have benefits for native forest bird conservation. Whatever its focus or form, adaptation is likely to be an iterative process that is needed over a continual basis to adjust management strategies as conditions shift. This is an expansion of the concept of adaptive management—that species and ecosystems should be managed in a way that is informed with experimentation and monitoring and management goals are continually updated to account for ecosystem responses and changing circumstances (Lawler, 2009).

6.2. DRAMATICALLY INCREASED INTEREST IN ADAPTATION

Over the past few years interest in climate adaptation has grown dramatically, accompanied by rapid advances in both the theory and practice of adaptation. Many Federal and State agencies and non-governmental organizations have begun shifting their focus to better deal with the impacts of climate change and plan for future impacts. For instance, at the Federal level, an interagency climate adaptation task force is providing high-level guidance to agencies for integrating adaptation into their planning and operations (CEQ, 2010, 2011a). Similarly, many State agencies and non-governmental conservation organizations are now explicitly incorporating adaptation into their strategies and work plans. The collapse in late 2010 of efforts to pass comprehensive Federal climate legislation, and consequent set-backs in efforts to achieve meaningful reductions in greenhouse gas emissions, has given added urgency and impetus to the field of climate change adaptation.

6.2.1. Trends in adaptation attention

Although climate change adaptation has been discussed in policy circles for nearly thirty years (Smith and Tirpak, 1989), over much of that time it was largely regarded as a taboo subject, with many scientists and activists concerned it would divert attention from addressing the underlying causes of climate change (Pielke and others, 2007). It has become increasingly clear, however, that no matter how vigorously greenhouse gas emissions are reduced, major shifts in climate will occur, necessitating aggressive action on climate adaptation as well as climate mitigation (IPCC, 2007a; NRC 2010). Consequently, over the past few years, interest in and acceptance of adaptation has increased sharply, within the biodiversity and ecosystem conservation community as well as among many other sectors.

One measure of the growth in interest in the field of adaptation is the level of attention to the issue in the news media. Moser (2009) explores how concern and public debate about adaptation have been changing over time through the lens of relevant media coverage. Charting what she refers to as an "explosive awakening to the need for adaptation" Moser (2009) documents a nearly four-fold increase in the number of articles mentioning adaptation from 2006 to 2007 alone, and suggests that this analysis likely under-represents the actual coverage of adaptation over time.

Glick and others (2011b) conducted a review of scientific literature to chart the rise in attention to climate change adaptation. Through a Web of Science query using broad climate change search terms first employed by Heller and Zavaleta (2009), along with subsequent filters to identify papers with keywords relevant to biodiversity or natural resource adaptation, Glick and others (2011b) identified approximately 600 "self-identified" adaptation papers. The number of such adaptation papers experienced a five-fold increase from 2007 to 2010. A topical analysis of these papers, however, revealed a heavy emphasis on papers focused on adaptation of human systems (including governance), followed by natural resource management and agricultural systems. Biodiversity and ecosystem conservation were the least well-represented topical area among the adaptation literature considered in this review.

Parallel to the increase in peer-reviewed papers on climate adaptation is the growth in gray literature reports, studies, plans, and resources on the topic. Although quantifying the increase in such gray literature is difficult, over the past few years there has been what seems like a rapid acceleration in the release of adaptation-specific reports and other gray literature. In addition, over the past few years an increasing number of web-based resources (for example, Climate Adaptation Knowledge Exchange [CAKE], Georgetown University's Adaptation

Clearinghouse, the Collaboratory for Climate Change Adaptation, and NOAA's Coastal Climate Adaptation) also reflects a greatly increased attention to the topic of climate adaptation. Finally, the development of formal adaptation policies and plans at Federal, State, tribal, and local levels, as well as in the private sector represent another indication of the rapid increase in attention to this field.

6.2.2. Adaptation at the Federal level

Since the last National Climate Assessment there has been a strong increase in adaptation activities within the Federal agencies. Of particular significance is the establishment in 2009 of an Interagency Climate Change Adaptation Task Force, co-chaired by the Council on Environmental Quality (CEQ), the Office of Science and Technology Policy (OSTP), and the National Oceanic and Atmospheric Administration (NOAA). This task force includes representatives from more than twenty Federal agencies and was directed by the President to develop recommendations for how the Federal government can strengthen policies and programs to better prepare the nation to adapt to the impacts of climate change (CEQ, 2010, 2011a).

In 2010 the Adaptation task force presented its initial recommendations to the President on a National Climate Adaptation Strategy (CEQ 2010) and adopted the following high-level policy goals to guide adaptation activities among the Federal agencies:

- Make adaptation a standard part of Agency planning to ensure that resources are invested wisely and services and operations remain effective in a changing climate.
- Ensure scientific information about the impacts of climate change is easily accessible so public and private sector decision-makers can build adaptive capacity into their plans and activities.
- Align Federal efforts to respond to climate impacts that cut across jurisdictions and missions, such as those that threaten water resources, public health, oceans and coasts, and communities.
- Develop a U.S. strategy to support international adaptation that leverages resources across the Federal Government to help developing countries reduce their vulnerability to climate change through programs that are consistent with the core principles and objectives of the President's new Global Development Policy.
- Build strong partnerships to support local, State, and tribal decision makers in improving management of places and infrastructure most likely to be affected by climate change.

In response to Presidential Executive Order 13514 Federal agencies are now evaluating how climate change is having an impact on their operations and services, and are beginning to integrate adaptation into agency planning processes. For instance, in 2011 Federal departments were required to adopt formal climate change adaptation policies, and by the summer of 2012 they are required to have developed and published adaptation plans (CEQ, 2011b).

In addition, three cross-cutting strategies have been developed or are in process that relate to the nation's biodiversity and ecosystem adaptation efforts: 1) the National Fish, Wildlife, and Plants Climate Adaptation Strategy (USFWS and NOAA, 2012); 2) the National Action Plan: Priorities for Managing Freshwater Resources in a Changing Climate (CEQ, 2011c; and 3) the National Ocean Policy Implementation Plan (CEQ, 2011d). Other important efforts that demonstrate the increasing attention to adaptation in the Federal government include the

National Climate Adaptation Summit, held in 2010, and the recent report by the President's Council of Science Advisors entitled "Sustaining Environmental Capital" (PCAST, 2011).

The National Fish, Wildlife, and Plants Climate Adaptation Strategy (USFWS and NOAA, 2012) is of particular relevance from the perspective of biodiversity, ecosystems, and ecosystem services. That strategy, released as a draft for public comment in January 2012, includes the following seven high-level goals:

- Conserve habitat to support healthy fish, wildlife and plant populations and ecosystem functions in a changing climate.
- Manage species and habitats to protect ecosystem functions and provide sustainable cultural, subsistence, recreational, and commercial use in a changing climate.
- Enhance capacity for effective management in a changing climate.
- Support adaptive management in a changing climate through integrated observation and monitoring and use of decision support tools.
- Increase knowledge and information on impacts and responses of fish, wildlife and plants to a changing climate.
- Increase awareness and motivate action to safeguard fish, wildlife and plants in a changing climate.
- Reduce non-climate stressors to help fish, wildlife, plants, and ecosystems adapt to a changing climate.

Several individual departments and agencies have been particularly active in promoting climate adaptation actions relevant to ecosystems and their inhabitants (Pew, 2010; C2ES, 2012). The Department of Interior (DOI), through Secretarial Order 3289, established a number of climate adaptation-oriented initiatives among its bureaus, including a network of Landscape Conservation Cooperatives, designed to promote cross-jurisdictional conservation and resource management planning and implementation, as well as a network of regional Climate Science Centers, designed to improve the integration of climate science into land and water management. Individual DOI bureaus, such as the U.S. Fish and Wildlife Service and National Park Service, have begun conducting climate vulnerability assessments of their land holdings and resource assets and in many cases developing unit-specific adaptation plans, while the U.S. Geological Survey, at the direction of Congress, has taken the lead in establishing a National Climate Change and Wildlife Science Center. The U.S. Forest Service, in the Department of Agriculture, is another major land management agency that has been actively integrating climate change into its planning and operations, and has instituted a climate change scorecard for measuring progress by each of its national forests and grasslands. The National Oceanic and Atmospheric Administration (NOAA) similarly has been active in providing a variety of products and services to support adaptation efforts of government and nongovernment entities, and incorporating climate change into its Federal-State coastal management programs, habitat conservation programs and endangered species programs.

Incorporation of climate adaptation into proposed Federal legislation is another measure of the increased interest in this topic. HR. 2454, the American Clean Energy and Security Act (ACES), was a comprehensive climate bill that passed the U.S. House of Representatives in June 2009, and included a title specifically devoted to advancing natural resources adaptation. A Senate version of the bill, passed out of the Environment and Public Works Committee but not taken up by the full Senate, also included strong natural resource adaptation provisions. In fall

2011, a stand-alone version of these natural resource adaptation provisions was introduced into the U.S. Senate as S.1881 the Safeguarding America's Future and Environment Act (SAFE Act). Similarly, a stand-alone bill was introduced in the U.S. House of Representatives in fall 2011 focusing on adaptation for water resources and infrastructure—HR. 2738, the Water Infrastructure Resiliency and Sustainability Act.

6.2.3. Adaptation at the State level

Attention to adaptation has also increased markedly at State levels. Responding to and planning for climate change provides State agencies an opportunity to promote the use of sound ecological conservation, not only for adaptation, but also as a potential component of mitigation strategies (for example, riparian restoration to reduce flood damage and increase carbon storage, and tidal wetland restoration to reduce impacts of sea level rise on human community infrastructure). Based on surveys conducted by the Association of Fish and Wildlife Agencies (Choudhury, 2012) as of March 2011 sixteen States (AK, CA, CT, FL, IL, MA, MD, ME, NC, NH, NY, OR, PA, VA, WA, and WI) have a State adaptation plan for biodiversity conservation or are in the process of completing their State adaptation plan. Within those sixteen States, seven (CT, FL, MA, ME, OR, PA, and WA) have a legislative mandate to create a sector-wide State adaptation plan. Those States that do not have a legislative mandate (for example, AK, CA, MD, NH, NY, VA, and WI) typically are guided by an Executive Order from their Governor, many of which involve a multi sector approach involving coastal, water resources, agriculture, forest and terrestrial ecosystems, bay and aquatic ecosystems, growth and land use, energy development, and public health.

Although many States do not yet have formal adaptation plans, State fish and wildlife agencies across the nation are beginning to create and implement strategies that address not only climate change, but conservation in the broader sense (that is, habitat restoration and landscape connectivity). Many State agencies are incorporating climate information into existing planning efforts, such as State Wildlife Action Plans, which currently are serving as a platform for climate change adaptation planning across the country. The first generation of these wildlife action plans were completed in 2005, and at that time very few explicitly addressed the impact of climate change on their wildlife resources, or proposed adaptation strategies to cope with these impacts. To be eligible for continued Federal funding for State and Tribal Wildlife Grants, fish and wildlife agencies are required to revise their State Wildlife Action Plans by 2015. Using guidance prepared by the Association of Fish and Wildlife Agencies (AFWA, 2009) States are incorporating climate change into their updated plans, and as of March 2011 all but twelve States have begun this process.

The growth in the number of climate change vulnerability assessments being carried out is another indication of the increased attention to adaptation among State natural resource agencies. As of March 2011 twenty-one State fish and wildlife agencies were undertaking a vulnerability assessment or had recently completed an assessment of either species, habitats, or both (Choudhury, 2012). These assessments will help State agencies restructure their management plans to incorporate strategies that will build resiliency, enhance ecosystem function and allow species and habitats to respond to change if possible.

Given the enormity of this issue and limited funds and workforce, collaboration with partners is critical (as it is for other pressing conservation issues). State agencies are actively engaging with a variety of traditional and non-traditional partners for adaptation planning and implementation such as DOI Landscape Conservation Cooperatives and Climate Science

Centers, Federal agencies such as USFWS, USFS, BLM, NRCS, NOAA, and EPA, tribes, national and local conservation organizations, academic institutions, local land trusts, outdoor organizations, and local governments to name a few. Partners provide technical expertise, serve on working groups, advocate congress for natural resource adaptation funding, collaborate on adaptation projects, assist in updating State Wildlife Action Plans, assist in vulnerability assessments as well as many other aspects of climate change adaptation work in support of State agencies.

6.2.4. Adaptation by tribal governments

Tribal governments and organizations also have increased efforts to plan for and respond to the impacts of climate change on species, habitats and ecosystems that are vital to their cultures and economies. For example, the Swinomish Tribe in the Pacific Northwest, which depends on salmon and shellfish, has developed a climate adaptation action plan (SITC, 2010). This effort seeks to assess local impacts, identify vulnerabilities, and prioritize planning areas and actions to address the impacts of climate change.

Other efforts like the Tribal Climate Change Project are designed to help understand the needs, lessons learned, and opportunities American Indians and Alaska Natives have in planning for the physical effects of climate change (http://tribalclimate.uoregon.edu/). This project is a collaborative effort between Tribes, the University of Oregon Environmental Studies Program and the USDA Forest Service Pacific Northwest Research Station. Of particular relevance is the efforts of many tribes to make use of traditional ecological knowledge (TEK) to understand climate change impacts on their lands and resources, and to develop adaptation responses (Salick and Ross, 2009, Vinyeta, 2012).

6.2.5. Challenges

Even as scientific evidence for climate change has been ever more robust and certain (IPCC, 2007a), for various reasons, public perception of climate change has become increasingly polarized (Ding and others, 2010). This polarization is becoming an obstacle to adaptation action for officials at all levels of government, including local levels. At the Federal level challenges include budget pressure on agencies and programs involved in wildlife and ecosystem conservation efforts. Additionally, many Federal laws and regulations are not well-prepared to address the challenges that will be posed by a shifting climatic baseline and its attendant ecological responses (Ruhl, 2010). Because of projected shifts in species and habitats with climate change, the need to address resource conservation and management across political and agency jurisdictions will challenge existing models for planning, regulation, and management. A number of Federal agencies, ranging from Department of Interior to the Department of Defense are taking more regional approaches in their ecosystem-based planning and management efforts.

At the State level, a lack of dedicated funding, an ever increasing work load for a downsized workforce, and waning political/public belief in climate change are sizable obstacles for State agencies to overcome in order to forge ahead with adaptation efforts. Not all efforts to respond to climate change will require new approaches. Many of the existing conservation efforts that State agencies currently are pursuing may go a long way in building resiliency in natural systems and helping agencies to plan for and minimize the impacts of climate change in the future. Many State fish and wildlife agencies are re-thinking existing programs and projects and integrating threats associated with climate change and appropriate adaptation responses. Responding and planning for climate change is requiring State agencies to think differently about how to manage species and habitats, and how to build on existing programs and projects to re-tool themselves to respond to climate change. It is important to emphasize that efforts to respond to climate change will not take away from the traditional management roles and responsibilities of State fish and wildlife agencies but may provide an opportunity to create more robust conservation responses to enhance and support existing conservation programs and projects.

6.3. THE NEED TO RECONSIDER CONSERVATION GOALS

Effective conservation and natural resource management relies on the articulation of clear goals, which make possible the development of specific management objectives and measures of success (Williams and others, 2009). At their core, goals are an expression of the desired condition of a landscape or other resource. Although conservation goals ideally are based on robust scientific analyses, fundamentally they are a reflection of human values. Conservation goals are not universal, and will be highly dependent on the particular values of the organizations, agencies, or people setting them. As such, multiple goals may apply to the same landscape or biological features, depending on the values or legal mandates of the organization or agency setting the goals. And while scientists can offer specific information to guide conservation actions, the choice of restoration or management goals is ultimately a process driven as much by societal values, economic constraints, and political feasibility as scientific knowledge (Lackey, 2004; Tear and others, 2005; Stein, 2009; Lindenmayer and Hunter, 2010).

Given the rate and magnitude of climate-mediated ecological changes, a reconsideration of conservation goals is both necessary and unsettling (Glick and others, 2011b), and a number of authors have touched on this issue over the past few years (for example, Pearson and Dawson, 2005; Simenstad and others, 2006; Hobbs and Cramer, 2008; Julius and West, 2008; Lawler, 2009). In this sense, goals articulate the *why* of adaptation, while strategies (discussed in section 6.4) describe the *how*. A particular dilemma facing many resource managers is the need to balance near-term goals for protecting and restoring species and ecosystems, with longer-term goals for sustaining functional ecological systems in the face of climate change.

6.3.1. Forward-looking goals: A key to successful adaptation

Conservation traditionally has been based on a paradigm of maintaining an existing desired condition, or restoring species or habitats to some desired historical state (Craig, 2010). Indeed, the widely used concept of reference condition generally refers to some presumed natural, pristine, or anthropogenically undisturbed condition (Stoddard and others 2006). Setting aside the philosophical and ecological issues associated with determining what historical time period to use as reference, the rapid changes now underway will make efforts to restore back to, and/or maintain, such historical conditions increasingly difficult. Choi (2007) for instance, has noted that although learning from the past will continue to be important, restoration efforts for the future should not be constrained by "historical-fidelity" (Higgs, 2003; Halvorson, 2004; Hobbs, 2004; Throop, 2004).

As climatic factors move outside the bounds of historical variability, they are expected to cause realignments and alterations in both the spatial and temporal patterns of biodiversity (Parmesan, 2006). Because plant and animal species will respond differentially to these climatic shifts, many ecosystems are expected to disaggregate, while novel, or non-analog, assemblages are expected to appear (Hobbs and others, 2006; Williams and Jackson, 2007; Seastedt and others, 2008). Such shifts and realignments will make protecting species and ecosystems in their current locations difficult and in many cases impossible. Given this challenge, one theme that emerges repeatedly in the adaptation literature is the need to move from a paradigm of protection and restoration, to one that is open to anticipating and actively managing change. Consequently, identifying forward-looking, rather than retrospective, goals will be key to successful climate change adaptation efforts.

6.3.2. Reconsidering Goals

Conservation and natural resource managers will not be faced with a choice of whether to reconsider many of our conservation and management goals; rather it will be a matter of when, how much, and in what ways should they change (Glick and others, 2011b). Julius and West (2008) noted that "for virtually every category of Federal land and water management, there will be situations where currently available adaptation strategies will not enable a manager to meet specific goals, especially where those goals are focused on keeping ecosystems unchanged or species where they are." Others have emphasized that managers will need to expand the definition of desirable ecosystems, accepting ecological processes as important goals in addition to species diversity and ecological patterns (Lemieux and others, 2011). Camacho and others (2010), summarized the range of possibilities by noting that climate change will force us to consider "whether we want to be curators seeking to restore and maintain resources for their historical significance; gardeners trying to maximize aesthetic or recreational values; farmers attempting to maximize economic yield; or trustees attempting to actively manage and protect wild species from harm even if that sometimes requires moving them to a more hospitable place."

Several key themes and issues emerge from the literature in the context of rethinking conservation goals. As these issues make clear, there are no easy answers and of necessity there will be trade-offs among many long-held values.

Among the most common suggestions for how conservation goals may need to shift is from those that focus on preserving current patterns of species compositions at particular locations towards goals focused on maintaining processes, both ecological and evolutionary (Harris and others, 2006; Pressey and others, 2007; Prober and Dunlop, 2011; Groves and others, In Press). It is worth noting that process-oriented goals are not novel in conservation. Resource managers have long known that managing ecological processes is key to achieving desired conditions. What is new is the notion that a focus on process might ensure the continuation of diverse and functioning ecosystems, even if the particular compositional and structural attributes may be strikingly different.

Compositional goals will still have relevance, but may need to be expressed at different spatial scales. For example, rather than retaining the full diversity of species at localized sites, such compositional goals may need to be restated as maintaining the full diversity across larger landscapes. Indeed, much of the emphasis on habitat corridors and landscape connectivity implicitly recognizes the need to take this broader geographic perspective as a means for sustaining broad-scale taxonomic and genetic diversity.

Although resource managers tend to focus more on ecological processes than evolutionary processes, the latter will be increasingly important in framing conservation goals under climate change. If managing biodiversity under climate change will largely be about "facilitating nature's response" (Prober and Dunlop, 2011), then having explicit goals for allowing adaptation in an evolutionary sense to proceed will be important, as will maintaining the distinctive evolutionary character of regional floras and faunas. One approach to retaining evolutionary potential that is gaining currency in the literature focuses on geophysical settings as promoters of future evolution (Anderson and Ferree, 2010; Beier and Brost, 2010). In this view, setting goals for the conservation of unique geophysical "stages" may play an important role in enabling the continuing adaptation and evolution of species in a region and in sustaining overall diversity even if the individual "actors" or species will be different.

An overarching conservation goal for many agencies and organizations focuses on the concept of "naturalness" (Cole and Yung, 2010). As species shift in response to climate change, existing ecosystems disaggregate, and novel ecosystems (composed of both native and non-native species) emerge, what will be viewed and accepted as the "new natural"? This question is of more than just theoretical interest. U.S. national parks and wilderness areas, for example, currently are managed to maintain "natural conditions," a term used to describe the condition of resources that would occur in the absence of human dominance over the landscape (NPS, 2006). The dilemma is most stark with regards to formally designated wilderness areas. Future trade-offs will be inevitable between two defining characteristics of wilderness—"untrammeled" quality and historical fidelity (Stephenson and Millar, 2012). The concept of untrammeled nature is embedded in the Wilderness Act of 1964, and refers to areas unencumbered by human influences. The concept of historical fidelity alludes to the primeval and natural character of these areas. Paradoxically, efforts designed to retain historical fidelity will likely require increased human intervention that contravenes the notion of untrammeled nature.

A particularly contentious aspect of reconsidering goals relates to the role of non-native and invasive species in climate-altered ecosystems (Scott and Lemieux, 2005; Green and Pearce-Higgins, 2010; Walther and others, 2009; Schlaepfer and others, 2011; Webber and Scott, 2012). Trade-offs are likely to be most pronounced where managers adopt goals focused on sustaining ecosystem processes and services rather than compositional patterns, which traditionally emphasize native species.

As climate change takes its toll on ecological resources, managers increasingly will be forced to grapple with the notion of triage as well. Bottrill and others (2008) define triage as "the process of prioritizing the allocation of limited resources to maximize conservation returns, relative to the conservation goals, under a constrained budget." In other words, conservation triage entails deciding which of one's conservation targets should be protected and how much intervention is necessary in a given place at a given time (Hagerman and others, 2010). As West and others (2009) stress, "even with substantial management efforts, some systems may not be able to maintain the ecological properties and services that they provide in today's climate. For other systems or species, the cost of adaptation may far outweigh the ecological, social, or economic returns it would provide. In such cases, resources may be better invested elsewhere."

6.3.3. Challenges

The questions that rapid climate change pose for reconsidering conservation goals are not easy to fathom, let alone answer. Even though the principles and practice of conservation have evolved in the past, there are formidable institutional and psychological barriers to shifting from current conservation paradigms and realigning goals (Jantarasami and others, 2010). Institutional barriers to reconsidering and recalibrating goals include those involving: 1) legislation and regulations; 2) management policies and procedures; 3) human and financial capital; and 4) information and science (Julius and West 2008). Legal mandates and policies, in particular, may present difficult barriers. Many existing laws, such as the Endangered Species Act, mandate particular approaches that constrain managers ability to modify goals and management

Impacts of Climate Change on Biodiversity, Ecosystems, and Ecosystem Services |
Technical Input to the 2013 National Climate Assessment

Chapter 6
Adaptation

objectives. Several recent legal reviews (for example, Fischman, 2007; Glicksman, 2009; Craig, 2010; Ruhl 2010), offer some insights on the nuances of Federal conservation laws as they relate to climate change adaptation.

Psychological barriers, however, can be equally challenging to overcome. Hagerman and others (2010) underscore the fact that many conservationists find it difficult to move beyond the familiar goals of restoring and protecting existing patterns of biodiversity and a priori-selected conservation targets due a strong resistance to making trade-offs—a concept described in the psychology literature as "protected values" (Gregory and others, 2006). Poiani and others (2011) similarly observed that a general reluctance of conservation practitioners to "give up on anything" may explain the relatively low number of adaptation strategies focused on transformation compared to maintaining status quo conditions.

6.4. CONVERGENCE ON ADAPTATION PRINCIPLES AND STRATEGIES

In concert with the dramatic increase in attention to adaptation over the past few years, considerable progress has occurred in the development of general principles and strategies for conserving biodiversity and ecosystems in the face of climate change. Although adaptation strategies vary by ecosystem type and depend on the biological levels being targeted, there has been a convergence on several general principles and strategies. Over this period there also has been a proliferation of adaptation planning approaches designed to facilitate the translation of those broad principles and strategies into specific adaptation actions at national, regional, and local scales. Even though these approaches may vary considerably in analytical techniques used, most can be characterized as containing a number of similar steps for moving from planning to implementation.

6.4.1. Overarching principles for adaptation

A large body of recent work has spurred the emergence of general principles for use in biodiversity and ecosystem adaptation (for example, Glick and others, 2009; Heller and Zavaleta, 2009; Mawdsley and others ,2009; West and others, 2009; Hansen and Hoffman, 2010; Kostyack and others, 2011; Game and others, 2011; Peterson and others 2011; Groves and others, In Press). Although many of these ideas resemble guidelines for conventional conservation practice, their novelty arises from the way in which the concepts and, ultimately, actions are applied (Mawdsley and others, 2009). For example, the timing and intensity of a given strategy (for example, prescribed fire) might be modified based on new knowledge of observed and projected impacts of climate change (for example, changes in optimal burning window). Here we distill recent guidance into five general principles that can be readily incorporated into most conservation and management planning processes. These general principles, in turn, apply to and support the application of particular adaptation strategies, which are addressed below (see 6.4.2.).

Link actions to climate impacts

Conservation actions should be designed specifically to address the ecological effects of climate change in concert with existing threats to conservation features and related management goals. As climate adaptation increases in prominence and becomes a primary lens for resource management, there is temptation for managers to re-label existing practices as adaptation, especially under an overly generous interpretation of "enhancing resilience." While some (or even many) such actions may indeed be climate-relevant, to truly be considered "climate adaptation" actions should be supported with an explicit understanding of how they are likely to reduce key vulnerabilities. To facilitate linking actions to climate impacts, managers are strongly encouraged to first conduct a climate change impacts or vulnerability assessment (West and others, 2009; Glick and others, 2011a; Rowland and others, 2011). This requires knowledge of both observed and projected changes in climate, as well as estimates of how these have and will affect physical processes. In turn, managers can use this information to evaluate impacts on ecological processes, vegetation dynamics, species populations and other management priorities. Such vulnerability assessments are a valuable tool for 1) identifying which species and systems are likely to be most strongly affected by projected changes, 2) understanding why they are likely to be vulnerable, and ultimately 3) setting priorities for conservation actions (Glick and others, 2011a). Such assessments can also provide transparency and credibility to strategic action planning and decision making efforts.

Embrace forward-looking goals

As discussed in Section 6.3., conservation goals should be framed to focus on future, rather than past, climatic and ecological conditions. Although embracing forward-looking goals may prove to be difficult in light of existing legislation, regulations, institutional cultures, and other barriers, historical reference points, particularly for restoration purposes, should be used with caution in the goal setting process (Millar and others, 2007). Managers also need to recognize that achievement of retrospective goals may be problematic given rapidly changing environments. Given this possibility, if an existing goal cannot be modified sufficiently, it may need to be abandoned (that is, triaged) and new goals established if ecological changes are particularly acute (West and others, 2009). Ideally, climate-informed management goals and priorities will incorporate a needed flexibility so they can take a long view (decades to centuries) yet also account for near-term conservation challenges and transition periods.

Consider broader landscape context

On-the-ground adaptation actions should be designed in the context of broader geographic scales for both ecological and socio-political considerations (Game and others, 2011; Groves and others In Press; Hilty and others In Press). From an ecological perspective, management of static, individual species populations in a given protected area may no longer be tenable under climate change, given the strong potential for dynamic range shifts, emergence of novel species assemblages, and changes in biotic interactions (Monzón and others, 2011). In addition to more readily accounting for such changes, a broader landscape context facilitates conservation and management approaches related to ecological processes and, in some cases, ecosystem services. For example, managers could more readily establish habitat buffer zones, wildlife refugia, and dispersal corridors using a broader approach (Mawdsley and others, 2009). This approach also provides flexibility for enabling transitions between ecological systems, which otherwise would be hindered by fine focused management practices alone (West and others, 2009). Moreover, from a socio-political perspective, a landscape approach promotes coordination and collaboration across management jurisdictions (Hansen and Hoffman, 2010). This will be necessary for turning institutional barriers, such as those related to management of shared resources, distrustful relationships, and management objectives that are at odds with one

Impacts of Climate Change on Biodiversity, Ecosystems, and Ecosystem Services |
Technical Input to the 2013 National Climate Assessment

Chapter 6
Adaptation

another, into opportunities for partnerships based on shared goals and trust (West and others, 2009).

Select strategies robust to an uncertain future

Managing under uncertain conditions is not new to conservation and management practitioners. However, managing for unknown future conditions using broad-scale climate projections and species distribution models as guidance is one of the greatest challenges managers currently face in an era of ongoing climate change (Lawler and others, 2010). Therefore, selection of strategies and actions should provide benefits across a range of plausible future conditions to account for uncertainties in future climatic conditions (West and others, 2009), and in ecological and human responses to climate shifts. Scenario planning approaches can greatly facilitate the selection of strategies under these conditions (Glick and others, 2011a, also see Section 6.3.4.). This approach is particularly effective when conducted in participatory, interactive settings with managers from multiple jurisdictions working together on common management objectives (Weeks and others, 2011; Cross and others, Accepted). Moreover, selection of strategies that potentially meet multiple objectives across different climate scenarios can provide an additional layer of confidence in support of decision making processes.

Employ agile and informed management

Given the pace of change confronting managers, it is becoming increasingly important to employ agile forms of management that can quickly respond to changing ecological and socio-economic conditions. Adaptive management is the best known approach for continuous learning and refinement of management practices, and can be an important component of climate change adaptation approaches (Williams and others, 2009). Central to this management approach is the need to pose actions as hypotheses that can be empirically tested. When employed in a proactive form, adaptive management can greatly facilitate the ability of conservation planners and resource managers to respond to the uncertainty associated with climate change (Lawler and others, 2010). This includes identifying actions that are directly tied to climate-informed management objectives, modeling potential outcomes, implementing strategies, monitoring their efficacy, and periodically adjusting strategies to better ensure achievement of goals (Heinz Center, 2008). Moreover, adaptive management facilitates a culture of continuous learning and dynamic adjustment to accommodate uncertainty, especially important when managers are attempting to cope with rapidly changing conditions across multiple fronts (for example, not only climatic and ecological, but also socio-economic) (Lawler and others, 2010). It is worth noting, however, that adaptive management is not the only approach for agile management and continuous learning. Indeed, adaptive management, in the formal sense, works best under conditions of high controllability and high uncertainty, while other approaches (for example, scenario planning) may have advantages when confronted by low controllability (Peterson and others, 2003).

Monitoring is essential to effective adaptive management practices and, likewise, should be considered as such to climate change adaptation practices (West and others, 2009; Lawler and others, 2010). This is particularly true relative to uncertainties in our understanding of ecological responses to climate change. Most existing monitoring frameworks and programs were not designed to capture climate change effects and will need to be adapted to do so (Heinz Center, 2008). This includes the identification of key indicators and metrics that track ecological response, including certain demographic parameters and the seasonal timing of life history events (phenology) across components of biodiversity (species, ecosystems, and biomes). In addition to monitoring leading indicators that provide advance warning of climate change effects, status indicators and strategy effectiveness measures will contribute to a comprehensive climate-informed monitoring strategy designed to support many critical aspects of climate adaptation for biodiversity conservation and management (Heinz Center, 2008).

6.4.2. Key adaptation strategies

Over the past few years considerable progress has been made in identifying potential adaptation strategies, and there has been broad convergence on a number of such strategies. Many of these strategies build on existing conservation approaches and conservation biology principles, but their application for climate adaptation often differs in when, where, and how they are applied.

Adaptation strategies for protecting ecosystems, biodiversity and ecosystem services can generally be grouped into three basic categories. First, several proposed strategies focus on improving the current conditions of systems with the stated goal of enhancing its resilience to climate change impacts. These strategies involve restoring ecosystem functioning and reducing other anthropogenic stresses. A second set of strategies involves protecting and managing large landscapes. These strategies include increasing the size of reserves, placing more reserves on the landscape, changing the way reserve networks are designed, and increasing connectivity among protected areas. The remainder of the strategies can generally be classified as site- and species-specific approaches. This last set is a diverse group of strategies that includes approaches such as assisted migration, supplemental watering, dam removal, habitat manipulations, intensive monitoring, and ecosystem engineering.

Improving current conditions

Many of the proposed strategies for addressing climate impacts on biodiversity involve ameliorating current conditions. These strategies include removing other, non-climate-induced, anthropogenic stressors and restoring ecosystem functions, disturbance regimes, or hydrological processes. The goal of each of these actions is to increase the resilience of a system to climate change, the theory being that an intact system, with fewer stressors, will be better able to absorb environmental and climatic stressors.

Many, if not all, recent reviews of the literature on adaptation strategies for protecting biodiversity in a changing climate highlight the need to reduce existing non-climatic stressors on species, populations, and ecosystems (for example, Glick and others, 2009; Heller and Zavaleta, 2009; Lawler, 2009; Mawdsley and others, 2009; Hansen and Hoffman, 2010). Other stressors, such as invasive species, land use, fragmentation, pathogens, and pollutants, can interact with climate change, further impacting species and systems (Schweiger and others, 2010; Walther, 2010). Reducing other such stressors has the potential to increase the resilience of a given system and hence its ability to withstand climate change (Glick and others, 2009). Actions designed to reduce the impacts of other stressors include reducing harvest, eliminating pathogens and pollutants, and reducing activities that cause habitat loss (Heller and Zavaleta, 2009). Although reducing these other threats may well increase the resilience of a species or system to climate change, the degree to which such efforts are successful will depend on the magnitude of the future climate impacts.

Similarly, adaptation strategies designed to restore ecosystem function, protect ecosystem features, or to maintain natural processes have all been suggested as approaches to increase the resilience of systems and species to climate change (Glick and others, 2009; Lawler, 2009; Mawdsley and others, 2009). Focusing on ecosystem functions and processes is likely to be more successful than focusing on suites of species because, although species assemblages will change as climate drives range shifts and alters interspecific interactions, major ecological functions, (for example, carbon cycling or fire regimes), will likely still play important roles even with system changes. More specific strategies aimed at restoring or maintaining functions and processes include restoring flood plains and riparian vegetation, fire management, and dam removal.

Landscape protection and connectivity

Addressing climate change impacts on biodiversity through protected areas or by connecting protected areas are some of the most often-cited adaptation strategies for biodiversity (Heller and Zavaleta, 2009). Protected areas are, arguably, the most important tool for protecting biodiversity today. However, climate change may compromise the ability of current protected lands and waters to protect biodiversity of the future. As climates change, species will likely move out of and into existing reserves. Consequently many suggestions have been made for how to locate, design, and connect reserves to best protect biodiversity as climates continue to change.

Many of the earliest recommendations for addressing climate change through reserve selection and reserve design focused on basic tenets or rules of thumb. For example, many have suggested increasing the total area protected and increasing the size of existing reserves. Larger reserves may provide the space to facilitate within-reserve range shifts. Thus, larger reserves may help to protect more species for longer periods of time in a changing climate (Peters and Darling, 1985). Similarly, reserves that span strong environmental gradients may facilitate range shifts within reserves by providing species access to areas where future suitable habitats may occur (for example, at higher elevations). To best protect species in a changing climate, new reserves can be placed at the core of species environmental distributions (Araújo and others, 2004), at elevational or poleward range-limit of key species (Peters and Darling 1985), or at transitions between major ecological systems or biomes (Halpin 1997).

More recently, more sophisticated approaches for locating new protected areas to address climate change have been suggested (Game and others, 2011). For example, projected shifts in species distributions can be used to identify areas that are likely to protect species today as well as into the future under multiple climate change scenarios (Hannah and others, 2007; Vos and others, 2008). The increase in specificity potentially makes this approach more effective for the targeted species; however, it incorporates higher levels of uncertainty inherent in forecasting future climatic changes and the biotic responses to those changes (Beier and Brost, 2010). Others have suggested placing reserves in areas that protect the underlying environmental gradients that largely determine patterns of biodiversity at broader scales (Anderson and Ferree, 2010; Beier and Brost, 2010). It is proposed that this approach will protect the elements of the landscape that are responsible for the distribution of regional biodiversity and that reserve networks based on them will be more robust to the uncertainties of climate change than will networks based on climate change projections. Although such approaches are theoretically justified, few studies have investigated their applicability (Schloss and others, 2011).

Additionally, the concept of climatic refugia has been invoked for the selection of reserves that will aid species in a changing climate (Groves and others, In Press). Climate refugia have been described both as areas that are projected to experience relatively small changes in climate and as areas projected to have suitable microclimates for given species (Saxon and others, 2005; Hansen and others, 2010; Dobrowski, 2011; Shoo and others, 2011). Ashcroft (2010) also distinguishes between climate refugia at macro versus micro spatial scales, and whether climate refugia are located within a species' current distribution (*in situ*) or outside the current distribution (ex situ). Protecting areas that are projected to change relatively little serves the purpose of protecting strongholds of current species, communities, and ecosystems. Protecting areas that are generally projected to have cooler (or moister or drier) climates than those in the surrounding landscape serves the purpose of giving species in a region a place to go. Locally cooler microclimates may occur at slightly higher elevations or be in areas with more vegetative cover or potential for vegetative cover.

Reserve networks will only be effective at protecting biodiversity in a changing climate if species are able to move among the reserves. In the past, as species moved in response to climatic changes, they often covered whole continents. Today's species, however, will likely have a much harder time tracking shifting climates. In contrast to past climate-driven range shifts, movement across today's landscapes will be limited by human land use and human activities. Fragmented habitats, roads, expansive agricultural fields, and other aspects of human dominated landscapes have the potential to limit species movements and (or) survival in regions of or between suitable climates. Not surprisingly, facilitating species movement by increasing the connectivity of landscapes is one of the most-often cited climate change adaptation strategies for biodiversity (Heller and Zavaleta, 2009).

Although increasing connectivity is likely to be a tool in the adaptation planners toolbox, traditional approaches to increasing connectivity will likely be insufficient for the purpose of addressing climate change (Cross and others, In Press). Traditional approaches for increasing connectivity for species have often included mapping corridors between habitat patches, through suitable habitat. This approach will allow species to move among areas that are currently climatically suitable, but won't likely let them move into areas that will become newly climatically suitable—areas where suitable habitat may develop in the future, but where it does not exist today. More recent studies have highlighted alternative approaches to connecting landscapes to address climate change. One such approach, uses climate projections to orient corridors and expand existing reserves in the direction of anticipated climatic changes (Ackerly and others, 2010). Others have suggested using projected shifts in species distributions to map potential routes that species might take to track shifting climates (Williams and others, 2005; Rose and Burton, 2009). Alternatively, Beier and Brost (2010) recommend using abiotic conditions or land facets (unique combinations of soil types, geologies, and topographies) to define movement corridors. Yet another approach involves connecting warmer areas to cooler areas (or drier areas to wetter areas) along routes that avoid lands that are more heavily impacted by humans and paths that do not follow more gentle climate gradients (Nuñez and others, written communication 2012).

Although much of the research and discussion of connectivity focuses on corridors for connecting landscapes, there are other, likely more cost effective approaches that are potentially more applicable to a wider range of species. For example, increasing the permeability of the landscape (that is, the degree to which a landscape is conducive to wildlife movement) by managing lands to facilitate species' movements will likely also be an effective method of increasing connectivity (Franklin and Lindenmayer, 2009). Selective harvest or retention cuts, tree-planting, alternative zoning, and rotational grazing may provide enough permeability to facilitate range expansions for many species (Kohm and Franklin, 1997; Manning and others, 2009).

Species and place-specific approaches

In addition to broad-scale actions such as reserve selection and promoting landscape connectivity, species- and place-specific approaches will also be important climate change adaptation strategies. These strategies are designed for a specific species or a specific place and as such may be implemented at relatively fine spatial scales. Species-based approaches may include altering microclimates of artificial nest-boxes for rare birds by painting boxes white or locating them on north-exposed slopes (Catry and others, 2011) and supplemental watering of key plant species in drought years (for example, Pavlik and others, 2002). Species-based adaption approaches could also include efforts to identify and manage for populations with higher genetic diversity, or promote populations with more plastic behaviors and morphologies. Place-based approaches, on the other hand could focus on microclimatic variations that provide refugia from large-scale changes in climate (Mosblech and others, 2011) or involve planting trees and other vegetation to provide shade and reduce temperatures (Wilby and others, 1998; Wilby and Perry, 2006).

Another specific example of a place-based adaptation strategy involves planting climate-resistant species or ecotypes (Glick and others, 2009). These species or ecotypes could be used to establish "neo-native forests"—forests consisting of species that existed in that location in the past, but that are not currently found there (Millar and others, 2007). In addition, it has been suggested that replanting efforts and restoration efforts in general make use of diverse genetic stocks instead of relying on local genotypes (Glick and others, 2009).

Assisted colonization, or managed translocation, is another species- and place-specific strategy for addressing climate change. This strategy involves the human movement of species outside their native range in order to facilitate their movement in response to climate change. Assisted colonization has become a hotly debated topic with proponents highlighting the need to move species that will be unable to keep pace with climate change and opponents highlighting the potential for negative, ecological, evolutionary, and economic impacts, as well as ethical concerns (Ricciardi and Simberloff, 2009; Sax and others, 2009; Schwartz and others, 2009). Frameworks for developing policies and making informed decisions for managed translocation have been developed (Hoegh-Guldberg and others, 2008; Richardson and others, 2009). Despite the ongoing debate over the appropriateness of the approach, at least some assisted colonization efforts have been undertaken (Lawler and Olden, 2011), and without policies regulating such actions, additional, uncontrolled translocations will likely occur in the future.

Ex-situ conservation may be a necessary alternative to assisted colonization or some species whose ranges or populations are dramatically reduced by climate impacts (Li and Pritchard, 2009). Such actions could include seed banking and captive breeding to ensure the long-term survival of a species.

6.4.3. Advances in Adaptation Planning

Although the general adaptation principles and strategies discussed above are useful for identifying the range of options for conserving biodiversity and ecosystems in light of climate change, these options need to be localized and translated into actionable recommendations (Heller and Zavaleta, 2009). Translating these principles and strategies into place- and target-specific actions, tied to explicit conservation goals, is often inhibited by complexity and uncertainty in projecting future climate and environmental conditions, by not knowing how to incorporate climate science to existing decision-making processes, and by the absence of readily apparent ways to respond to those changes (Cross and others, Accepted; Lawler and others, 2010). Fortunately, a growing number of adaptation planning approaches are being designed to help practitioners integrate climate change into local conservation decisions. Although these approaches vary in their details, many share a number of key elements. The Climate-Smart Conservation workgroup convened by the National Wildlife Federation has identified the following phases in a generalized adaptation planning and implementation cycle (**Figure 6.2**) (Stein and others, written communication 2012):

1. Identify existing conservation goals and objectives
2. Assess climate change impacts and vulnerabilities
3. Review conservation goals and objectives in light of climate vulnerabilities and revise as necessary
4. Identify adaptation options (that is, strategies and actions capable of reducing vulnerabilities to achieve stated goals)
5. Evaluate and prioritize adaptation options
6. Implement priority actions
7. Track effectiveness of actions and ecological responses (that is, review and refine actions, strategies, and goals based on monitoring and other new information)

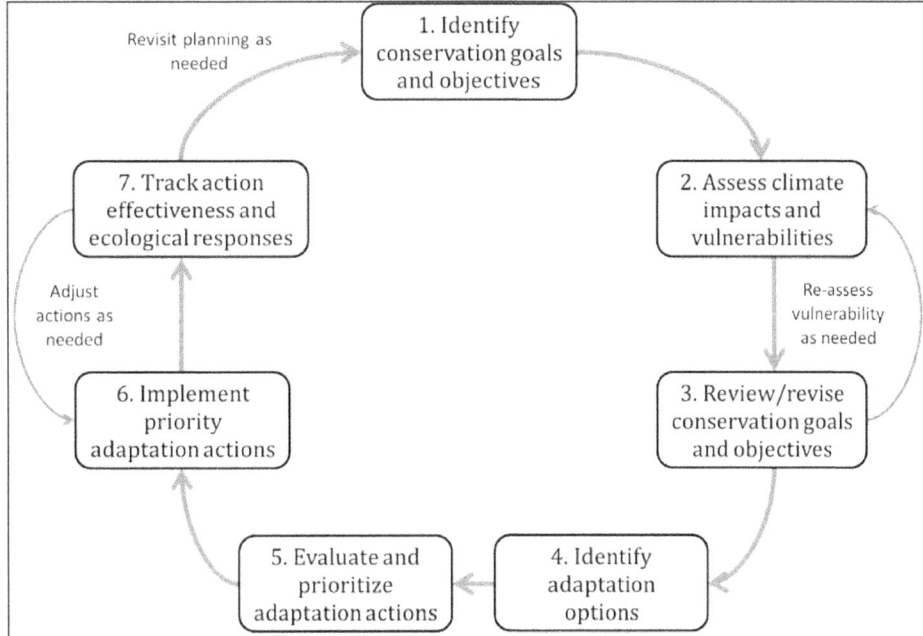

Figure 6.2. Generalized framework for climate change adaptation planning and implementation (Stein and others, written communication 2012).

This planning and implementation framework mirrors standard conservation planning processes (for example, Groves, 2003), and follows the "plan-act-check-adjust" approach of many adaptive management cycles (for example, Williams and others, 2009). Yet these steps are designed specifically to incorporate climate considerations into the planning process, particularly through an emphasis on reconsidering goals in light of climate-related impacts and vulnerabilities. The iterative nature of climate-adaptation planning recognizes the need to incorporate new monitoring information, updated climate and ecological projections, and changing societal values and regulatory constraints into decisions. Although the ultimate goal may be to mainstream climate change into comprehensive decision-making processes that simultaneously address multiple stressors (IPCC, 2007b), targeted biodiversity and ecosystem adaptation planning may be necessary in the near term as decision-makers become more familiar with understanding climate change threats and identifying adaptation strategies.

Assessing climate change impacts and vulnerabilities (Step 2 in **Figure 6.2**) is the most obviously climate-centric part of adaptation planning. Climate change vulnerability is defined as a function of a species' or ecosystem's 1) exposure to climate changes, 2) sensitivity to those changes, and 3) adaptive capacity to cope with those changes (see Section 6.1.2. and *Chapter 2*) for further discussion of vulnerability) (Dawson and others, 2011; Glick and others, 2011a; IPCC 2007b). There are a growing number of tools available for assessing climate change vulnerabilities and impacts, each with its own set of benefits and limitations (Rowland and others, 2011). Climate change vulnerability and impact assessments can explore the relative level of vulnerability of a set of species or ecosystems within a planning region, in addition to highlighting the direct and indirect ways that a species or ecosystem may be affected by climate change (Glick and others, 2011a).

Vulnerability assessments ideally should inform a review and reconsideration of existing goals and management objectives to ensure they continue to be feasible and appropriate in light of climate change (Step 3 in **Figure 6.2**, and as discussed in section 6.3). Such a review may either validate existing goals, or indicate that revisions are needed to address changing conditions and values. In either instance, the intent is to articulate "climate-informed" conservation goals that enable short-term actions to be carried out within a longer term context, rather than to create a set of stand-alone "climate change goals" that may be perceived by managers as disconnected from their day-to-day challenges. Finally, these assessments help identify conservation actions and management interventions (that is, adaptation actions) designed to reduce specific vulnerabilities through either reducing sensitivity or exposure, or increasing adaptive capacity (Step 4 in **Figure 6.2**) (Rowland and others, 2011; Glick and others, 2011a).

Adaptation planning approaches must explicitly consider uncertainties involved in projecting future climate changes and associated impacts to human and natural systems, only some of which can be quantified (Morgan and others, 2009). Planners can fortunately turn to a number of familiar tools for making management decisions in light of uncertainty. Risk management (for example, Willows and Connell, 2003), structured decision-making (for example, Ohlson and others, 2005), scenario-based planning (for example, Peterson and others, 2003), and adaptive management (for example, Conroy and others, 2011) are commonly-used tools for addressing uncertainty that increasingly are being applied to climate change planning. Scenario-based planning has received increasing attention as an important tool for adaptation planning because of its usefulness in situations where uncertainties are high and uncontrollable (Peterson and others, 2003). In this context, the IPCC (2007b) defines a scenario as "a coherent, internally consistent, and plausible description of a possible future state of the world". Scenarios are not meant to be a forecast or prediction of the future, but rather are intended to describe alternate, plausible trajectories for the future (Mahmoud and others, 2009). A key goal of scenario planning is to identify those conservation actions that are recommended across all or most future scenarios. These actions—sometimes called "no regrets" or "low regrets" actions (Willows and Connell, 2003)—are then considered relatively robust to uncertainty in how climate change will play out in a given location.

In the last few years, there has been a proliferation of guidance on adaptation planning and the application of specific adaptation planning approaches to the conservation and management of biodiversity and ecosystems in the United States (for example, EPA, 2009; NOAA, 2010; Halofsky and others, 2011; Peterson and others, 2011; Weeks and others, 2011; Poiani and others, 2011). These approaches tend to follow the generalized adaptation planning steps outlined in **Figure 6.2**, although some rely on specific tools, such as scenario-based planning (Weeks and others, 2011), or the Open Standards for the Practice of Conservation (Poiani and others, 2011). Guidance for adaptation planning also has been developed for Federal agencies (CEQ, 2011b), particularly Federal land management units such as National Forests (Peterson and others, 2011), as well as State fish and wildlife (AFWA, 2009) and coastal management agencies (NOAA, 2010). However, there is broad recognition of the need for adaptation planning to cross jurisdictional boundaries (Heller and Zavaleta, 2009). Many adaptation planning approaches emphasize the value of collaborative dialogue between scientists and managers, focusing on local interpretation of climate change projections and ecological responses, in helping produce practical science-based strategies for climate change adaptation (Halofsky and others, 2011; Peterson and others, 2011; Cross and others, Accepted). There is also a growing recognition that successful adaptation planning requires sustained engagement of science-management partnerships.

6.4.4. Challenges

Despite good progress over the past few years on developing principles, strategies and planning approaches for climate change adaptation, action lags in implementing these plans and putting these strategies to work on the ground. Indeed, moving from planning to implementation is perhaps the greatest current challenge. An additional challenge relates to the application of adaptive management in the practice of adaptation. Although the general concept of adaptive management—entailing an iterative cycle of learning and doing—is widely acknowledged, adaptive management in the formal sense has proven challenging to put widely into practice. As a result, there is a need both to improve the capacity of managers to incorporate adaptive management into their existing decision and management processes, as well as the need for developing new approaches for incorporating agile and informed management into climate adaptation efforts.

As society moves from adaptation planning to implementation there will be an increasing need to consider unintended consequences of adaptation actions. Reducing the potential for unintended and counterproductive consequences will require the development and application of risk assessment methods for selecting among possible adaptation responses. Risk analysis will be needed to better distinguish among those things that we can do, from those that we should do.

Even the most widely cited adaptation strategy, establishment of corridors to facilitate species migrations and range shifts (Heller and Zavaleta, 2009), may have unintended negative effects. Some authors (for example, Proches and others, 2005; Simberloff and others, 1992), have raised the concern that such corridors could open pathways for invasive species to colonize

Impacts of Climate Change on Biodiversity, Ecosystems, and Ecosystem Services |
Technical Input to the 2013 National Climate Assessment

Chapter 6
Adaptation

core habitat areas, although others suggest this may not be the case (for example, Damschen and others, 2006), or that the conservation benefits far outweigh the potential negative effects (Levey and others, 2005). The vigorous debate within the scientific and conservation community over the merits or dangers of managed translocations or "assisted migration" is an example of where risk assessment frameworks are needed to evaluate the potential positive and negative effects of proposed actions (Hoegh-Goldberg, 2008; Richardson and others, 2009).

Ecosystems are highly complex, and integrated and stochastic systems are impossible to control. These complexities will generate surprises in response to natural and human-caused adjustments of ecosystems to climate change, even in the case of adaptation actions that have positive effects in terms of their stated goals (for example, conservation of a threatened species). As a result, assessing the options and risks of intervention will become increasingly important. Dawson and others (2011), for instance, have summarized the range of possible adaptation responses in terms of intensity of intervention, ranging from preparedness, to low intensity intervention, to high intensity intervention (**Figure 6.3**). They note that management decisions will depend on judgments of potential risks and benefits, balanced against costs and the availability of resources. Evaluation of specific adaptation responses will also need to incorporate socioeconomic considerations in order to have a more complete understanding of the potential risks as well as opportunities for implementation (Adger and others, 2005).

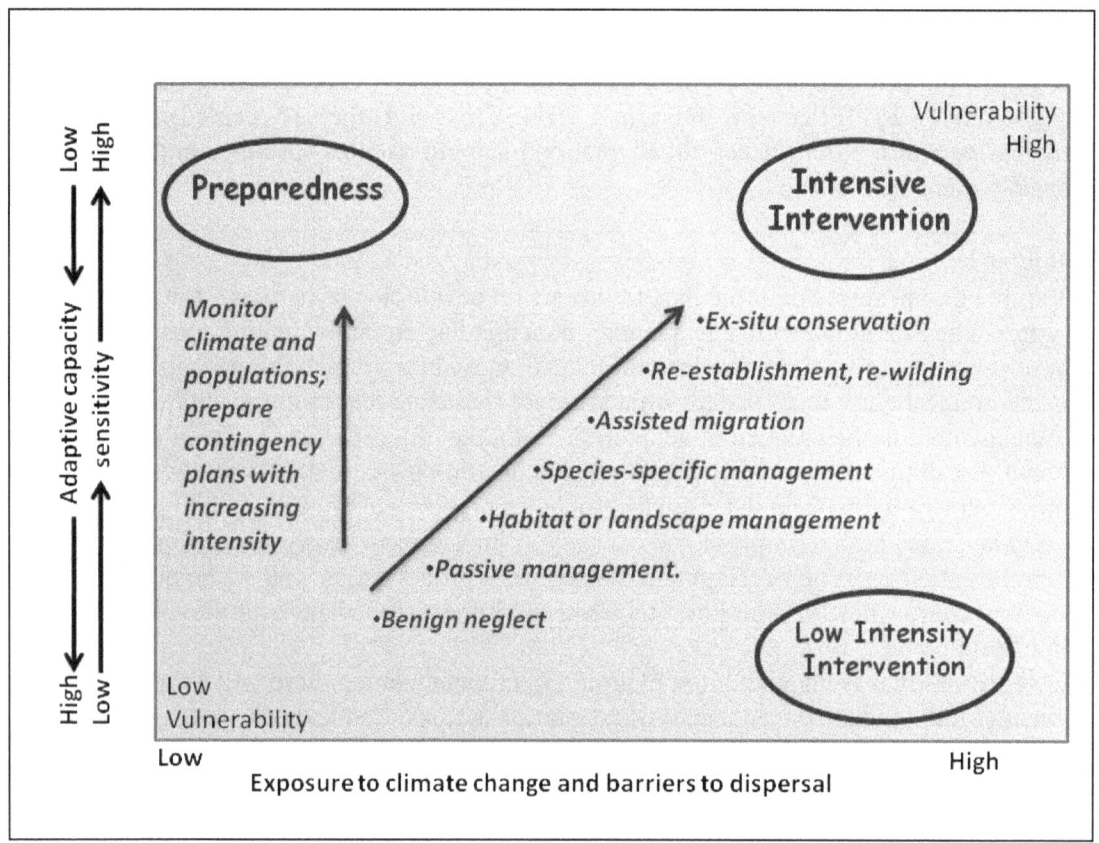

Figure 6.3. *Framework for assessing the relative intensity of conservation interventions based on the vulnerability of a species or ecosystem (from Dawson and others, 2011).*

Impacts of Climate Change on Biodiversity, Ecosystems, and Ecosystem Services |
Technical Input to the 2013 National Climate Assessment

Chapter 6
Adaptation

6.5. MAINSTREAMING ADAPTATION

Early efforts to develop climate adaptation strategies to improve outcomes for biodiversity and ecosystems began as fairly isolated endeavors, often independent from even broader conservation and management activities. Over the past decade, however, it has become increasingly apparent that good biodiversity and ecosystem adaptation not only will be important in the general practice of conservation, but that biodiversity and ecosystem adaptation should be integrated into cross-sectoral planning efforts (Hansen and Hoffman, 2010). Successful ecosystem adaptation will bridge the natural/built environment divide.

Mainstreaming biodiversity and ecosystem adaptation, both within the natural resource community and into larger societal efforts to address the climate change is just beginning; however, it seems vital for improving biodiversity and ecosystem outcomes over the long-term. Indeed, integrating adaptation planning into existing agency planning efforts is a key tenet of the Federal interagency adaptation task force (see Section 6.2.1). Although eliminating sectoral management and planning silos to allow for better integration of cross-sectoral solutions will be necessary (CEQ, 2010), for the most part this is not part of current management and planning governance structures.

In order to develop a model for how to begin this integration, the concept of Ecosystem-Based Adaptation (EBA) has been proposed (CBD, 2009; Colls and others, 2009; Vignola and others, 2009; World Bank, 2010). Ecosystem-based approaches to adaptation (EBA) have been defined by the Convention on Biological Diversity as "the use of biodiversity and ecosystem services as part of an overall adaptation strategy to help people adapt to the adverse effects of climate change" (CBD, 2009) Ecosystem-based adaptation has not been widely used by name in the United States, but has been growing in recognition internationally. This approach aims to demonstrate the cost savings and mutual benefits to social, economic, cultural and biological systems that can occur when biodiversity and ecosystem service adaptation is integrated into broader societal adaptation efforts. An example of ecosystem-based adaptation is the role of intact ecosystems (for example, mangroves or wetlands) in contributing to coastal storm attenuation and protection of human communities. Such a role was in evidence during the 2004 tsunami in Southeast Asia, where coastal regions with intact mangroves fared better than deforested coastline (Mascarenhas and Jayakumar, 2008; Kaplan and others, 2009), as well as in coastal Louisiana post-Hurricane Katrina (Shaffer and others, 2009). Other examples include improving agriculture through ecosystem-based rather than agrochemical approaches, and water quality protection along the lines of New York City's protection of upstate watersheds.

As with all environmental issues, societal priorities tend to focus on social systems and the built infrastructure—things that have easily recognized economic and social value. Adaptation is no different. However there is a growing understanding that using ecosystem attributes and services can be a more sustainable and affordable path to social and built environment adaptation. For example many regions are exploring floodplain protection and restoration instead of seawalls and levies as tools for addressing the challenges of sea level rise and flooding (Kershner, 2010). Increasing our understanding of the utility of these natural systems for long-term protection will be necessary to garner better long-term outcomes for their own successful adaptation endeavors.

Impacts of Climate Change on Biodiversity, Ecosystems, and Ecosystem Services |
Technical Input to the 2013 National Climate Assessment

Chapter 6
Adaptation

6.6. ILLUSTRATIVE CASE STUDY: CLIMATE ADAPTATION IN THE SAN FRANCISCO BAY

The San Francisco Bay is home to approximately seven million people and a rich diversity of fish, wildlife, and other ecological resources. Ecosystem services—including fishing, hunting, recreation, and salt harvesting—are valued at $240-293 million annually for the Bay (Battelle Memorial Institute, 2008). Recognizing the significant threat of sea level rise to infrastructure, ecosystems and the services they provide, and cultural and recreational resources, the region undertook a process to assess the vulnerability of coastal resources and develop adaptation strategies. Their experience provides a useful case study of the recent advances in adaptation, as well as some of the remaining challenges.

The adaptation planning process was initiated in November 2008 by the San Francisco Bay Conservation and Development Commission (SFBCDC). Having a conservation-oriented, cross-jurisdictional entity in place was quite fortuitous for facilitating an effective adaptation process. For one thing, SFBCDC leadership of this effort enabled a landscape approach to sea-level rise adaptation, ensuring that individual communities in the Bay had a shared vision and strategy. Furthermore, during the last several decades, the SFBCDC had already established a broad-based recognition that conservation objectives should be a significant consideration in making development decisions for the Bay. The rationale for considering ecological impacts, both for conserving wildlife and for sustaining ecosystem services, had already been articulated in the San Francisco Bay Plan, first authored in 1968 (SFBCDC, 2008). Thus, when addressing the implications and response options for sea level rise, it was already natural to mainstream biodiversity and ecosystem considerations in the adaptation decisions relevant to the built infrastructure.

The SFBCDC began the adaptation planning process by conducting a vulnerability assessment, culminating in the publication of "Living with a Rising Bay: Vulnerability and Adaptation in San Francisco Bay and on the Shoreline" (SFBCDC, 2011a), first released in April 2009, and a revised version released in September 2011. The assessment examined two sea-level rise scenarios: a 16-inch rise by mid-century and a 55-inch rise by the end of the century. They found that coastal development valued at $62 billion would be at increased risk of flooding by the end of the century. Areas at risk include about 128-square miles of residential development, 93 percent of the San Francisco and Oakland airports, and 87 percent of the waterfront areas available for public access. The multiple ways that climate change is likely to affect the Bay's ecosystem are also discussed, including the inundation or erosion of coastal habitats, altered species composition, changing freshwater flow regimes, and degraded water quality.

Using the results of the vulnerability assessment, the SFBCDC initiated a collaborative process to engage scientists, managers, and stakeholders in amending the San Francisco Bay Plan to include consideration of climate change (SFBCDC, 2011b). Preliminary recommendations for amendments to the Bay Plan, included in the vulnerability assessment, and two subsequent drafts of amendment language were offered for public comment. The SFBCDC held a total of 36 public hearings, workshops, and meetings as the amendment language evolved. Although the process took longer than initially anticipated, the amendments were officially adopted on October 6, 2011. The amendments recommend a next step of formulating a regional sea level rise adaptation strategy that specifically maps out where and how coastal development and conservation projects should proceed.

The amendments to the Bay Plan feature many of the key findings highlighted in this chapter. For example, they explicitly identify the need to reconsider the previously adopted

Impacts of Climate Change on Biodiversity, Ecosystems, and Ecosystem Services |
Technical Input to the 2013 National Climate Assessment

Chapter 6
Adaptation

conservation goal of restoring 65,000 acres of tidal marsh (Goals Project, 1999). This goal was based on the conditions observed in the 1990s, and the plan now recognizes that climate change may require new regional targets. The amendments also highlight the need to use adaptive management and to consider a range of strategies in vulnerable shoreline areas, from armoring the shoreline to discouraging and ultimately removing shoreline development to allow for managed retreat of coastal habitats. Another example is the identification of several existing stressors—including modification of sedimentation rates, introduction of invasive species, and pollution— and the potential for sea level rise to interact with these stressors, thereby magnifying their effects on the Bay ecosystem.

Despite the significant progress in assessing the vulnerability of the Bay to sea level rise and identifying potential adaptation opportunities, much work remains to translate this guidance into specific, on-the-ground development and conservation actions. Nonetheless, the amendments state that the region must "begin now to adapt to the impacts of climate change" and identifies the sorts of projects that should be encouraged even as more detailed guidance is being developed. These preferred projects do not negatively impact the Bay, do not increase risks to public safety, or have regional benefits that outweigh the risks from climate change. This risk-based approach to weighing the costs and benefits of different development choices and the trade-offs between different shared goals will be a necessary element of adaptation efforts, particularly when the limits of adaptation might be approached.

6.7. INTO THE FUTURE: ADAPTATION AND ITS LIMITS

Over the past few years the field of climate adaptation has experienced considerable progress and has begun to identify opportunities for society to prepare for and cope with the impacts of rapid climate change. There are, however, limits to the amount of change that can be accommodated without major ecosystem disruptions and loss of biodiversity. Limits to adaptation tend to revolve around thresholds of an ecological, economic, or technological nature (Adger and others, 2009). There are, for instance, ecological or physical thresholds beyond which adaptation responses are unable to prevent climate change impacts (for example, temperature thresholds for organisms, such as thermal stress in corals or cold-water salmonids). Economic thresholds exist where the costs of adaptation exceeds the costs of averted impacts (that is, it is more expensive to adapt than to experience the impacts). Finally, there are technological thresholds beyond which available technologies cannot avert the climate impacts (for example, limits to engineered solutions to avoid extreme flooding). In practice, however, at least the latter two thresholds are dependent on societal constructs, and influenced by attitudes to risk, values, and ethics (Adger and others, 2009).

The rate, magnitude, and character of climatic changes will influence whether and when these limits are exceeded. As an example, a system may be capable of accommodating a level of change that occurs gradually, but may not be capable of accommodating that same amount of change if it takes place more rapidly. If faced with enough external change, species and systems will exceed their adaptive capacity (even with the benefit of targeted adaptation actions), cross ecological thresholds, and undergo regime shifts (CCSP, 2009). Although development of indicators to predict regime shifts is an active area of research (for example, Brock and Carpenter, 2006, Scheffer and others, 2009), ecological thresholds are notoriously difficult to forecast, and tend to be recognized only once they have been exceeded (Groffman and others, 2006; CCSP, 2009). Depending upon the extent of future climatic change, even with the most aggressive adaptation strategies, society may be unable to prevent irreversible losses of biodiversity or serious degradation of ecosystems and their services.

It looks increasingly likely that projected global average temperature increases will exceed the 2 degrees Celsius target that scientists and policymakers had identified as a threshold for avoiding dangerous interference with the climate system (IEA, 2011). Accordingly, the need to cope with increasing climate impacts (that is, adaptation) will only become more acute with higher levels of warming. The paradox, however, is that even as the need for adaptation becomes more intense, the success of adaptation efforts increasingly will be tested and possibly compromised as ecological, economic, and technological thresholds are reached (Hansen and others, 2010). This paradox highlights the importance of viewing adaptation as a process rather than an outcome, and as fundamentally about managing rather than resisting change. Indeed, for adaptation to be successful over the long term it will need to promote what the eminent conservationist Aldo Leopold referred to as the "capacity for self renewal" (Leopold, 1949).

6.8. LITERATURE CITED

Ackerly DD, Loarie SR, Cornwell WK, Weiss SB, Hamilton H, Branciforte R, and Kraft NJB. 2010. The geography of climate change: implications for conservation biogeography. *Diversity and Distributions* **16**: 476-487.

Adger WN, Arnell NW, and Tompkins EL. 2005. Successful adaptation to climate change across scales. *Global Environmental Change* **15**: 77-86.

Adger WNS, Dessai S, Goulden M, Hulme M, Lorenzoni I, Nelson DR, Naess LO, Wolf J, and Wreford A. 2009. Are there limits to adaptation to climate change? *Climatic Change* **93**: 335-354.

AFWA (Association of Fish and Wildlife Agencies). 2009. Voluntary guidance for States to incorporate climate change into State Wildlife Action Plans and other management plans. Washington, DC: AFWA.

Anderson MG and Ferree CE. 2010. Conserving the stage: climate change and the geophysical underpinnings of species diversity. *PLoS ONE* **5**: e11554.

Araújo MB, Cabeza M, Thuiller W, Hannah L, and Williams PH. 2004. Would climate change drive species out of reserves? An assessment of existing reserve-selection methods. *Global Change Biology* **10**: 1618-1626.

Ashcroft MB. 2010. Identifying refugia from climate change. *Journal of Biogeography* **37**: 1407-1413.

Battelle Memorial Institute. 2008. San Francisco Bay Subtidal Habitat Goals Report. Appendix 1-2: Economic Valuation of San Francisco Bay Natural Resources Services. Available at: www.sfbaysubtidal.org/PDFS/Ap1-2%20Econ%20Evaluation.pdf

Beier P and Brost B. 2010. Use of land facets to plan for climate change: conserving the arenas, not the actors. *Conservation Biology* **24**: 701-710.

Bottrill MC, Joseph LN, Carwardine J, Bode M, Cook C, Game ET, Grantham H, Kark S, Linke S, McDonald-Madden E, Pressey RL, Walker S, Wilson KA, and Possingham HP. 2008. Is conservation triage just smart decision making? *Trends in Ecology and Evolution* **23**: 649-654.

Brock, WA and Carpenter SR. 2006. Variance as a leading indicator of regime shift in ecosystem services. *Ecology and Society* **11**: 9. Available at: www.ecologyandsociety.org/vol11/iss2/art9/

C2ES (Center for Climate and Energy Solutions). 2012. Climate change adaptation: what Federal agencies are doing? February 2012 update. Arlington, VA: Center for Climate and Energy Solutions.

Camacho AE, Doremus H, McLachlan JS, and Minteer BA. 2010. Reassessing conservation goals in a changing climate. *Issues in Science and Technology Online*. Available at: www.issues.org/26.4/p_camacho.html

Carey PD., Fitter AH, and Watkinson AR. 1992. A field study using the fungicide benomyl to investigate the effect of mychorrhizal fungi on plant fitness. *Oecologia* **90**: 550-555.

Carpenter S, Walker B, Anderies JM, and Abel N. 2001. From metaphor to measurement: Resilience of what to what? *Ecosystems* **4**: 765-781.

Catry I, Franco AMA, and Sutherland WJ. 2011. Adapting conservation efforts to face climate change: modifying nest-site provisioning for lesser kestrels. *Biological Conservation* **144**: 1111-1119.

CBD (Convention on Biological Diversity). 2009. Connecting biodiversity and climate change mitigation and adaptation: key messages from the report of the second ad hoc technical

Impacts of Climate Change on Biodiversity, Ecosystems, and Ecosystem Services |
Technical Input to the 2013 National Climate Assessment

Chapter 6
Adaptation

expert group on biodiversity and climate change. Technical Series No. 41. Montreal: Secretariat of the Convention on Biological Diversity.

CCSP (Climate Change Science Program). 2009. Thresholds of climate change in ecosystems. A report by the U.S. Climate Change Science Program and the Subcommittee on Global Change Research. Reston, VA: U.S. Geological Survey. Available at: www.climatescience.gov/Library/sap/sap4-2/final-report/

CEQ (Council on Environmental Quality). 2010. Progress report of the Interagency Climate Change Adaptation Task Force: recommended actions in support of a national climate change adaptation strategy. Washington, DC: White House Council on Environmental Quality. Available at: www.whitehouse.gov/sites/default/files/microsites/ceq/Interagency-Climate-Change-Adaptation-Progress-Report.pdf

CEQ (Council on Environmental Quality). 2011a. Federal actions for a climate resilient nation: progress report of the Interagency Climate Change Adaptation Task Force. Washington, DC: White House Council on Environmental Quality. Available at: www.whitehouse.gov/sites/default/files/microsites/ceq/2011_adaptation_progress_report.pdf

CEQ (Council on Environmental Quality). 2011b. Instructions for implementing climate change adaptation planning in accordance with Executive Order 13514. http://www.whitehouse.gov/sites/default/files/microsites/ceq/adaptation_final_implementing_instructions_3_3.pdf

CEQ (Council on Environmental Quality). 2011c. Draft national action plan: priorities for managing freshwater resources in a changing climate. Washington, DC: White House Council on Environmental Quality. Available at: www.whitehouse.gov/sites/default/files/microsites/ceq/napdraft6_2_11_final.pdf

CEQ (Council on Environmental Quality). 2011d. Draft national ocean policy implementation plan. Washington, DC: White House Council on Environmental Quality. Available at: www.whitehouse.gov/sites/default/files/microsites/ceq/national_ocean_policy_draft_implementation_plan_01-12-12.pdf

Chen I., Hill JK, Ohlemüller R, Roy DB, Thomas CD. 2011. Rapid range shifts of species associated with high levels of climate warming. *Science* **333**: 1024-1026.

Choi YD. 2007. Restoration ecology to the future: a call for new paradigm. *Restoration Ecology* 15: 351-353.

Choudhury A. (Ed). 2012. AFWA 2011 State Climate Adaptation Summary Report. Washington, DC: Association of Fish and Wildlife Agencies. Available at: www.fishwildlife.org/files/AFWA2011StateClimateAdaptationSummaryReport.pdf

Cole DN. and Yung L (Eds). 2010. Beyond naturalness: rethinking park and wilderness stewardship in an era of rapid change. Washington, DC: Island Press.

Colls A, Ash N, and Ikkala N. 2009. Ecosystem-based adaptation: a natural response to climate change. Gland, Switzerland: IUCN.

Conroy MJ, Runge MC, Nichols JD, Stodola KW, Cooper RJ. 2011. Conservation in the face of climate change: the roles of alternative models, monitoring, and adaptation in confronting and reducing uncertainty. *Biological Conservation* **144**: 1204-1213

Craig RK. 2010. "Stationarity is dead" – long live transformation: five principles for climate change adaptation law. *Harvard Environmental Law Review* **34**: 9-75.

Cross MS, Hilty JA, Tabor GM, Lawler JJ, Graumlich LJ, and Berger J. In Press. From connect-the-dots to dynamic networks: Maintaining and enhancing connectivity as a strategy to

Impacts of Climate Change on Biodiversity, Ecosystems, and Ecosystem Services | Technical Input to the 2013 National Climate Assessment

Chapter 6
Adaptation

address climate change impacts on wildlife. *In* Brodie J, Doak D, and Post E (Eds), Wildlife conservation in a changing climate. University of Chicago Press, Chicago, IL.

Cross MS, McCarthy PD, Garfin G, Gori D, and Enquist CAF. Accepted. Accelerating climate change adaptation for natural resources in southwestern United States. *Conservation Biology.*

Damschen EI, Haddad NM, Orrock JL, Tewksbury JJ, and Levey DJ. 2006. Corridors increase plant species richness at large scales. *Science* **313**: 1284-1286.

Dawson TP, Jackson ST, House JI, Prentice IC, and Mace GM. 2011. Beyond predictions: Biodiversity conservation in a changing climate. *Science* **332**:53-58.

Ding D, Maibach EW, Zhao X, Roser-Renouf C, and Leiserowitz A. 2010. Support for climate policy and societal action are linked to perceptions about scientific agreement. *Nature Climate Change* **1**: 462-466.

Dobrowski SZ. 2011. A climatic basis for microrefugia: the influence of terrain on climate. *Global Change Biology* **17**: 1022-1035.

Drever CR, Peterson G, Messier C, Bergeron Y, and Flannigan MD. 2006. Can forests management based on natural disturbances maintain ecological resilience? *Canadian Journal of Forest Research* **36**: 2285-2299.

Egerton FN. 1973. Changing concepts of the balance of nature. *Quarterly Review of Biology* **48**: 322-350.

EPA (Environmental Protection Agency). 2009. Synthesis of adaptation options for coastal areas. EPA 430-F-08-024. Washington, DC: U.S. Environmental Protection Agency, Climate Ready Estuaries Program.

Fazey I, Gamarra JGP, Fischer J, Reed MS, Stringer LC, and Christie M. 2010. Adaptation strategies for reducing vulnerability to future environmental change. *Frontiers in Ecology and Environment* **8**: 414-422.

Fischman RL. 2007. From words to action: The impact and legal status of the 2006 National Wildlife Refuge System Management Policies. *Stanford Environmental Law Journal* 77. http://ssrn.com/abstract=921073. Viewed May 25, 2011.

Folke C. 2006. Resilience: the emergence of a perspective for social-ecological systems analyses. *Global Environmental Change* **16**: 253-267.

Franklin JF and Lindenmayer D. 2009. Importance of matrix habitats in maintaining biological diversity. *Proceedings of the National Academy of Science* **106**: 349-350.

Gallopín GC. 2006. Linkages between vulnerability, resilience, and adaptive capacity. *Global Environmental Change* **16**: 293-303.

Game ET, Lipsett-Moore G, Saxon E, Peterson N, and Sheppard S. 2011. Incorporating climate change adaptation into national conservation assessments. *Global Change Biology* **17**: 3150-3160.

Glick P, Staudt A, and Stein B. 2009. A new era for conservation: review of climate change adaptation literature. Washington DC: National Wildlife Federation. Available at: http://www.nwf.org/~/media/PDFs/Global-Warming/Reports/NWFClimateChangeAdaptationLiteratureReview.ashx

Glick P, Stein BA, and Edelson N. 2011a. Scanning the conservation horizon: a guide to climate change vulnerability assessment. Washington, DC: National Wildlife Federation. Available at: www.nwf.org/vulnerabilityguide

Glick P, Chmura H, and Stein BA. 2011b. Moving the conservation goalposts: a review of climate change adaptation literature. Washington, DC: National Wildlife Federation.

Impacts of Climate Change on Biodiversity, Ecosystems, and Ecosystem Services |
Technical Input to the 2013 National Climate Assessment

Chapter 6
Adaptation

Available at: www.nwf.org/~/media/PDFs/Global-Warming/Reports/Moving-the-
Conservation-Goalposts-2011.ashx

Glicksman RL. 2009. Ecosystem resilience to disruptions linked to global climate change: an adaptive approach to Federal land management. *Nebraska Law Review* 87: 833. Available at: http://works.bepress.com/robert_glicksman/8

Goals Project. 1999. Baylands ecosystem habitat goals. A report of habitat recommendations prepared by the San Francisco Bay Area Wetlands Ecosystem Goals Project. San Francisco, CA and Oakland, CA: U.S. Environmental Protection Agency and S.F. Bay Regional Water Quality Control Board. Available at: http://baeccc.org/pdf/sfbaygoals031799.pdf

Green RE and Pearce-Higgins J. 2010. Species management in the face of a changing climate. *In* Baxter JM and Galbraith CA (Eds), Species management: challenges and solutions for the 21st Century. HMSO, Edinburgh.

Gregory R, Ohlson D, and Arvai J. 2006. Deconstructing adaptive management: criteria for applications to environmental management. *Ecological Applications* 16: 2411-2425.

Groffman PM, Baron JS, Blett T, Gold AJ, Goodman I, Gunderson LH, Levinson BM, Palmer MA, Paerl HW, Peterson GD, Poff NL, Rejeski DW, Reynolds JF, Turner MG, Weathers KC, and Wiens J. 2006. Ecological thresholds: the key to successful environmental management or an important concept with no practical application? *Ecosystems* 9: 1-13

Groves CR. 2003. Drafting a conservation blueprint: a practitioner's guide to planning for biodiversity. Washington, DC: Island Press.

Groves CR, Game ET, Anderson MG, Cross M, Enquist C, Ferdana Z, Girvetz E, Gondor A, Hall KR, Higgins J, Marshall R, Popper K, Schill S, and Shafer SL. In Press. Incorporating climate change into systematic conservation planning. *Biodiversity and Conservation.*

Hagerman S, Sowlatabadi H, Satterfield T, and McDaniels T. 2010. Expert views on biodiversity conservation in an era of climate change. *Global Environmental Change* 20: 192-207.

Halofsky JE, Peterson DL, O'Halloran KA, and Hawkins-Hoffman C (Eds). 2011. Adapting to climate change at Olympic National Forest and Olympic National Park. Gen. Tech. Rep. PNW-GTR-844. Portland, OR: U.S. Department of Agriculture, Forest Service, Pacific Northwest Research Station.

Halpin PN. 1997. Global climate change and natural-area protection: management responses and research directions. *Ecological Applications* 7: 828-843.

Halvorson W. 2004. A response to the article (Hobbs 2004) "Restoration ecology: the challenge of social values and expectation." *Frontiers in Ecology and the Environment* 2: 46-47.

Hannah L, Midgley GF, Andelman S, Araujo MB, Hughes G, Martinez-Meyer E, Pearson RG, and Williams P. 2007. Protected area needs in a changing climate. *Frontiers in Ecology and the Environment* 5: 131-138.

Hansen LJ, Biringer JL, and Hoffman JR. 2003. Buying time: a user's manual for building resistance and resilience to climate change in natural systems. Washington, DC: World Wildlife Fund.

Hansen LJ and Hoffman JR. 2010. Climate savvy: Adapting conservation and resource management to a changing world. Washington, DC: Island Press.

Hansen L, Hoffman J, Drews C, and Mielbrecht E. 2010. Designing climate-smart Conservation: Guidance and case studies. *Conservation Biology* 24: 63-69.

Harris JA, Hobbs RJ, Higgs E, and Aronson J. 2006. Ecological restoration and global climate change. *Restoration Ecology* 14: 170-176.

Heinz Center. 2008. Strategies for managing the effects of climate change on wildlife and ecosystems. Washington, DC: Heinz Center.

Heller NE and Zavaleta ES. 2009. Biodiversity management in the face of climate change: a review of 22 years of recommendations. *Biological Conservation* **142**: 14-32.

Higgs E. 2003. Nature by design, people, natural processes, and ecological restoration. Cambridge, MA: MIT Press.

Hilty JA, Chester CC, and Cross MS (Eds). In Press. Climate and conservation: landscape and seascape science, planning and action. Washington, DC: Island Press.

Hobbs RJ. 2004. Restoration ecology; the challenge of social values and expectations. *Frontiers in Ecology and the Environment* **2**: 43-44.

Hobbs RJ, Arico S, Aronson J, Baron JS, Bridgewater P, Cramer VA, Epstein PR, Ewel JJ, Klink CA, Lugo AE, Norton D, Ojima D, Richardson DM, Sanderson EW, Valladares F, Vilà M, Zamora R, and Zobel M. 2006. Novel ecosystems: theoretical and management aspects of the new ecological world order. *Global Ecology and Biogeography* **15**: 1-7.

Hobbs R. and Cramer V. 2008. Restoration ecology: interventionist approaches for restoring and maintaining ecosystem function in the face of rapid environmental change. *Annual Review of Environment and Resources* **33**: 39-61.

Hoegh-Guldberg O, Hughes L, McIntyre S, Lindenmayer DB, Parmesan C, Possingham HP, and Thomas CD. 2008. Assisted colonization and rapid climate change. *Science* **321**: 345-346.

Hoffmann AA, and Sgrò CM. Climate change and evolutionary adaptation. *Nature* **470**: 479-485.

Holling CS.1996. Engineering resilience versus ecological resilience. *In* Schulze PC (Ed), Engineering within ecological constraints. National Academy Press, Washington, DC.

Holling CS. 1973. Resilience and stability of ecological systems. *Annual Review of Ecology and Systematics* **4**: 1-23.

IEA (International Energy Agency). 2011. World energy outlook: executive summary. Paris, France: IEA.

IPCC (Intergovernmental Panel on Climate Change). 2001. Climate change 2001: synthesis report. Cambridge, UK: Cambridge University Press.

IPCC (Intergovernmental Panel on Climate Change). 2007a. Climate change 2007: synthesis report. Cambridge, UK and New York, NY: Cambridge University Press.

IPCC (Intergovernmental Panel on Climate Change). 2007b. Climate change 2007: working group II: impacts, adaptation and vulnerability. Cambridge, UK and New York, NY Cambridge University Press.

IPCC (Intergovernmental Panel on Climate Change). 2012. Managing the risks of extreme events and disasters to advance climate change adaptation. A special report of working groups I and II of the Intergovernmental Panel on Climate Change. Cambridge, UK, and New York, NY: Cambridge University Press.

Jantarasami LC, Lawler JJ, and Thomas CW. 2010. Institutional barriers to climate change adaptation in U.S. national parks and forests. *Ecology and Society* **15**: 33. Available at: www.ecologyandsociety.org/vol15/iss4/art33

Julius SH and West JM (Eds). 2008. Preliminary review of adaptation options for climate sensitive ecosystems and resources. A report by the U.S. Climate Change Science Program and the Subcommittee on Global Change Research. Washington, DC:U.S. Environmental Protection Agency. Available at: http://downloads.climatescience.gov/sap/sap4-4/sap4-4-final-report-all.pdf

Kaplan M, Renaud FG, and Luchters G. 2009. Vulnerability assessment and protective effects of coastal vegetation during the 2004 tsunami in Sri Lanka. *Natural Hazards and Earth System Sciences* **9**: 1479-1494.

Kelly AE, and Goulden ML. 2008. Rapid shifts in plant distribution with recent climate change. *Proceeding of the National Academy of Science* **105**: 11823-11826.

Kershner J. 2010. South Bay Salt Pond Restoration Project [Case study on CAKE website]. Available at: http://www.cakex.org/case-studies/2876

Kohm KA and Franklin JF (Eds). 1997. Creating a forestry for the 21st Century: the science of ecosystem management. Washington, DC: Island Press,

Kostyack J, Lawler JJ, Goble DD, Olden JD, and Scott JM. 2011. Beyond reserves and corridors: policy solutions to facilitate the movement of plants and animals in a changing climate. *Bioscience* **61**: 713-719.

Landres PB, Morgan P, and Swanson FJ. 1999. Overview of the use of natural variability concepts in managing ecological systems. *Ecological Applications* **9**: 1179-1188.

Lackey RT. 2004. Restoration ecology: the challenge of social values and expectations. *Frontiers in Ecology and the Environment* **2**: 45-46.

Lawler JJ. 2009. Climate change adaptation strategies for resource management and conservation planning. *Annals of the New York Academy of Sciences* **1162**: 79-98.

Lawler JJ, Tear TH, Pyke C, Shaw MR, Gonzalez P, Kareiva P, Hansen L, Hannah L, Klausmeyer K, Aldous A, Bienz C, and Pearsall S. 2010. Resource management in a changing and uncertain climate. *Frontiers in Ecology and the Environment* **8**: 35-43.

Lawler JJ and Olden JD. 2011. Reframing the debate over assisted colonization. *Frontiers in Ecology and the Environment* **9**: 569–574. doi:10.1890/100106.

Lemieux CJ, Beechey TJ, and Gray PA. 2011. Prospects for Canada's protected areas in an era of rapid climate change. *Land Use Policy* **28**: 928-941.

Leopold A. 1949. *A Sand County almanac.* New York: Oxford University Press.

Levey DJ, Bolker BM, Tewksbury JJ, Sargent S, and Haddad NM. 2005. Response to Proches and others *Science* **310**: 782-783.

Li D and Pritchard HW. 2009. The science and economics of ex situ plant conservation. *Trends in Plant Science* **14**: 614-621.

Lindenmayer D and Hunter M. 2010. Some guiding concepts for conservation biology. *Conservation Biology* **24**: 1459-1468.

Link JS, Nye JA, and Hare JA. 2010. Guidelines for incorporating fish distribution shifts into a fisheries management context. *Fish and Fisheries* **12**: 461-469.

Magness DR, Morton JM, Huettmann F, Chapin III FS, and McGuire AD. 2011. A climate-change adaptation framework to reduce continental-scale vulnerability across conservation reserves. *Ecosphere* **2**: article 112.

Mahmoud M, Liu Y, Hartmann H, Stewart S, Wagener T, Semmens D, Stewart R, Gupta H, Dominguez D, Dominguez F, Hulse D, Letcher R, Rashleigh B, Smith C, Street R, Ticehurst J, Twery M, v. Delden H, Waldick R, White D, and Winter L. 2009. A formal framework for scenario development in support of environmental decision-making. *Environmental Modeling and Software* **24**: 798-808.

Manning AD, Gibbons P, and Lindenmayer DB. 2009. Scattered trees: a complementary strategy for facilitating adaptive responses to climate change in modified landscapes? *Journal of Applied Ecology* **46**: 915-919.

Mascarenhas A and Jayakumar S. 2008. An environmental perspective of the post-tsunami scenario along the coast of Tamil Nadu, India: role of sand dunes and forests. *Journal of Environmental Management* **89**: 24-34.

Mawdsley JR, O'Malley R, and Ojima DS. 2009. A review of climate-change adaptation strategies for wildlife management and biodiversity conservation. *Conservation Biology* **23**: 1080-1089.

Metzger MJ, Leemans R, and Schröter D. 2005. A multidisciplinary multi-scale framework for assessing vulnerabilities to global change. *International Journal of Applied Earth Observations and Geoinformation* **7**: 253-267.

Millar C, Stephenson NL, and Stephens SL. 2007. Climate change and forests of the future: managing in the face of uncertainty. *Ecological Applications* **17**: 2145-2151.

Milly PCD, Betancourt J, Falkenmark M, Hirsch RM, Kundzewicz ZW, Lettenmaier DP, and Stouffer RJ. 2008. Stationarity is dead: whither water management? *Science* **319**: 573-57.

Monzón J, Moyer-Horner L, and Palamar MB. 2011. Climate change and species range dynamics in protected areas. *BioScience* **61**: 752-761.

Morgan G, Dowlatabadi H, Henrion M, Keith D, Lempert R, McBrid S, Small M, and Wilbanks T. 2009. Best practice approaches for characterizing, communicating, and incorporating scientific uncertainty in decisionmaking. A report by the U.S. Climate Change Science Program and the Subcommittee on Global Change Research. Synthesis and Assessment Product 5.2. Washington, DC: NOAA.

Mosblech NAS., Bush MB, and van Woesik R. 2011. On metapopulations and microrefugia: palaeoecological insights. *Journal of Biogeography* **38**: 419-429.

Moser SC. 2009. Good morning America: the explosive awakening to the need for adaptation. Sacramento, CA and Charleston, SC: California Energy Commission and NOAA Coastal Services Center. Available at: www.csc.noaa.gov/publications/need-for-adaptation.pdf

Moser SC and Eckstrom JA. 2010. A framework to diagnose barriers to climate change adaptation. *Proceedings of the National Academy of Sciences* **107**: 22026-22031.

NOAA (National Oceanic and Atmospheric Administration). 2010. Adapting to climate change: a planning guide for State coastal managers. Silver Spring, MD: NOAA, Office of Ocean and Coastal Resource Management.

NPS (National Park Service). 2006. Management Policy 2006. Washington, DC: U.S. Department of the Interior, National Park Service. Available at: www.nps.gov/policy/mp2006.pdf

NRC (National Research Council). 2010. Adapting to the impacts of climate change. Washington, DC: National Academies Press.

NRC (National Research Council). 2011. America's climate choices. Washington, DC: National Academies Press.

Nye JA, Link JS, Hare JA, and Overholtz WJ. 2009. Changing spatial distribution of fish stocks in relation to climate and population size in the Northeast United States continental shelf. *Marine Ecology Progress Series* **393**: 111-139.

Ohlson DW, McKinnon GA, and Hirsch KG. 2005. A structured decision-making approach to climate change adaptation in the forest sector. *The Forestry Chronicle* **81**: 97-103.

Palmer MA, Reidy Liermann CA, Nilsson C, Florke M, Alcamo J, Lake PS, and Bond N. 2008. Climate change and the world's river basins: anticipating management options. *Frontiers in Ecology and the Environment* **6**: 81-89.

Parmesan C. 2006. Ecological and evolutionary responses to recent climate change. *Annual Reviews of Ecology, Evolution, and Systematics* **37**: 637-669.

Pavlik B, Murphy D, and Tahoe Yellow Cress Technical Advisory Group. 2002. Conservation strategy for Tahoe yellow cress (*Rorippa subumbellata*). Zephyr Cove, NV: Tahoe Regional Planning Agency.

PCAST (President's Council of Advisors on Science and Technology). 2011. Sustaining environmental capital: protecting society and the economy. Washington, DC: White House Office of Science and Technology Policy.

Pearson RG and Dawson TP. 2005. Long-distance plant dispersal and habitat fragmentation: identifying conservation targets for spatial landscape planning under climate change. *Biological Conservation* **123**: 389-401.

Peters RL and Darling JDS. 1985. The greenhouse effect and nature reserves: global warming would diminish biological diversity by causing extinctions among reserve species. *BioScience* **35**: 707-717.

Peterson DL, Millar CI, Joyce LA, Furniss MJ, Halofsky JE, Neilson RP, and Morelli TL. 2011. Responding to climate change in national forests: a guidebook for developing adaptation options. General Technical Report PNW-GTR-855. Portland OR: U.S. Department of Agriculture, Forest Service, Pacific Northwest Research Station.

Peterson GD, Cumming GS, and Carpenter SR. 2003. Scenario planning: a tool for conservation in an uncertain world. *Conservation Biology* **17**: 358-366.

Pew (Pew Center on Global Climate Change). 2010. Climate change adaptation: what Federal agencies are doing. Washington, DC: Pew Center on Global Climate Change.

Pielke Jr R, Prins G, Rayner S, and Sarewitz D. 2007. Climate change 2007: lifting the taboo on adaptation. *Nature* **445**: 597-598.

Poiani KA, Goldman RL, Hobson J, Hoekstra JM, and Nelson KS. 2011. Redesigning biodiversity conservation projects for climate change: examples from the field. *Biological Conservation* **20**: 185-201.

Pressey RL, Cabeza M, Watts ME, Cowling RM, and Wilson KA. 2007. Conservation planning in a changing world. *Trends in Ecology and Evolution* **22**: 583-592.

Prober SM and Dunlop M. 2011. Climate change: A cause for new biodiversity conservation objectives but let's not throw the baby out with the bathwater. *Ecological Management and Restoration* **12**: 2-3.

Proches S, Wilson JRU, Veltman R, Kalwij JM, Richardson DM, and Chown SL. 2005. Landscape corridors: possible dangers? *Science* **310**: 778-779.

Reusch TBH and Wood TE. 2007. Molecular ecology of global change. *Molecular Ecology* **16**: 3973-3992.

Ricciardi A and Simberloff D. 2009. Assisted colonization is not a viable conservation strategy. *Trends in Ecology and Evolution* **24**: 248-253.

Richardson DM., Hellmann JJ, McLachlan JS, Sax DF, Schwartz MW, Gonzalez P, Brennan EJ, Camacho A, Root TL, Sala OE, Schneider SH, Ashe DM, Clark JR, Early R, Etterson JR, Fielder ED, Gill JL, Minteer BA, Polasky S, Safford HD, Thompson AR, and Vellend M. 2009. Multidimensional evaluation of managed relocation. *Proceedings of the National Academy of Sciences* **106**: 9721-9724.

Richter BD, Mathews R, Harrison DL, and Wigington R. 2003. Ecologically sustainable water management: managing river flows for ecological integrity. *Ecological Applications* **13**: 206-224.

Rose NA and Burton PJ. 2009. Using bioclimatic envelopes to identify temporal corridors in support of conservation planning in a changing climate. *Forest Ecology and Management* **258**: S64-S74.

Rowland EL, Davison JE, and Graumlich LJ. 2011. Approaches to evaluating climate change impacts on species: a guide to initiating the adaptation planning process. *Environmental Management* **47**: 322-337.

Ruhl JB. 2010. Climate change adaptation and the structural transformation of environmental law. *Environmental Law* **40**: 363-431.

Salick J and Ross N. 2009. Traditional peoples and climate change. *Global Environmental Change* **19**: 137-139.

Sax DF, Smith KF, and Thompson AR. 2009. Managed relocation: a nuanced evaluation is needed. *Trends in Ecology and Evolution* **24**: 472-473.

Saxon E, Baker B, Hargrove W, Hoffman F, and Zganjar C. 2005. Mapping environments at risk under different global climate change scenarios. *Ecological Letters* **8**: 53-60.

Schlaepfer MA, Sax DF, and Olden JD. 2011. The potential conservation value of non-native species. *Conservation Biology* **25**: 428-437.

Schloss CA, Lawler JJ, Larson ER, Papendick HL, Case MJ, Evans DM, DeLap JH, Langdon JGR, McRae BH, and Hall SH. 2011. Systematic conservation planning in the face of climate change: bet-hedging on the Columbia Plateau. *PLoS ONE* **6**:e28788. doi:28710.21371/journal.pone.0028788.

Schwartz MW, Hellmann JJ, and McLachlan JS. 2009. The precautionary principle is misguided advice. *Trends in Ecology and Evolution* **24**: 747.

Schweiger O, Biesmeijer JC, Bommarco R, Hickler T, Hulme PE, Klotz S, Kuhn I, Moora M, Nielsen A, Ohlemuller R, Petanidou T, Potts SG, Pysek P, Stout JC, Sykes MT, Tscheulin T, Vila M, Walther GR, Westphal C, Winter M, Zobel M, and Settele J. 2010. Multiple stressors on biotic interactions: how climate change and alien species interact to affect pollination. *Biological Reviews* **85**: 777-795.

Scott D and Lemieux C. 2005. Climate change and protected area policy and planning in Canada. *The Forestry Chronicle* **81**: 696-703.

Seastedt TR, Hobbs RJ, and Suding KN. 2008. Management of novel ecosystems: are novel approaches required? *Frontiers in Ecology and the Environment* **6**: 547-553.

SFBCDC (San Francisco Bay Conservation and Development Commission). 2008. *San Francisco Bay plan.* Available at: www.bcdc.ca.gov/pdf/planning/plans/bayplan/bayplan.pdf

SFBCDC (San Francisco Bay Conservation and Development Commission). 2011a. Living with a rising bay: vulnerability and adaptation in San Francisco Bay and on the shoreline. Available at: www.bcdc.ca.gov

SFBCDC (San Francisco Bay Conservation and Development Commission). 2011b. Resolution No. 11-08: Adoption of Bay Plan Amendment No. 1-08 adding new climate change findings and policies to the Bay Plan; and revising the Bay Plan tidal marsh and tidal flats; safety of fills; protection of the shoreline; and public access findings and policies. Available at: www.bcdc.ca.gov/proposed_bay_plan/bp_amend_1-08.shtml

Shaffer GP, Day Jr JW, Mack S, Kemp GP, van Heerden I, Poirrier MA, Westphal KA, FitzGerald D, Milanes A, Morris CA, Bea R, and Penland PS. 2009. The MRGO navigation project: a massive human-induced environmental, economic, and storm disaster. *Journal of Coastal Research* **54**: 206-224.

Impacts of Climate Change on Biodiversity, Ecosystems, and Ecosystem Services | Technical Input to the 2013 National Climate Assessment

Chapter 6
Adaptation

Scheffer M, Bascompte J, Brock WA, Brovkin V, Carpenter SR, Dakos V, Held H, van Nes EH, Rietkerk M. and Sugihara G. 2009. Early-warning signals for critical transitions. *Nature* **461**: 53-59

Shoo LP, Storlie C, Vanderwal J, Little J, and Williams SE. 2011. Targeted protection and restoration to conserve tropical biodiversity in a warming world. *Global Change Biology* **17**: 186-193.

Simberloff D, Farr JA, Cox J, Mehlman DW. 1992. Movement corridors: conservation bargains or poor investments? *Conservation Biology* **6**: 493-502.

Simenstad C, Reed D, Ford M. 2006. When is restoration not? incorporating landscape-scale processes to restore self-sustaining ecosystems in coastal wetland restoration. *Ecological Engineering* **26**: 27-39.

SITC (Swinomish Indian Tribal Community). 2010. Swinomish climate change initiative climate adaptation action plan. La Conner, WA: Swinomish Indian Tribal Community, Office of Planning. Available at: www.swinomish-nsn.gov/climate_change/Docs/SITC_CC_AdaptationActionPlan_complete.pdf

Smith JB, and Tirpak D (Eds). 1989. The potential effects of global climate change on the United States: report to Congress. Washington, DC:U.S. EPA, Office of Policy, Planning and Evaluation, Office of Research and Development.

Smith B and Wandel J. 2006. Adaptation, adaptive capacity, and vulnerability. *Global Environmental Change* **16**: 282-292.

Stein BA. 2009. Bridging the gap: incorporating science-based information into land use planning. *In* Kihslinger R and McElfish J (Eds), Nature-friendly land use practices at multiple scales. ELI Press, Washington, DC.

Stephensen NL and Millar CI. 2012. Climate change: wilderness's greatest challenge. *Park Science* 28. Available at: www.nature.nps.gov/ParkScience/index.cfm?ArticleID=538&Page=1

Stoddard JL, Larson DP, Hawkins CP, Johnson RK, and Norris RH. 2006. Setting expectations for the ecological condition of streams: the concept of reference condition. *Ecological Applications* **16**: 1267-1276.

Tear TH, Kareiva P, Angermeier PL, Comer P, Czech B, Kautz R, Landon L, Mehlman D, Murphy K, Ruckelshaus M, Scott JM, and Wilhere G. 2005. How much is enough? The recurrent problem of setting measurable objectives in conservation. *BioScience* **55**: 835-849.

Throop W. 2004. A response to the article (Hobbs 2004) "Restoration ecology: the challenge of social values and expectation." *Frontiers in Ecology and the Environment* **2**: 47-48.

USFWS (U.S. Fish and Wildlife Service) and NOAA (National Oceanic and Atmospheric Administration). 2012. National fish, wildlife, and plants climate adaptation strategy. Washington, DC: USFWS and NOAA.

Vignola R, Locatelli B, Martinez C, and Imbach P. 2009. Ecosystem-based adaptation to climate change: what role for policy-makers, society and scientists? *Mitigation and Adaptation Strategies for Global Change* **14**: 691-696.

Vinyeta T. 2012. A synthesis of literature on traditional ecological knowledge and climate change. Available at: http://tribalclimate.uoregon.edu/files/2010/11/TEK_CC_Draft_3-13-2012.pdf

Vos CC, Berry P, Opdam P, Baveco H, Nijhof B, O'Hanley J, Bell C, and Kuipers H. 2008. Adapting landscapes to climate change: examples of climate-proof ecosystem networks and priority adaptation zones. *Journal of Applied Ecology* **45**: 1722-1731.

Walker B, Holling CS, Carpenter SR, and Kinzig A. 2004. Resilience, adaptability and transformability in social–ecological systems. *Ecology and Society* **9**: 5.

Walther GR, Roques A, Hulme PE Sykes MT, Pysek P, Kuhn I, Zobel M, Bacher S, Botta-Dukat Z, Bugmann H, Czucz B, Dauber J, Hickler T, Jarosık V, Kenis M, Klotz S, Minchin D, Moora M, Nentwig W, Ott J, Panov VE, Reineking B, Robinet C, Semenchenko V, Solarz W, Thuiller W, Vila M, Vohland K, and Settele J. 2009. Alien species in a warmer world: risks and opportunities. *Trends in Ecology and Evolution* **24**: 686-693.

Walther GR. 2010. Community and ecosystem responses to recent climate change. *Philosophical Transactions of the Royal Society B-Biological Sciences* **365**: 2019-2024.

Webber BL and Scott JK. 2012. Rapid global change: implications for defining natives and aliens. *Global Ecology and Biogeography* **21**: 305-311.

Weeks D, Malone P, and Welling L. 2011. Climate change scenario planning: a tool for managing parks into uncertain futures. *Park Science* **28**: 26-33.

West JM, Julius SH, Kareiva P, Enquist C, Lawler JJ, Petersen B, Johnson AE, and Shaw MR. 2009. U.S. natural resources and climate change: concepts and approaches for management adaptation. *Environmental Management* **44**: 1001-1021.

Wilby RL, Wigley TML, Conway D, Jones PD, Hewitson BC, Main J, and Wilks DS. 1998. Statistical downscaling of general circulation model output: a comparison of methods. *Water Resources Research* **34**: 2995-3008.

Wilby RL and Perry GLW. 2006. Climate change, biodiversity and the urban environment: a critical review based on London, UK. *Progress in Physical Geography* **30**: 73-98.

Wilcove DS., Rothstein D, Dubow J, Phillips A, and Losos E. 1998. Quantifying threats to imperiled species in the United States. *BioScience* **48**: 607-615.

Williams PH, Hannah L, Andelman SJ, Midgley GF, Araújo MB, Hughes G, Manne L, Martinez-Meyer E, and Pearson RG. 2005. Planning for climate change: identifying minimum-dispersal corridors for the Cape Proteaceae. *Conservation Biology* **19**: 1063-1074.

Williams JW and Jackson ST. 2007. Novel climates, no-analog communities, and ecological surprises. *Frontiers in Ecology and the Environment* **5**: 475-482.

Williams SE, Shoo LP, Isaac JL, Hoffmann AA, and Langham G. 2008. Towards an integrated framework for assessing the vulnerability of species to climate change. *PLoS Biology* **6**: 2621-2626.

Williams BK, Szaro RC, and Shapiro CD. 2009. Adaptive management: the U.S. Department of the Interior technical guide. Washington, DC: U.S. Department of the Interior, Adaptive Management Working Group. Available at: www.doi.gov/initiatives/AdaptiveManagement/TechGuide.pdf

Willows RI, and Connell RK (Eds). 2003. Climate adaptation: risk, uncertainty and decision-making. UKCIP Technical Report. Oxford, UK: United Kingdom Climate Impacts Program.

Wilson GWT and Hartnett DC. 1997. Effects of mychorrhizae on plant growth and dynamics in an experimental tall grass prairie microcosms. *American Journal of Botany* **84**: 478-482.

World Bank. 2010. Convenient solutions to an inconvenient truth : ecosystem-based approaches to climate change. Washington, DC: International Bank for Reconstruction and Development/The World Bank.

Chapter 7. Proposed Actions for the Sustained Assessment of Biodiversity, Ecosystems, and Ecosystem Services

Lead Authors: Shawn L. Carter, Amanda Staudt, Gary Geller, and Woody Turner

Key Proposed Actions

- Identify a core set of widely recognized, policy-relevant questions about impacts on biodiversity and ecosystem services.
- Establish a broader ecosystem assessment process and framework.
- Align monitoring, modeling, and assessment activities for climate with those for biodiversity and ecosystem services.
- Identify and convey clear connections between biodiversity loss, reduced ecosystem services, and societal benefits.

7.1. PROPOSED ACTIONS FOR THE SUSTAINED ASSESSMENT

The impacts of climate change are manifest at all scales of biological organization – from genes to ecosystems – and are occurring at rates unprecedented in recent history. These impacts are expected to increase during the coming decades, as this report shows. Thus, regular and ongoing assessments are essential to track changes, anticipate future changes, and inform response strategies. The assessment process provides information that allows implementation of rapid management responses under conditions of high uncertainty.

At present, there is no ongoing assessment process for the United States that exclusively focuses on biodiversity, ecosystems, and ecosystem services. Only a handful of past assessment efforts have focused primarily on biodiversity (for example, MA, 2005; CBD, 2010), while large national and international climate assessments have addressed ecosystems or biodiversity as one of many different affected sectors (for example, IPCC, 2007; USGCRP, 2009). A recent White House report (PCAST, 2011) has charged the Federal interagency community to conduct a Quadrennial Ecosystems Services Trends (QuEST) Assessment, which will provide a comprehensive evaluation of the condition of United States ecosystems. We laud this recommendation and note that the significant policy-relevant information collected in this technical input report clearly supports the need for a focused assessment of this type.

In this concluding section, we highlight several proposed actions relevant to establishing a sustained assessment process in the United States that addresses the effects of climate change and other stressors on biodiversity and ecosystem services. It is important to recognize that an assessment of climate impacts cannot be divorced from the need for a holistic ecosystem assessment, especially given the broad range of stressors affecting ecosystems. A major challenge will be to craft coordinated assessment processes – including monitoring, modeling, research, and synthetic efforts – that work across the different levels of diversity and address the decision-making needs arising from climate change impacts on ecosystems.

Identify a core set of widely recognized, policy-relevant questions about impacts on biodiversity and ecosystem services.

Successful assessments should be initiated with a clear mandate and set of guiding questions stemming from interactive discussions among policy-makers, other stakeholders, and the research community (NRC 2007). **Figure 7.1** illustrates how the guiding questions for the NCA can build on the Global Change Research Act of 1990 (GCRA, 1990), address the needs of stakeholders, and build on past assessments.

In the case of the National Climate Assessment, the GCRA requires a quadrennial assessment ██ment, agriculture, energy production and use, land and water resources, transportation, human health and welfare, human ██-related nature of environmental research by calling for an ██ assessments conducted under this mandate have focused primarily on climate change and given limited attention to the effects of other environmental stressors (NAST, 2001; USGCRP, 2009). As such, the treatment of ecosystems and biodiversity in these past assessments does not reflect the full complexity of how they are being affected by human activities. Further efforts are needed to identify the core set of policy-relevant questions for future ecosystem assessments in the United States.

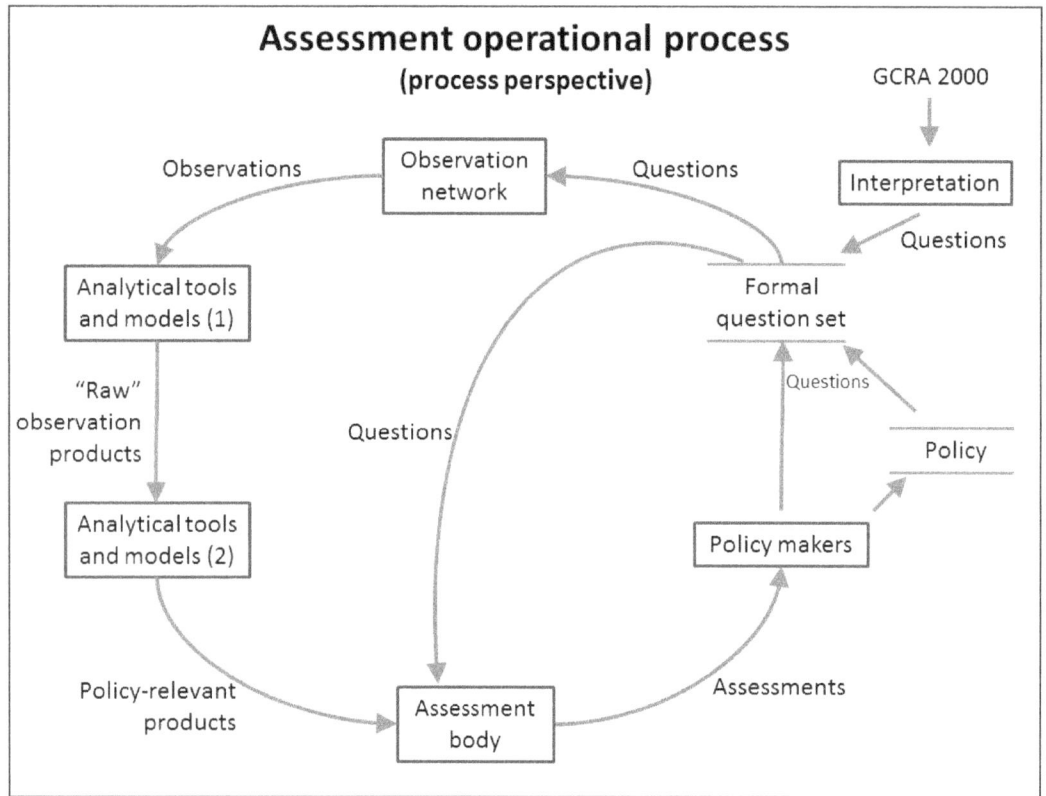

Figure 7.1. An operational process for conducting regular assessments of biodiversity and ecosystem services. Policy requirements and legal mandates are interpreted into assessment questions that determine the required observations, models, and tools to inform assessment products.

The management and policy needs of stakeholders should help guide a more integrated research approach to predicting, understanding, and adapting to the impacts of climate change on biodiversity and ecosystem services. Important stakeholders for this sector include Federal and State policy makers; natural resource managers at Federal, State, local and tribal scales; non-governmental organizations engaged in conserving natural resources; and outdoor recreation and other industries that depend on ecosystem services for their livelihood. Significant advances in effective ways to engage stakeholders can be used to guide these efforts (for example, Glick and others, 2011).

International assessment efforts can provide a starting point for identifying the policy-relevant questions. The Intergovernmental Science-Policy Platform on Biodiversity and Ecosystem Services (IPBES) provides an IPCC-like science and policy framework (Larigauderie and Mooney, 2010) that could be used to inform a national assessment framework. Furthermore, assessment targets specific to biodiversity (for example, the 2010 targets established via the Convention on Biological Diversity) that have already been established by the international community can also inform the initial questions (Jones and others, 2011).

Establish a broader ecosystem assessment process and framework.

The assessment of climate impacts on biodiversity and ecosystem services in the United States may be nested within a broader ecosystem assessment framework with relevance to other international activities. A widely recognized, credible forum is suggested to solicit and capture stakeholder needs and expectations. Currently, the National Climate Assessment process, which is coordinated by the U.S. Global Change Research Program, plays an important role in coordinating Federal research activities related to global change. Future QuEST or related United States assessments, with oversight from the Sustainability Task Force of the National Sc▨▨▨▨▨▨▨▨▨▨▨▨▨▨▨▨▨▨▨▨▨▨▨▨▨▨▨▨▨▨ ▨▨▨ ▨▨▨▨ ▨ ▨▨▨▨▨▨▨▨▨▨ ▨▨▨▨▨ ▨▨▨▨▨▨▨▨▨▨▨▨▨▨▨ Sustainability (CENRS), could offer a broader biodiversity and ecosystems services-focused assessment framework and harness IPBES efforts to assess global biosphere change. A critical first step to establishing this broader framework is to build capacity for research activities that improve policy relevance and catalyze external funding (Perrings and others, 2011).

Align monitoring, modeling, and assessment activities for climate with those for biodiversity and ecosystem services.

Monitoring and modeling frameworks for climate change and ecological change have typically been designed and implemented with different objectives, underlying assumptions, and with little regard for interoperability. Consequently, access to monitoring data and modeling output that can provide information on biotic interactions across different spatial domains is limited (McMahon and others, 2011; Bellard and others, 2012). As mentioned in *Chapter 5: Other Stressors* and *Chapter 2*: *Biodiversity*, a lack of integrated data networks limits our ability to investigate the interaction of multiple stressors under changing climate regimes. Information technology needs associated with the integration of observational data, models, and assessments are extensive, and a coordinated effort across Federal, State, local, academic, non-governmental, and other spheres is needed to ensure that data are accessible, discoverable and useful for assessments.

Multiple observation networks capable of integrating data on multiple spatial scales offer flexible solutions for responding to the formal set of assessment questions. Multi-tiered monitoring approaches that combine long-term monitoring networks with spatially intensive, *in situ* monitoring and remotely sensed data provide good examples for informing broad-scale assessments (Jones and others, 2010). Data from observation networks drive 1st-order model projections of species and ecosystem responses, which then inform model frameworks capable of generating scenarios (Pereira and others, 2010). These can then be processed into indicators of direct use to the assessment body (**Figure 7.1**).

Identify and convey clear connections between biodiversity loss, reduced ecosystem services, and societal benefits.

    Clear connections between biodiversity loss, changes in ecosystem services (whether reduced or enhanced), and societal benefits or detriments should be articulated in assessment findings (Carpenter and others, 2009). Recent assessments of biodiversity and ecosystems have adopted frameworks that include ecosystem services and human well-being (MA, 2005; UKNEA, 2011) and our understanding of the tangible links between biodiversity, ecological processes, and ecosystem services is rapidly expanding (Loreau, 2010).

    The intersecting pressures among energy development, land use change, provision of food and water, and climate change require that future assessments make clear the tradeoffs associated with biodiversity conservation (MA, 2005; Dale and others, 2011). The Millennium Ecosystem Assessment, for example, explicitly recognized that present-day use of ecosystem services should be weighed against the potential loss of benefits in the future (MA, 2005). The ability to convey the relative tradeoffs associated with ecosystem service protection has a critical role to play in helping the American public, key policymakers, and resource managers make well-informed decisions about our natural resources.

7.2. LITERATURE CITED

Bellard C, Bertelsmeier C, Leadley P, Thuiller W, and Courchamp F. 2012. Impacts of climate change on the future of biodiversity. *Ecology Letters* **15**: 365-377.

Carpenter SR, Mooney HA, Agard J, Capistrano D, DeFries RS, Diaz S, Dietz T, Duraiappah AK, Oteng-Yeboah A, Pereira HM, Perrings C, Reid WV, Sarukhan J, Scholes RJ, and Whyte A. 2009. Science for managing ecosystem services: Beyond the Millennium Ecosystem Assessment. *Proceedings of the National Academy of Sciences* **106**: 1305-1312.

CBD (Secretariat of the Convention on Biological Diversity). 2010. Biodiversity Scenarios: projections of 21st century change in biodiversity and associated ecosystem services. Available at: http://www.cbd.int/

Dale VH, Efroymson RA, and Kline KL. 2011. The land use-climate change-energy nexus. *Landscape Ecology* **26**: 755-773.

GCRA (Global Change Research Act). 1990. Public Law 101-606(11/16/90) 104 Stat. 3096-3104.

Glick P, Stein BA, and Edelson N. 2011. Scanning the Conservation Horizon: A Guide to Climate Change Vulnerability Assessment. National Wildlife Federation, Washington, DC.

IPCC (Intergovernmental Panel on Climate Change). 2007. Climate Change 2007: The Physical Science Basis. Contribution of Working Group I to the Fourth Assessment Report of the Intergovernmental Panel on Climate Change, Cambridge , UK.

Jones JPG, Collen B, Atkinson G, Baxter PWJ, Bubb P, Illian JB, Katzner TE, Keane A, Loh J, McDonald-Madden E, Nicholson E, Pereira HM, Possingham HP, Pullin AS, Rodrigues ASL, Ruiz-Gutierrez V, Sommerville M, and Milner-Gulland EJ. 2011. The why, what, and how of global biodiversity indicators Beyond the 2010 target. *Conservation Biology* **25**: 450-457.

Jones KB, Bogena H, Vereecken H, and Weltzin JF. 2010. Design and importance of multi-tiered ecological monitoring networks in long-term ecological research. *In* Müller F, Schubert H, and Klotz S. (Eds), Long-term ecological research: Between theory and application. Springer, Berlin, Germany. 355-374 p.

Larigauderie A, and Mooney HA. 2010. The intergovernmental science-policy platform on biodiversity and ecosystem services: moving a step closer to an IPCC-like mechanism for biodiversity. *Current Opinion in Environmental Sustainability* **2**: 9-14.

Loreau M. 2010. Linking biodiversity and ecosystems: toward a unifying ecological theory. *Philosophical Transactions of the Royal Society* B **365**: 49-60.

MA (Millennium Ecosystem Assessment). 2005. Ecosystems and Human Well-being: Biodiversity Synthesis. World Resources Institute, Washington, D.C.

McMahon SM, Harrison SP, Armbruster WS, Bartlein PJ, Beale CM, Edwards ME, Kattge J, Midgley G, Morin X, and Prentice IC. 2011. Improving assessment and modelling of climate change impacts on global terrestrial biodiversity. *Trends in Ecology and Evolution* **26**: 249-259.

NAST (National Assessment Synthesis Team). 2001. Climate Change Impacts on the United States: The Potential Consequences of Climate Variability and Change, Report for the U.S. Global Change Research Program. Cambridge University Press, Cambridge, UK.

NRC (National Research Council). 2007. Analysis of Global Change Assessments: Lessons Learned. National Academies Press, Washington, DC.

environmental capital: protecting society and the economy. 145 p. Available at: http://www.whitehouse.gov/administration/eop/ostp/pcast/docsreports

Pereira HM, Leadley PW, Proenca V, Alkemade R, Scharlemann JPW, Fernandez-Manjarres JF, Araujo MB, Balvanera P, Biggs R, Cheung WWL, Chini L, Cooper HD, Gilman EL, Guenette S, Hurtt GC, Huntington HP, Mace GM, Oberdorff T, Revenga C, Rodrigues P, Scholes RJ, Sumaila UR, and Walpole M. 2010. Scenarios for global biodiversity in the 21st century. *Science* **330**: 1496-1501.

Perrings C, Duraiappah A, Larigauderie A, and Mooney H. 2011. The Biodiversity and Ecosystem Services Science-Policy Interface. *Science* **331**: 1139-1140.

UKNEA (UK National Ecosystem Assessment). 2011. The UK National Ecosystem Assessment Technical Report. UNEP-WCMC, Cambridge, UK. Available at: http://uknea.unep-wcmc.org/Resources/tabid/82/Default.aspx

USGCRP (U.S. Global Change Research Program). 2009. Global Climate Change Impacts in the United States. Cambridge University Press, Cambridge, UK.

Appendix A: List of contributors to the National Climate Assessment Technical Input on Biodiversity, Ecosystems, and Ecosystem Services

Name	Affiliation
Shere Abbott	Syracuse University
Katie Arkema	Stanford University
Doug Beard	USGS National Climate Change and Wildlife Science Center
Rosina Beirbaum	University of Michigan
Elena Bennett	McGill University
Michael J. Bernstein	Arizona State University
Britta Bierwagen	U.S. Environmental Protection Agency
Micheal Bernstein	Arizona State University
Kate A. Brauman	University of Minnesota Institute on the Environment
Shawn L. Carter	USGS National Climate Change and Wildlife Science Center
F. Stuart Chapin III	University of Alaska, Fairbanks
Molly S. Cross	Wildlife Conservation Society
Amy Daniels	US Forest Service, Research
Cynthia J. Decker	NOAA
Natalie S. Dubois	Defenders of Wildlife
J. Emmett Duffy	Virginia Institute of Marine Science, The College of William and Mary
Jeff Dukes	Purdue University
Gary Geller	NASA/JPL
Evan Girvetz	The Nature Conservancy
Dave Goodrich	USDA-Agricultural Research Service
Patrick Gonzalez	National Park Service
Roger Griffis	NOAA Fisheries Service
Nancy B. Grimm	National Science Foundation, Arizona State University, National Climate Assessment
Peter M. Groffman	Cary Institute of Ecosystem Studies
Carolyn Enquist	USA National Phenology Network and The Wildlife Society National Coordinating Office, School of Natural Resources & the Environment, The University of Arizona
Lara Hansen	EcoAdapt
Bruce Hayden	University of Virginia
Jessica Hellmann	Department of Biological Sciences, University of Notre Dame
Jennifer Howard	NOAA
Kurt Johnson	U.S. Fish and Wildlife Service
Peter Kareiva	The Nature Conservancy
Josh Lawler	University of Washington

Allison Leidner	NASA/AAAS
Yiqi Luo	University of Oklahoma
Virginia Matzek	Santa Clara University
Forrest Melton	NASA Ames / CSU Monterey Bay
Scott A. Morrison	The Nature Conservancy
Knute Nadelhoffer	University of Michigan/LTER
Erik Nelson	Bowdoin College
Daniel Nover	EPA (AAAS)
John O'Leary	Massachusetts Division of Fisheries and Wildlife
Amber Pairis	California Department of Fish and Game
Craig Paukert	USGS and Univ of Missouri
Malin Pinsky	Princeton University
Pete Raymond	Yale FES
Walt Reid	David and Lucile Packard Foundation
Jennifer Riddell	EPA (AAAS)
Mary Ruckelshaus	NaturalCapital Project
John Sabo	Arizona State University
Susan Ruffo	Council on Environmental Quality
Marin Saunders	Santa Clara University
Josh Schimel	UC Santa Barbara
Darius Semmens	USGS
Joel Smith	Stratus Consulting
Lesley Sneddon	NatureServe
Luis A. Solórzano	Gordon and Betty Moore Foundation
Michelle D. Staudinger	USGS National Climate Change & Wildlife Science Center; University of Missouri
Amanda Staudt	National Wildlife Federation
Bruce A. Stein	National Wildlife Federation
Heather Tallis	Natural Capital Project, Stanford University
Laura Thompson	USGS National Climate Change and Wildlife Science Center
Woody Turner	Earth Science Division, NASA Headquarters
Craig E. Williamson	Miami University, Ohio
Joanna Whittier	University of Missouri
Elda Varela-Acevedo	Michigan State University

National Climate Assessment
Biodiversity/Ecosystems/Ecosystem Services Chapter
Technical Input Workshop

January 17-19, 2012
Moore Foundation, Palo Alto, CA
1661 Page Mill Road

AGENDA

Monday January 16, 2012

5:00 p.m.   Steering Committee meeting/dinner

Tuesday January 17, 2012

8:15 a.m.   Continental Breakfast Available at Moore Foundation

9:00 a.m.   PLENARY MEETING

- Welcome and introductions (Michelle Staudinger) (10 min)
- Peter Kareiva to introduce Moore Foundation
- Welcome by Moore Foundation / Importance of Assessments to Science: Making a difference (Steve McCormic and Guillermo Castilleja) (20 min)
- Process and products: Q&A (Nancy Grimm) (30 min)
- Overview of meeting: Outline and agenda (Peter Kareiva) (15 min)

10:15 a.m. BREAK

10:30 a.m. OVERVIEW PRESENTATIONS

- Working group leads focusing on overall approach and current status; objective is to ensure coordination as separate working groups move forward over next 3 days (three PPT slides: (i) overview of outline, (ii) gaps/issues/questions in our section, (iii) intersections with other groups) – 5 minutes presentations; 15 Q&A and feedback

  o Biodiversity
  o Ecosystems
  o Ecosystem Services
  o Other stressors

- Feedback on what is and / or is not being done
- Facilitated discussion of any issues that came up during working group overview presentations
- Charge to working groups for afternoon + tomorrow

Noon   LUNCH

1:15 p.m.   BREAKOUT SESSIONS

            Objectives:
            o Update on progress of pre-workshop work
            o What still needs to be done
            o Development of figures, tables, and graphics
            o Development of key messages/findings of subgroup activities for presentations at end of day

4: 00 p.m.   PLENARY SESSION (presentations from each group)

            o Key findings
            o Cool Science (emerging science, what is relevant to the public and policy makers)
            o Results from new analyses

4:50 p.m.

5:00 p.m.   RECEPTION WITH MOORE STAFF: MEET OUR COLLEAGUES

Wednesday January 18, 2012

8:15 a.m.   Continental Breakfast Available at Moore Foundation

9:00 a.m.   PLENARY MEETING

- Status check and marching orders for breakout sessions

9:15 a.m.   BREAKOUT SESSIONS

Biodiversity	Ecosystems	Ecosystem Services	Other Stressors
Room: A Facilitators: Michelle	Room: B Facilitators: Nancy	Room: C Facilitators: Peter, Mary	Room: D Facilitators: Amanda

11:00 a.m.   PLENARY SESSION

- Presentation of breakout groups and discussion of other themes for afternoon session

Noon   LUNCH

1:15 p.m.   BREAKOUT SESSIONS

- Research gaps/sustaining the assessment
- Adaptation
- Policy implications
- Risk portraits
- Integration

4:00 p.m.   PRESENTATIONS

- Major findings from breakout sessions
- Discussion and feedback on presentations

5:00 p.m.   ADJOURN FOR THE DAY

Thursday January 19, 2012

8:15 a.m.   Continental Breakfast Available at Moore Foundation

9:00 a.m.   PLENARY MEETING

- How to make our assessment impactful (Walt Reid, Doug Beard)

9:30 a.m.   DISCUSSION

10:00 a.m.   BREAK

10:30 a.m.   BREAKOUT SESSIONS

Objectives for working groups:
- o Incorporate feedback from plenary discussions
- o Key findings/impactfulness
- o Continue writing as time allows

Biodiversity	**Ecosystems**	**Ecosystem Services**	**Other Stressors**	**Integration**
Room: A Facilitators: Shawn, Michelle	Room: B Facilitators: Nancy	Room: C Facilitators: Mary	Room: D Facilitators: Amanda	Room: E Facilitators: Peter

Noon   LUNCH

1:00 p.m.   BREAKOUT SESSION

- 4 workgroups
- Next steps and assigning final responsibilities
- Writing activities

2:45 p.m.   CONCLUDING REMARKS FOR THE ENTIRE TECHNICAL INPUT

3:00 p.m.   ADJOURN